Reconstructing Mobility

Kristian J. Carlson • Damiano Marchi
Editors

Reconstructing Mobility

Environmental, Behavioral, and Morphological Determinants

Springer

Editors
Kristian J. Carlson
Evolutionary Studies Institute
University of the Witwatersrand
Johannesburg, South Africa

Damiano Marchi
Department of Biology
University of Pisa
Pisa, Italy

ISBN 978-1-4899-7459-4 ISBN 978-1-4899-7460-0 (eBook)
DOI 10.1007/978-1-4899-7460-0
Springer New York Heidelberg Dordrecht London

Library of Congress Control Number: 2014942689

Printed on acid-free paper

Springer is part of Springer Science+Business Media (www.springer.com)

Preface

For decades, scientists have relied on the concept of mobility in describing activity patterns of past and present human populations. Population-level comparisons have traditionally sought to demonstrate differential mobility (e.g., logistical or residential) amongst Pleistocene or Holocene *Homo* groups, using this as a basis for inferring convergent or contrasting adaptive behavior. For example, shifting from a hunter-gatherer to a more sedentary agricultural subsistence strategy generally has been associated with a relative decline in mobility associated with the latter. Substantial efforts have been devoted towards inferring which musculoskeletal adaptations best reflect such a potential shift in mobility. The central role of bipedalism in human locomotion predisposes lower limb musculoskeletal anatomy to feature prominently in these inferences, although it is important to note that expressions of mobility in other areas of the postcranium (e.g., the upper limb) are gaining traction in the field when studying select populations (e.g., coastal or island groups). It is problematic that often mobility is not defined a priori in precise enough terms to facilitate comparability of results across studies. Typically, some derivation of an ethnographic definition of mobility is adopted, whether explicitly recognized or not (e.g., populations with greater mobility travel farther than populations with lesser mobility). Usually, in applying the ethnographic definition, unstated motivations for travel focus on resource acquisition or intergroup relationships (e.g., trading).

On the other hand, an excessively narrow application of the concept of mobility, such as a mechanically focused one, equally limits comparisons of results across studies. Not all studies would (nor should) integrate experimental approaches in order to quantify mobility. Resources necessary for the requisite acquisition of ground reaction force and kinematic data are not equally available to all researchers, and there are ethical and logistical constraints when studying human subjects. Rather, the optimal solution for defining mobility, or fully capturing its essence, should embrace a multidisciplinary approach in how the concept is applied. Despite such a long-standing and widespread reliance on the concept of mobility for reconstructing and comparing activity patterns and life histories of human populations, such an inclusive attempt at defining mobility has not yet been made.

To address this notable absence, in the spring of 2011, we organized a symposium on mobility at the 80th Annual Meeting of the American Association of Physical Anthropologists held in Minneapolis, MN. The symposium assembled an array of experts using different approaches for quantifying and comparing the effects of mobility on postcranial musculoskeletal anatomy. The symposium and subsequent discussions were aimed at embracing current perspectives and stimulating new ones that emphasized a holistic view of the interaction among intrinsic (i.e., skeletal) and extrinsic (i.e., environmental) factors relevant for quantifying and studying the differential expression of mobility. Moreover, the symposium highlighted the importance of disentangling environmental effects some of which transcend traditional categorical groupings, such as coastal versus inland and/or mountainous versus flat terrain environments.

This volume emanates from the original symposium. It is not intended to be the final word on the concept of mobility, but we hope that it will serve as a suitable starting point from which new discussion and future work can begin (or continue), perhaps with a renewed focus on critical issues identified herein or to be expanded laterally. We also hope that this volume represents a useful advance by articulating a consensus working definition of mobility that can be widely applied in anthropological studies in order to overcome the lack of consistency in explicitly defining the concept of mobility that currently cripples the comparisons of results across studies.

There are a number of people we would like to thank, for this volume would not have materialized without the substantial efforts of many. First, we would like to acknowledge the original participants in the 2011 symposium, not all of whom were able to contribute chapters to the volume for one reason or another. The discussions that took place leading up to, during, and following the symposium helped shape this volume considerably. Thank you for your contributions in driving this effort forward. We also would like to thank contributors to the volume who did not participate in the 2011 symposium for one reason or another. Your contributions to the collective effort have broadened its scope in new, exciting ways. Chapters were reviewed by a mix of fellow contributors, co-editors, and additional colleagues. We are indebted to everyone who assisted with reviewing the individual chapters. Thank you for your time and willingness to offer constructive suggestions. Finally, we would like to thank Janet Slobodien, Jacob Gallay, and others at Springer Press for encouraging the efforts that ultimately led to this volume. We are extremely grateful for this unwavering support throughout the entire process.

Johannesburg, South Africa — Kristian J. Carlson
Pisa, Italy — Damiano Marchi

Contents

Contributors

Kristian J. Carlson Evolutionary Studies Institute, University of the Witwatersrand, Johannesburg, South Africa

School of Geosciences, University of the Witwatersrand, South Africa

Department of Anthropology, Indiana University, Bloomington, IN, USA

Steven E. Churchill Department of Evolutionary Anthropology, Duke University, Durham, NC, USA

Evolutionary Studies Institute, University of the Witwatersrand, Johannesburg, South Africa

Libby W. Cowgill Department of Anthropology, University of Missouri, Columbia, Columbia, MO, USA

Shamsi R. Daneshvari Department of Anthropology, University of New Mexico, Albuquerque, NM, USA

Health Sciences Library and Informatics Center, University of New Mexico, Albuquerque, NM, USA

Thomas G. Davies PAVE Research Group, Department of Archaeology & Anthropology, University of Cambridge, Cambridge, UK

Churchill College, Cambridge, UK

Frederick E. Grine Departments of Anthropology and Anatomical Sciences, Stony Brook University, Stony Brook, NY, USA

Ryan W. Higgins Center for Functional Anatomy and Evolution, Johns Hopkins School of Medicine, Baltimore, MD, USA

Clark Spenser Larsen Department of Anthropology, The Ohio State University, Columbus, OH, USA

Damiano Marchi Department of Biology, University of Pisa, Pisa, Italy

Evolutionary Studies Institute, University of the Witwatersrand, Johannesburg, South Africa

Osbjorn M. Pearsen Department of Archaeology, Durham University, Durham, UK

Timothy R. Petersen Department of Anthropology, University of New Mexico, Albuquerque, NM, USA

Department of Anesthesiology & Critical Care Medicine, University of New Mexico Hospital, Albuquerque, NM, USA

Christopher B. Ruff Center for Functional Anatomy and Evolution, Johns Hopkins University School of Medicine, Baltimore, MD, USA

Timothy M. Ryan Department of Anthropology, Pennsylvania State University, State College, PA, USA

Center for Quantitative X-Ray Imaging, EMS Energy Institute, Pennsylvania State University, State College, PA, USA

Laura L. Shackelford Department of Anthropology, University of Illinois at Urbana-Champaign, Urbana, IL, USA

Colin N. Shaw PAVE Research Group, Department of Archaeology & Anthropology, University of Cambridge, Cambridge, UK

McDonald Institute for Archaeological Research, Department of Archaeology and Anthropology University of Cambridge, Cambridge, UK

Vitale S. Sparacello Department of Archaeology, Durham University, Durham, UK

Department of Anthropology, University of New Mexico, Albuquerque, NM, USA

Jay T. Stock PAVE Research Group, Department of Archaeology & Anthropology, University of Cambridge, Cambridge, UK

Kelli H. Tamvada Department of Anthropology, University at Albany, State University of New York, Albany, NY, USA

Christopher S. Walker Department of Evolutionary Anthropology, Duke University, Durham, NC, USA

Cara M. Wall-Scheffler Department of Biology, Seattle Pacific University, Seattle, WA, USA

Department of Anthropology, University of Washington, Seattle, WA, USA

Elizabeth Weiss Department of Anthropology, SJSU, San Jose, CA, USA

Daniel J. Wescott Department of Anthropology, FACTS, Texas State University, San Marcos, TX, USA

Chapter 1
Introduction: Towards Refining the Concept of Mobility

Kristian J. Carlson and Damiano Marchi

Abstract Bone has an ability to model and remodel itself such that its distribution and material properties reflect factors occurring during the lifetime of an individual. Known factors influencing bone properties range from nonmechanical (e.g., age, sex, diet, health, and hormones) to mechanical ones (e.g., activity level and patterns). A lifetime accumulation of these inputs, therefore, should be reflected in the structure of bone diaphyses at the death of an individual. Inferring the inputs of these factors from long bone diaphyses of long dead individuals, whether Holocene agriculturalists or hunter-gatherers, or earlier human ancestors, depends in part on modern analogues being used to help identify and isolate the contributions of these factors. This chapter is both an introduction to and a synthesis of the collaborative effort that is recounted within the volume, and that is aimed at understanding the impact of human mobility as one such input to diaphyseal form.

Keywords Bone functional adaptation • Activity patterns • Hunter-gatherer • Travel distance • Terrain complexity

---

K.J. Carlson (✉)
Evolutionary Studies Institute, University of the Witwatersrand, WITS 2050, Johannesburg, South Africa

School of Geosciences, University of the Witwatersrand, WITS 2050 Johannesburg, South Africa

Department of Anthropology, Indiana University, Bloomington, IN, USA
e-mail: kristian.carlson@wits.ac.za

D. Marchi
Department of Biology, University of Pisa, Pisa, Italy
e-mail: damiano.marchi@unipi.it

Evolutionary Studies Institute, University of the Witwatersrand, Johannesburg, South Africa

K.J. Carlson and D. Marchi (eds.), *Reconstructing Mobility: Environmental, Behavioral, and Morphological Determinants*, DOI 10.1007/978-1-4899-7460-0_1,

Over the course of an individual's lifetime, bone, as a material, has an ability to model (deposition) or remodel (resorption followed by deposition), with this ability impacted by a number of mechanical (e.g., activity patterns) and nonmechanical factors including age, sex, diet, health, and hormonal fluctuations (Martin et al. 1998; Carter and Beaupré 2001). For example, bone modeling and remodeling processes appear to be age sensitive, in that bone responses to mechanical loading appear to be stronger during growth than during adulthood (see review in Ruff et al. 2006). Hormonal fluctuations or dietary factors may mediate these responses (e.g., Devlin and Lieberman 2007; Devlin et al. 2010). In order to prevent structural deficiency, or injury, a bone must resist deformations (strains) that occur during activities, accomplishing this through redistributive and reparative (re)modeling processes that are mediated by nonmechanical factors, such as those mentioned above, guided by the principle of material economy, and ultimately aimed at minimizing internal forces (stresses) within the bone. A lifetime's accumulation of these processes, therefore, should be reflected in the structure of a bone diaphysis preceding death.

Wolff (1892) was the first to coalesce some of these governing rules into what became known as Wolff's Law. Originally, Wolff envisaged only trabecular strut arrangements, having observed similarities between the internal structure of the proximal femur and lines of maximum internal stress in a Fairbairn crane (Roesler 1981). Corroborating work of contemporaries (e.g., Meyer 1867; Roux 1881), and subsequent studies that built upon preceding work (e.g., Kummer 1959; Pauwels 1968, 1980; Amtmann 1971), eventually extended Wolff's Law to cortical bone. More recently, the term "Bone Functional Adaptation" has been coined for the modern evolution of this concept (Ruff et al. 2006), with substantial work clearly remaining before its complexities can be fully understood (Pearson and Lieberman 2004; Judex and Carlson 2009; Robling 2009).

The concept of bone functional adaptation uses inverse dynamics to reconstruct loading profiles over the lifetime of an individual. Importantly, as bone functional adaptations during pre-adulthood and adulthood may differ, for example in response magnitude and rate (see review in Ruff et al. 2006), the sum total lifetime response to mechanical loading is probably not linear. In order to estimate how an individual may have loaded their limb bones (e.g., humerus, femur, tibia) over the course of their lifetime, where frequency (activity level) and magnitude (activity pattern) of these loads are amongst the most crucial determinants, the structure of a long bone diaphysis is modeled using engineering concepts, such as beam theory (Mott 1996). Estimating age-equivalent activity profiles in this manner (e.g., by calculating and comparing cross-sectional geometric properties) is a particularly powerful means of inferring adaptive strategies of individuals whose activities are no longer physically observable (e.g., Pleistocene and Holocene human groups characterized by different subsistence strategies). The earliest comparisons of bone functional adaptations focused on intra-group sources of structural variability (e.g., age and sex) rather than true activity profile differences between populations (see review by Ruff and Larsen, Chap. 2). The first detailed comparison of population-level activity profiles, in what would initiate a framework for subsequent attempts to interpret "mobility" differences between populations, assessed femoral cross-sectional properties from preagricultural and agricultural archeological samples of the Georgia Coast of North America (Ruff et al. 1984). Following this trailblazing approach instigated by Ruff et al. (1984), contrasting population mobility became an increasingly popular aspect of studies examining bone functional adaptations in populations characterized by different adaptive strategies (e.g., subsistence activities and life histories) (see review by Ruff and Larsen, Chap. 2).

1.1 Bone Functional Adaptation and Quantifying Mobility

Studies of bone functional adaptations incorporating mobility comparisons often utilize ethnographic accounts (e.g., Bridges 1989; Ruff and Larsen 2001; Stock and Pfeiffer 2001; Carlson et al. 2007) or archeological data (Marchi et al. 2011; Stock et al. 2011; Walker and Churchill 2014), when available, in order to corroborate the inferred activity profiles of the study populations generated from long bone diaphyseal structure. Unfortunately, behavioral resolution in such datasets rarely facilitates quantification of mobility, instead typically offering only comparative qualitative characterizations, such as high versus low distance traveled (e.g., White 1985; Williams 1988). This creates a disparity between the qualitative extent to which mobility is superficially characterized versus the subtleties that can be quantified in the structure of the postcranium. Greater refinement of the concept of mobility is needed.

Efforts to differentiate between logistical (individual) and residential (group) mobility (Kelly 1995; Binford 2001; see also Walker and Churchill, Chap. 12), and between broad substrate differences (e.g., terrestrial versus aquatic/marine: Stock and Pfeiffer 2001; Weiss 2003), have improved resolution in quantifying mobility somewhat. Additional studies (Ruff 1999; Carlson et al. 2007; Marchi 2008) have proposed that substrate complexity, particularly terrain unevenness (e.g., Sparacello et al., Chap. 6; Higgins, Chap. 13; Carlson, Chap. 14; but see Shackelford, Chap. 9), could add explanatory power (in terms of bone functional adaptations) to the concept of mobility, irrespective of distance traveled. This body of work suggests that there is a crucial need, therefore, to link substrate complexity with the behavioral complexity it evokes. Experimental studies have quantified the ground reaction forces and kinematics of gait responses resulting from perturbations created by substrate complexity (e.g., Demes et al. 2006; Daley et al. 2006; Voloshina et al. 2013). A limited number of bone strain studies describe diaphyseal surface strains during gaits where substrate complexity was introduced as well (Burr et al. 1996; Demes et al. 2001; Moreno et al. 2008). Athlete studies (e.g., Jones et al. 1977; Shaw and Stock 2009a, b; Shaw et al., Chap. 4) that document bone functional adaptations associated with varying activities also help clarify form–function relationships that are relevant to the concept of mobility. Ultimately, greater integration of experimental and comparative approaches may offer the best way to continue refining the concept of mobility. In this regard, a recent edited volume dedicated to linking field and laboratory research with respect to nonhuman primate locomotion (D'Août and Vereecke 2011) could be worth consulting.

1.2 Rationale and Organization of the Book

The primary motivation behind this edited volume is to assemble a diverse range of specialists in the study of form–function relationships manifested in the human postcranium, particularly those emphasizing approaches useful for reconstructing human mobility patterns. The goals are to (1) demonstrate the importance of the concept of mobility to understanding bone functional adaptations in the postcranium, (2) compile various factors worthy of consideration in defining the concept of mobility, and (3) provide a consensus, working definition consisting of the factors that are most integral. There are undoubtedly important contributors to the study of mobility and bone functional adaptations whose contributions have been omitted from this volume. We hope that these colleagues understand the space constraints encountered in producing this volume, and we eagerly anticipate their continued contributions in advancing issues raised herein.

This edited volume can be divided into several themes, unequal in representation by chapter counts, but which collectively fulfill the first stated goal: (1) contributions of musculoskeletal markers (MSMs) to studies of mobility (Weiss, Chap. 3); (2) comparisons of human adult morphologies from populations characterized by different activity profiles, e.g., mobility levels (Shaw et al., Chap. 4; Davies and Stock, Chap. 5; Sparacello et al., Chap. 6; Wescott, Chap. 7; Pearson et al., Chap. 8; Shackelford, Chap. 9; Wall-Scheffler, Chap. 10); (3) the role of ontogeny in differentiating lower limb morphologies of human populations characterized by different activity profiles, e.g., mobility levels (Cowgill, Chap. 11); (4) nonhuman models used to enlighten the relationship between bone functional adaptations and mobility (Walker and Churchill, Chap. 12; Higgins, Chap. 13; and Carlson, Chap. 14); and (5) a final chapter (Tamvada, Chap. 15) illustrating a promising future direction for the field.

Ruff and Larsen (Chap. 2) begin the volume with an historical account of studies that use postcranial bone functional adaptations to reconstruct mobility differences, touching on major issues such as the influence of sex-specific activities, terrain, and body shape. The authors recount how the field has responded to challenges in the past (e.g., limitations in data acquisition, scaling of cross-sectional geometric properties, and competing structural influences of activity and body proportions), and indicate future directions of critical importance (e.g., incorporating population history and genetics into evaluations and critically assessing techniques that enable larger sample sizes).

Weiss (Chap. 3), unlike other contributors, focuses on the role of muscles in reconstructing behavioral profiles. She reviews current literature on the expression of MSMs in human upper and lower limbs, and how these features are used for evaluating activity patterns (e.g., mobility) in groups characterized by differences in subsistence strategies, sexual division of labor, and home range terrain. She compares variability in MSM expression amongst upper and lower limbs, noting that the latter display as much or more variation than the former, which is opposite the expectation, since bipedalism, the author reasons, should result in more uniformity in the human lower limb. The author notes that age and body size are known confounding variables with activity levels. The author ends by suggesting that MSMs may be useful for characterizing hominin behavior, and by calling for more research into their etiology.

Two studies, Shaw and colleagues (Chap. 4) and Davies and Stock (Chap. 5), evaluate bone functional adaptations along the entire length of the human lower limb, which is seldom attempted due to constraints in acquiring such extensive datasets. Shaw and colleagues evaluate long bones in proximal and distal segments of the human lower limb in order to assess whether their strength and variability taper proximo-distally in response to tissue economy constraints and energetic trade-offs that appear to drive proximal mass concentration in the limb. Davies and Stock use a solid section model to compare diaphyseal rigidity of these long bones, including the location of minimum bending rigidity, across several human groups characterized by different mobility patterns, body sizes, and body shapes. Shaw and colleagues observe larger section moduli and cortical area in more proximal regions (e.g., proximal femoral diaphysis) and smaller values in more distal regions (e.g., distal tibial diaphysis), with exceptions around the knee joint, corroborating the notion of distal tapering. Variability in these properties, however, does not change along diaphyses, leading the authors to suggest that morphological plasticity is constant along diaphyses, and that morphological constraints (canalization) do not appear to drive the observed tapering. Davies and Stock note that Australian aborigines do not exhibit relatively high bending rigidity, or robusticity, corroborating results of earlier structural studies (e.g., Pearson 2000; Carlson et al. 2007), and contradicting the common perception that Australian aborigines are a highly mobile group. While Ruff and Larsen (Chap. 2) point out a few limitations in using the solid section model adopted by Davies and Stock (Chap. 5), it is worth noting that Davies and Stock corroborate results reported by Shaw and colleagues (Chap. 4) where both studies suggest that the tibial diaphysis may be a better indicator of group-wide mobility differences than the femoral diaphysis.

Two studies, Sparacello and colleagues (Chap. 6) and Shackelford (Chap. 9), emphasize terrain effects, specifically comparing relatively flat versus mountainous terrain. While Sparacello and colleagues examine fibular properties, including relative ratios of fibular/tibial properties, Shackelford focuses attention on femoral and tibial midshaft properties. Sparacello and colleagues report high fibular rigidity, including relative rigidity, in Late Upper Paleolithic, Neolithic, and Iron Age groups, despite the latter group being considered more sedentary than the former groups. The authors suggest that despite relative sedentism (i.e., low mobility) of the latter group, all three groups inhabited areas characterized by uneven terrain, possibly indicating similarly enhanced leg strength from frequent inversion/eversion of the foot while moving on uneven terrain. Shackelford, on the other hand, reports comparatively gracile femoral and tibial diaphyses in a Late Pleistocene Asian sample, which inhabited areas of more uneven terrain than the more robust northern African samples. Shackelford suggests that comparatively greater mechanical efficiency at the hips and knees of the Asian sample may be an alternative mechanism for counteracting elevated loading of the lower limbs associated with high mobility on uneven terrain.

Three studies, Wescott (Chap. 7), Pearson and colleagues (Chap. 8), and Cowgill (Chap. 11), examine several potential factors that influence the shape ratio (I_{max}/I_{min})

and mobility index (I_x/I_y) at selected sites on human femoral and tibial diaphyses (i.e., midshafts). All make a case for the problem of equifinality[1] in shape ratios and mobility indices. Wescott compares femoral diaphyses of ambulatory individuals with those of impaired (disabled) individuals and also documents secular trends in femoral properties of modern North American populations. Ultimately, Wescott suggests that one potential solution for addressing the problem of equifinality in these ratios is to study mobility differences using a whole limb approach, incorporating comparisons of multiple cross-sectional properties when possible. Pearson and colleagues (Chap. 8) and Cowgill (Chap. 11), among other issues, examine confounding effects of body shape (i.e., bi-iliac breadth) and activity patterns (i.e., mobility) expressed in femoral and tibial midshaft shape and mobility indices. Pearson and colleagues find inconsistent (weak) correlations between femoral midshaft shape and bi-iliac breadth, while Cowgill observes stronger ties, though it is important to note that the two studies use different samples. Cowgill, in particular, observes evidence of mobility indices differentiating earlier (less than 6 years of age) in some populations (Point Hope) than others, which she suggests is attributable to the cold-adapted body proportions specific to the former. Pearson and colleagues observe low correlations between femoral and tibial midshaft shapes, suggesting that these two locations may record different activities (i.e., fast gaits preferentially affect the former location and slow gaits preferentially affect the latter location). Both chapters call for investigating midshaft shape ratios and mobility indices using mobility, body shape, and other determinants.

Wall-Scheffler (Chap. 10) reviews the literature on burden carrying, inclined walking, and energetics. Multiple lines of evidence support links between energetic savings or performance during specific behaviors and morphological variation expressed in populations. For example, a shorter tibia is correlated with energy efficiency during uphill walking, while a longer tibia is correlated with increased speed along flat terrain. Predominantly through her own work, Wall-Scheffler has documented that a wider pelvis for a given mass (e.g., typically a female trait) provides more energy savings during front and back burden carrying. The author also points out that during burden carrying a wider pelvis allows an individual to vary walking speed without substantially adjusting cost of transport (i.e., incurring a metabolic penalty). The evolutionary relevance to mobility could be profound, as the author points out, since women with broad pelves would be able to walk together in groups during burden carrying, but still adhere to their own individual optimal speed.

Three studies, Walker and Churchill (Chap. 12), Higgins (Chap. 13), and Carlson (Chap. 14), use nonhuman models in order to address questions relevant to linking human mobility and bone functional adaptations. Walker and Churchill use ranging data and group aggregate mass of social carnivores (grey wolves) to build a model for predicting home range areas of variably-sized Neandertal social groups. The authors suggest that even small groups of Neandertals (less than 33 individuals)

[1] Ludwig von Bertalanffy (1956) defined equifinality as the same final state arising from different initial states. He originally used the term in helping found general systems theory. More recently, the term has been frequently applied in the study of taphonomic processes (see Lyman 2004).

would have required large territories (1,400–5,400 km^2), which they point out corroborates ranging estimates using lithic raw material procurement patterns. Agreement between the two lines of evidence suggests, according to the authors, that lithic raw material procurement was embedded in subsistence mobility during the European Mousterian. Higgins (Chap. 13) compares metacarpal structure of similarly-mobile bovid species characterized by flat terrain, mountainous terrain, and mixed terrains in order to assess the effect of sloped terrain on bone functional adaptations. The author notes that bovid species characterized by mountainous terrain routinely exhibit elevated anteroposterior (AP) and mediolateral (ML) bending rigidity, with perhaps greater relative increases in the ML direction. In contrast, human tibiae from sampled individuals occupying non-flat terrain, according to the author, usually exhibit relatively more enhanced AP bending rigidity than ML bending rigidity. Higgins suggests this is probably because of lateral buttressing by the fibula in humans, though this is not directly assessed in the study. As Sparacello and colleagues (Chap. 6) demonstrated, a structural analysis of fibulae from these human groups could prove to be enlightening. Carlson (Chap. 14) uses C57BL/6J mice to assess the effect of a specific behavior (i.e., change in direction), resulting from terrain complexity (i.e., obstacle avoidance), on bone functional adaptations in the femoral diaphysis. Structural differences in femoral diaphyses corresponding to presumed greater ML orientation of loading regimes in the experimental group engaging in more turning are observed. The author suggests that these experimental data support the notion that greater terrain complexity not only in the vertical direction, but in the horizontal direction (e.g., obstacle frequency on a landscape) would be worth evaluating when comparing bone functional adaptations of variably-mobile human groups.

Finally, Tamvada (Chap. 15) provides a glimpse of an exciting new application for finite element (FE) modeling. The author uses a finite element analysis (FEA) to explore structural integrity of the human femur. While validating an FE model presents a few logistical obstacles, the opportunity to calculate stresses and strains arising during an array of specific behaviors, or associated with specific kinematic variables (e.g., excursion angles), offers refreshing opportunities for understanding bone functional adaptations at a level that is seldom obtainable. For example, an FEA approach could permit assessment of stresses or strains associated with specific behavioral (gait) responses to elements of terrain complexity. This could offer a particularly powerful means of insight into documenting and understanding bone functional adaptations of human populations characterized by even the subtlest differences in mobility.

1.3 Conclusions and Future Directions

In summarizing the contributed chapters in this edited volume, the second and third goals stated above are fulfilled. The most commonly-adopted criteria in defining mobility are: (1) cumulative behavior over an individual's lifetime, (2) a focus on the lower limb, and (3) overall distance traveled (e.g., see Shaw et al., Chap. 4; Sparacello et al., Chap. 6; Pearson et al., Chap. 8; Cowgill, Chap. 11; and Higgins, Chap. 13). Other chapters offer definitions that include two of these three criteria, often excluding the lifetime cumulative behavior focus (e.g., see Shackelford, Chap. 9; Carlson, Chap. 14; and Tamvada, Chap. 15), or the lower limb focus (e.g., see Davies and Stock, Chap. 5). Wescott (Chap. 7) uses a logistic mobility definition, while Walker and Churchill (Chap. 12) employ both a logistic and residential mobility definition to examine different scales of mobility within their study. The concept of logistic mobility would seem to be consistent with the three most frequently adopted criteria noted above, but perhaps with the potential exclusion of the lifetime cumulative behavior criterion. It is worth noting that the chapter by Walker and Churchill stands apart from others in the volume due to its focus on both a unit of comparison at the population level (residential mobility) and at the individual level (logistic mobility). Bone functional adaptations are perhaps less useful in informing about residential mobility than in informing about logistic mobility, as individual variability in bone functional adaptations is better suited to association with individual variability in the latter. Other chapters, such as those by Shaw and colleagues (Chap. 4), Wall-Scheffler (Chap. 10), and Carlson (Chap. 14), incorporate additional criteria in defining mobility, for example, terrain complexity (e.g., elevation change and lateral movements).

Parallels between defining human mobility and defining positional behavior in observational studies of free-ranging primates offer a few intriguing points worthy of consideration. Prost (1965:1202) originally defined positional behavior as the "study of how and when an animal establishes particular spatial relations between his body mass and his physical environment" in response to existing disorder in classifications of primate locomotion. Prost argued that positional behaviors should be exhaustively categorized into dynamic (locomotor) and static (postural) states, with the former being most relevant to the concept of mobility adopted throughout this edited volume, and defined by Prost as the summary displacement of body mass. Subsequent attempts to standardize and refine classifications of positional (locomotor) behaviors into more discrete categories (i.e., Hunt et al. 1996) focused on spatial relationships between body segments, the center of body mass, and substrates/superstrates. Even this exhaustive attempt, however, has required additional fine-tuning, often due to species-specific locomotor habits (Walker 1998; Thorpe and Crompton 2006). Despite these additional efforts, there are still behavioral subcategories (e.g., turning) that remain underemphasized in classification schemes of positional behavior. Despite the continual need for adjusting its categories and subcategories, the adoption of the standardized positional behavior classificatory scheme ensured emphasis was placed on interactions between the body, its segments, and the environment, which allowed field behavioralists and morphologists to begin examining broader issues by comparing positional (locomotor) behavior profiles across groups of primates, and across different studies. It would seem that parallel benefits could come from standardizing the concept of human mobility in a similar fashion.

To this end, we suggest that standardizing an explicit definition of the concept of mobility (just as positional behavior eventually became explicitly defined) could be fruitful for strengthening behavioral inferences (e.g., activity patterns) originating from quantification of bone functional adaptations. Chapters in this edited volume converge on three primary criteria, which we suggest would be a good starting point for such a working definition of mobility: (1) cumulative behavior over a lifetime, though not necessarily implicating a life-long linear response to mechanical loading (i.e., age-equivalent samples should be compared when possible; see Cowgill, Chap. 11; Ruff et al. 2006); (2) overall distance traveled, though clearly incorporating terrain complexity is of growing importance; and (3) priority placed on use of the lower limb, which predominates as the instrument of human movement. Notably, amalgamation of the whole limb rather than consideration of discrete segments in isolation may help overcome the problem of equifinality for specific properties (e.g., diaphyseal midshaft shape and mobility indices).

While standardizing the concept of mobility going forward may benefit cross-study comparisons, just as the creation of a positional behavior classification scheme (Prost 1965; Hunt et al. 1996) enhanced opportunities to compare behavioral repertoires of different free-ranging primate species, it is important to leave open the possibility for study-specific adjustments. For example, comparisons of aquatic/marine mobility require incorporating upper limb comparisons. Also, substrate complexity clearly offers an additional, rich source of information for interpreting bone functional adaptations, particularly as separate elements of this complexity continue to be assessed. Our hope is that this edited volume stimulates further attempts to refine the concept of mobility, and of equal importance, to determine how mobility can inform on the activity patterns and substrate use visible through comparisons of bone functional adaptations. The value of experimental determinism/modeling for linking morphology and behavior (mobility), as opposed to "common sense" arguments, should not be overlooked, nor should the importance of integrating naturalistic conditions of environments occupied by the study populations be underappreciated. Unquestionably, much exciting work lies ahead!

References

Amtmann E (1971) Mechanical stress, functional adaptation and the variation structure of the human femur diaphysis. Ergeb Anat Entwicklungsgesch 44:1–89

Binford LR (2001) Constructing frames of reference—and analytical method for archaeological theory building using hunter-gatherer and environmental data sets. University of California Press, Berkeley, CA

Bridges PS (1989) Changes in activities with the shift to agriculture in the southeastern United States. Curr Anthropol 30:385–394

Burr DB, Milgrom C, Fyhrie D, Forwood M, Nyska M, Finestone A, Hoshaw S, Saiag E, Simkin A (1996) *In vivo* measurement of human tibial strains during vigorous activity. Bone 18:405–410

Carlson KJ, Grine FE, Pearson OM (2007) Robusticity and sexual dimorphism in the postcranium of modern hunter-gatherers from Australia. Am J Phys Anthropol 134:9–23

Carter DR, Beaupré GS (2001) Skeletal function and form—mechanobiology of skeletal development, aging, and regeneration. Cambridge University Press, Cambridge

D'Août K, Vereecke EE (eds) (2011) Primate locomotion—linking field and laboratory research. Springer, New York

Daley MA, Usherwood JR, Felix G, Biewener AA (2006) Running over rough terrain: guinea fowl maintain dynamic stability despite a large unexpected change in substrate height. J Exp Biol 209:171–187

Demes B, Qin Y, Stern JT Jr, Larson SG, Rubin CT (2001) Patterns of strain in the Macaque tibia during functional activity. Am J Phys Anthropol 116:257–265

Demes B, Carlson KJ, Franz TM (2006) Cutting corners: the dynamics of turning behaviors in two primate species. J Exp Biol 209:927–937

Devlin MJ, Lieberman DE (2007) Variation in estradiol level affects cortical bone growth in response to mechanical loading in sheep. J Exp Biol 210:602–613

Devlin MJ, Cloutier AM, Thomas NA, Panus DA, Lotinun S, Pinz I, Baron R, Rosen CJ, Bouxsein ML (2010) Caloric restriction leads to high marrow adiposity and low bone mass in growing mice. J Bone Miner Res 25:2078–2088

Hunt KD, Cant JGH, Gebo DL, Rose MD, Walker SE (1996) Standardized descriptions of primate locomotor and postural modes. Primates 37:363–387

Jones HH, Priest JD, Hayes WC, Tichenor CC, Nagel DA (1977) Humeral hypertrophy in response to exercise. J Bone Joint Surg 59A:204–208

Judex S, Carlson KJ (2009) Is bone's response to mechanical signals dominated by gravitational loading? Med Sci Sports Exerc 41:2037–2043

Kelly RL (1995) The foraging spectrum—diversity in hunter-gatherer lifeways. Smithsonian Institution Press, Washington, DC

Kummer B (1959) Bauprinzipien des Säugetierskeletes. Georg Thieme, Stuttgart

Lyman RL (2004) The concept of equifinality in taphonomy. J Taphonomy 2:15–26

Marchi D (2008) Relationships between lower limb cross-sectional geometry and mobility: the case of a Neolithic sample from Italy. Am J Phys Anthropol 137(2):188–200

Marchi D, Sparacello VS, Shaw CN (2011) Mobility and lower limb robusticity of a pastoralist Neolithic population from North-Western Italy. In: Pinhasi R, Stock JT (eds) Human bioarchaeology of the transition to agriculture. Wiley, New York, pp 317–346

Martin RB, Burr DB, Sharkey NA (1998) Skeletal tissue mechanics. Springer, New York

Meyer GH (1867) Die Architektur der Spongiosa. Anat Physiol Wiss Med 34:615–628

Moreno CA, Main RP, Biewener AA (2008) Variability in forelimb bone strains during non-steady locomotor activities in goats. J Exp Biol 211:1148–1162

Mott RL (1996) Applied strength of materials, 3rd edn. Prentice Hall, Englewood Cliffs, NJ

Pauwels F (1968) Beitrag zur funktionellen Anpassung der Corticalis der Röhrenknochen. Untersuchung an drei rachitisch deformierten Femora. Z Anat Entwicklungsgesch 127:121–137

Pauwels F (1980) Biomechanics of the locomotor apparatus: contributions on the functional anatomy of the locomotor apparatus. Springer, Berlin

Pearson OM (2000) Activity, climate, and postcranial robusticity—implications for modern human origins and scenarios of adaptive change. Curr Anthropol 41:569–607

Pearson OM, Lieberman DE (2004) The aging of Wolff's "Law": ontogeny and responses to mechanical loading in cortical bone. Yearb Phys Anthropol 47:63–99

Prost JH (1965) A definitional system for the classification of primate locomotion. Am Anthropol 67:1198–1214

Robling AG (2009) Is bone's response to mechanical signals dominated by muscle forces? Med Sci Sports Exerc 41:2044–2049

Roesler H (1981) Some historical remarks on the theory of cancellous bone structure (Wolff's law). In: Cowin SC (ed) Mechanical properties of bone. ASME, New York, pp 27–42

Roux W (1881) Der zuchtende Kampf der Teile, oder die 'Teilauslese' im Organismus (Theorie der 'funktionelle Anpassung'). Wilhelm Englemann, Leipzig

Ruff CB (1999) Skeletal structure and behavioral patterns of prehistoric Great Basin populations. In: Hemphill BE, Larsen CS (eds) Prehistoric lifeways in the Great Basin wetlands: bioarchaeological reconstruction and interpretation. University of Utah Press, Salt Lake City, pp 290–320

Ruff CB, Larsen CS (2001) Reconstructing behavior in Spanish Florida: the biomechanical evidence. In: Larsen CS (ed) Bioarchaeology of Spanish Florida: the impact of colonialism. University Press of Florida, Gainesville, pp 113–145

Ruff CB, Larsen CS, Hayes WC (1984) Structural changes in the femur with the transition to agriculture on the Georgia coast. Am J Phys Anthropol 64:125–136
Ruff CB, Holt BH, Trinkaus E (2006) Who's afraid of the big bad Wolff? Wolff's law and bone functional adaptation. Am J Phys Anthropol 129:484–498
Shaw CN, Stock JT (2009a) Intensity, repetitiveness, and directionality of habitual adolescent mobility patterns influence the tibial diaphysis morphology of athletes. Am J Phys Anthropol 140:149–159
Shaw CN, Stock JT (2009b) Habitual throwing and swimming correspond with upper limb diaphyseal strength and shape in modern human athletes. Am J Phys Anthropol 140:160–172
Stock JT, Pfeiffer S (2001) Linking structural variability in long bone diaphyses to habitual behaviors: foragers from the southern African Later Stone Age and the Andaman Islands. Am J Phys Anthropol 115:337–348
Stock JT, O'Neill MC, Ruff CB, Zabecki M, Shackelford L, Rose JC (2011) Body size, skeletal biomechanics, mobility and habitual activity from the late Palaeolithic to the mid-dynastic Nile valley. In: Pinhasi R, Stock JT (eds) Human bioarchaeology of the transition to agriculture. Wiley, Chichester, pp 347–370
Thorpe SK, Crompton RH (2006) Orangutan positional behavior and the nature of arboreal locomotion in Hominoidea. Am J Phys Anthropol 131:384–401
Voloshina AS, Kuo AD, Daley MA, Ferris DP (2013) Biomechanics and energetics of walking on uneven terrain. J Exp Biol 216:3963–3970
von Bertalanffy L (1956) General system theory. Gen Syst 1:1–10
Walker SE (1998) Fine-grained differences within positional categories: a case study of Pithecia and Chiropotes. In: Strasser E, Feagle J, Rosenberger A, McHenry HM (eds) Primate locomotion—recent advances. Plenum, New York, pp 31–43
Walker CS, Churchill SE (2014) Territory size in canis lupus: implications for neandertal mobility. In: Carlson KJ, Marchi D (eds) Reconstructing Mobility: Environmental, Behavioral, and Morphological Determinants. Springer, New York
Weiss E (2003) The effects of rowing on humeral strength. Am J Phys Anthropol 121:293–302
White NG (1985) Sex differences in Australian Aboriginal subsistence: possible implications for the biology of hunter-gatherers. In: Ghesquiere J, Martin RD, Newcombe F (eds) Human Sexual Dimorphism. Taylor and Francis: London, pp. 323–361
Williams E (1988) Complex Hunter-Gatherers—a Late Holocene example from temperate Australia. British Archaeological Reports International Series 423, Oxford
Wolff J (1892) Das Gesetz der Transformation der Knochen. Springer, Berlin

Chapter 2
Long Bone Structural Analyses and the Reconstruction of Past Mobility: A Historical Review

Christopher B. Ruff and Clark Spencer Larsen

Abstract The use of long bone structural analysis to reconstruct past human behavior had its origins in the 1970s, although it was only in the last 30 years that true population-level comparisons began to be carried out. Since then, several dozen studies of archaeological and paleontological samples have been completed, illustrating the complexity as well as some consistencies in the relationship between bone morphology and mobility. Bone cross-sectional shape rather than relative size appears to be more clearly related to mobility differences. This is particularly true in comparisons between males and females within the same population. Terrain has a strong effect on relative strength of the lower limb bones. Body shape differences also have an effect on structural properties, and must be factored into comparisons. New methods of noninvasively acquiring structural properties promise even greater accessibility of information and larger samples in the future, although caution must be applied when extrapolating from approximations to true section properties.

Keywords Biomechanics • Long bone • Mobility • Sexual dimorphism • Body size • Body shape

C.B. Ruff (✉)
Center for Functional Anatomy and Evolution, Johns Hopkins University School of Medicine, 1830 E. Monument Street, Baltimore, MD 21205, USA
e-mail: cbruff@jhmi.edu

C.S. Larsen
Department of Anthropology, The Ohio State University, 4034 Smith Laboratory, 174 W. 18th Avenue, Columbus, OH 43210-1106, USA
e-mail: larsen.53@osu.edu

K.J. Carlson and D. Marchi (eds.), *Reconstructing Mobility: Environmental, Behavioral, and Morphological Determinants*, DOI 10.1007/978-1-4899-7460-0_2,

2.1 Earlier Studies

Assessing long bone strength through the analysis of diaphyseal cross-sectional geometry can be traced back as far as Galileo (Galilei 1638). The first full-scale analysis of an actual human long bone (a femur) is probably that of Koch (1917), almost 100 years ago. Endo and Kimura first applied this technique in a human paleoanthropological or archaeological context in 1970, in their comparative analysis of the Amud 1 Neandertal tibia. Several mechanically oriented analyses of human samples rather than individuals soon followed (Amtmann 1971; Kimura 1971; Minns et al. 1975; Lovejoy et al. 1976; Martin and Atkinson 1977; Miller and Piotrowski 1977; Lovejoy and Trinkaus 1980). With the exception of Lovejoy and coworker's studies of modern and Neandertal tibias, none of these were explicitly comparative, being mainly concerned with general mechanical modeling and/or age and sex differences within samples. A number of other investigations of archaeological samples were carried out during the same general time period (Dewey et al. 1969; Van Gerven et al. 1969; Van Gerven and Armelagos 1970; Van Gerven 1973; Carlson et al. 1976) but with a focus on cortical thickness and areal measurements and systemic age-related bone loss (osteoporosis) rather than mechanical effects.

Most of these early studies were limited in size and scope by two factors: the need to destructively sample (i.e., section) specimens in order to obtain cross sections, and the use of manual point-counting methods to input bone distribution information. The development and wider availability of computed tomography (CT) for anthropological research (e.g., Jungers and Minns 1979; Tate and Cann 1982; Sumner et al. 1985; Ruff and Leo 1986) helped to alleviate the first problem. Other new noninvasive techniques, including external molding combined with multiplane radiography (Trinkaus and Ruff 1989; O'Neill and Ruff 2004), provided additional ways to accurately reconstruct cross sections without physical sectioning of specimens. Biplanar radiographs alone are not adequate for reconstructing human lower limb long bone cross sections (O'Neill and Ruff 2004), even when corrected for eccentricity of the endosteal and periosteal contours (Ohman 1993). At the same time, new automated and semi-automated methods for deriving section properties from images were developed (Nagurka and Hayes 1980; Sumner et al. 1985). These made possible, for the first time, truly demographic studies of large samples (Ruff and Hayes 1983a, b; Sumner 1984).

The first controlled population-level comparison of long bone structural properties was carried out by the present authors on archaeological samples from the Georgia coast (Ruff et al. 1984). Cross-sectional properties of femora from preagricultural (2200 B.C.–A.D. 1150) and agricultural (A.D. 1150–1550) groups were compared. The agricultural group showed a decline in all properties (see Fig. 2.1), with many of the declines remaining significant even after adjustment for different body sizes (bone lengths). The agricultural group also showed an increase in circularity (Fig. 2.1). Both results were attributed to a decline in activity levels in the agricultural group. Interestingly, when compared to similar data from the Pecos Pueblo, New Mexico sample (Ruff and Hayes 1983a, b), the Georgia coast preagricultural

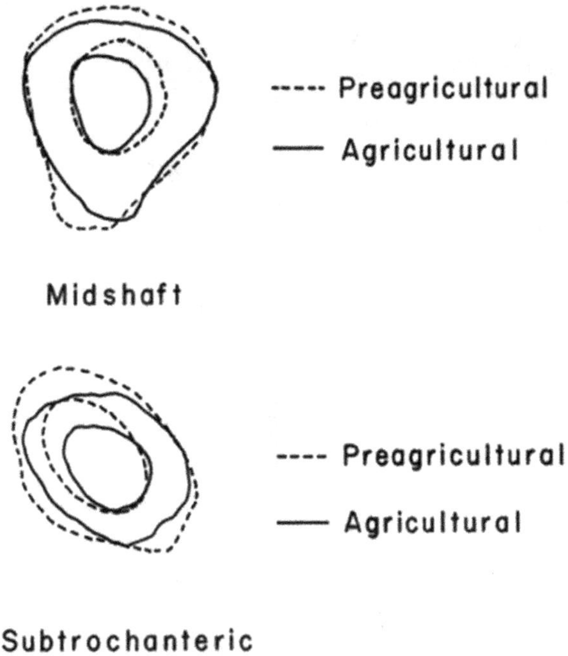

Fig. 2.1 Cross sections of femora from pre-contact preagricultural and agricultural Georgia coast samples, scaled to equal bone lengths. (Reproduced with permission from Ruff et al. 1984)

group was more similar in overall robusticity, or relative size, to Pecos (which was agricultural), while the Georgia coast agricultural group was more similar to Pecos in bone shape, i.e., relative bending rigidity in different planes. This was interpreted to reflect relatively high overall mechanical loadings at Pecos, due to the rugged terrain (a hypothesis later confirmed by broader comparisons: Ruff 1999), but more similarity in *types* of activities between the two agricultural samples, specifically, lower levels of overall mobility. There was also some evidence for more diachronic change in bone shape among Georgia coast males than among females, suggesting greater effects of subsistence strategy on activities among males. Sex differences in femoral and tibial structure had also been noted within the Pecos sample (Ruff and Hayes 1983b), and tentatively attributed to a combination of sexual dimorphism in body shape (wider hips in females) and activity differences between the sexes (males more mobile).

This theme was taken up again in broader comparisons between the Pecos sample and a modern US autopsy sample, as well as a number of other femoral and tibial archaeological samples (Ruff 1987). Males were shown to have relatively greater anteroposterior (AP) strength in the region about the knee, and females to have relatively greater mediolateral (ML) strength in the region near the hip. The AP/ML strength difference near the knee declined from hunter-gatherers to agriculturalists to modern industrial samples, while sexual dimorphism near the hip showed no trend. External breadth measurements, available for a wider sampling of populations, exhibited similar patterns. The decline in sexual dimorphism near the knee (including the femoral and tibial midshafts) was due to a decrease in male AP/ML

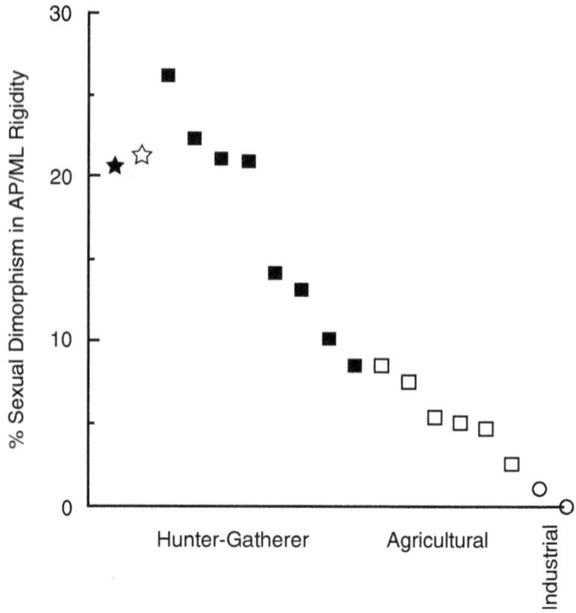

Fig. 2.2 Sexual dimorphism in femoral midshaft AP/ML bending rigidity in relation to subsistence strategy [(male–female)/female × 100]. *Filled* and *open squares*: Native North American hunter-gatherers and agriculturalists, respectively; *open circles*: industrial samples; *filled* and *open stars*: Neandertals and Upper Paleolithic humans, respectively. Each point represents a population (or in the case of the two earliest groups, sample). (Reproduced with permission from Ruff 2005)

strength, with no change in females. AP bending rigidity or strength in this region is probably related to the degree of flexion of the knee and applied force of the knee flexors and extensors during locomotion, all of which should increase with more rapid locomotion over longer distances, i.e., increased mobility (Ruff 1987, 2005). These results were thus consistent with ethnographic data indicating a decline in sexual division of labor through the same subsistence changes, specifically, a decline in male mobility. Later comparisons incorporating a larger number of population samples further supported this conclusion (Ruff 1999, 2005) (see Fig. 2.2). Interestingly, Neandertal and early anatomically modern (Upper Paleolithic) humans showed similar levels of sexual dimorphism in bone shape to modern hunter-gatherers (Fig. 2.2), suggesting a similar division of labor. The relatively constant sexual dimorphism in bone shape near the hip is consistent with observed sexual dimorphism in pelvic shape and predictions based on biomechanical models of this region (Ruff 1995).

The Georgia coast study was later extended to include more population samples in the region, including several living during the Spanish contact period, and the humerus as well as the femur, with a total sample size of 168 femora and 189 humeri (Fresia et al. 1990; Ruff and Larsen 1990, 2001). This broader sampling revealed several interesting patterns and trends, with implications for reconstructing mobility and other behavioral characteristics. First, the reduction in overall femoral robusticity (strength relative to body size) and midshaft AP/ML bending rigidity observed earlier between preagricultural and agricultural samples did not continue in a uniform manner in the contact period. Both sexes actually increased slightly in overall femoral rigidity in missionized Guale samples, and in humeral rigidity among males, although not among females. This was interpreted to reflect an overall increase in workload as well as possible increases in relative body mass under mission conditions. However, AP/ML bending rigidity of the midshaft femur continued to decline in contact period females, while showing a slight initial increase in males followed by a decrease. The early missionized males also showed a large increase in variability in this index suggestive of a bimodal distribution. Together this evidence suggested that some males from the early mission period became less mobile and some increased in mobility, while females as a whole became less mobile. These observations are consistent with historical evidence for enforced long-distance travel by some male Guale as part of the Spanish *repartimiento* labor system, with other males (and all females) not involved. Thus, sexual dimorphism in bone shape was quite high on average among mission period Guale, similar to that of many hunter-gatherer populations, possibly because sexual "division of labor" (in this case, forced labor) also increased (on average). Interestingly, a contemporary, geographically adjacent but less acculturated sample of Timucua did not show this same increase in male diversity or sexual dimorphism, as would be predicted given the less drastic effects of missionization in this population (they also had relatively lower overall robusticity).[1] The somewhat disparate temporal patterns for the upper and lower limbs in male and female Guale also suggests different behaviors, i.e., work requirements, during the mission period, with males engaging in heavier or more frequent lifting activities. Again, the less acculturated Timucua did not show this pattern.

Several other comparative studies of Native North American archaeological samples were carried out during this time period, from geographic regions ranging from the Delaware coast (Robbins et al. 1989) to the Tennessee River Valley (Bridges 1989), Great Plains (Ruff 1994a), New Mexico (Brock and Ruff 1988), and the Great Basin (Ruff 1999). The effect of subsistence strategy and relative mobility on long bone cross-sectional geometry was a major theme of each of these studies. One of the most consistent patterns observed was a decline in sexual dimorphism in midshaft femoral shape with increased sedentism, supporting the general model presented above (Fig. 2.2). Wescott (2006) reported similar findings for a number of other North American archaeological and modern samples. In some other respects, these various studies showed a fair degree of heterogeneity in results. For example, Bridges (1989) found an increase in relative strength at some skeletal locations in the femur and humerus between preagricultural and agricultural samples in the Tennessee River Valley, unlike the temporal decline found in the Georgia coast samples (and see Larsen and Ruff 2011). However, she did find that circularity of sections increased with the adoption of agriculture, which is similar to the Georgia coast findings. As with the original comparisons with the Pecos sample

[1] Our earlier study (Ruff and Larsen 2001) assumed that the "Yamassee" sample represented a group who had emigrated to Amelia Island, Florida, from South Carolina. New biodistance and mortuary evidence indicates the likelihood that the series is from an early Timucua population, the descendants of a native tribe indigenous to Amelia Island (see Stojanowski 2013).

(Ruff et al. 1984), then, bone shape appeared to be a better indicator of types of activity (including mobility) than overall cross-sectional size. The combined effects of nutrition and behavior on cross-sectional morphology were emphasized in a study of three Great Basin samples (Ruff 1999). These samples had relatively thin, but expanded long bone cortices, leading to high levels of bending rigidity. This morphology may reflect the combination of a relatively poor diet, leading to endosteal resorption of bone (Garn et al. 1969), combined with a very vigorous lifestyle, which would favor periosteal expansion (Ruff and Hayes 1988). In broader comparisons with other Native North American samples, this study also demonstrated a marked effect of terrain on relative rigidity of the femur, with femora from mountainous regions (including the Pecos as well as Great Basin samples) exhibiting greater rigidity relative to body size than those from plains or coastal regions.

2.2 Other Factors: Terrain and Body Shape

The effects of different types of "terrain" on long bone robusticity were also addressed in two later studies (Stock and Pfeiffer 2001; Weiss 2003). Stock and Pfeiffer (2001) compared relative rigidity of the femur and humerus in Andamanese Islanders and Later Stone Age South Africans. Both groups were highly "mobile," but in different ways: via marine transport (canoeing) in the Andamanese and via long-distance terrestrial travel in the South Africans. Consistent with these behavioral differences, the Andamanese showed greater relative rigidity in the humerus and clavicle, and the South Africans in the femur, tibia, and metatarsal (see Fig. 2.3). AP/ML rigidity was not assessed directly, but the Andamanese did show increased circularity

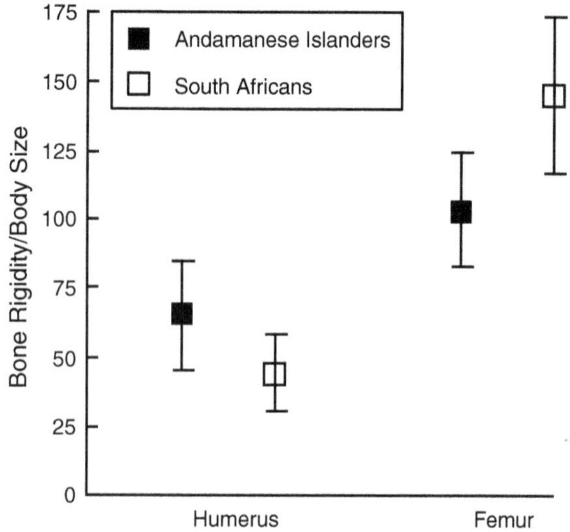

Fig. 2.3 Humeral and femoral overall rigidity (polar second moment of area) relative to body size in Andamanese Island and South African Stone Age samples (mean ± SD). (Data from Stock and Pfeiffer 2001)

(reduced maximum/minimum bending rigidity) in the midshaft femur compared to the South African sample, with a much more marked difference among males.

Weiss (2003) compared a combined measure of humeral diaphyseal robusticity in five different Native North American samples with varying degrees of dependence on water transport, including ocean-rowing Aleut and British Columbian samples, river-rowing Georgia coast samples, and the non-rowing Pecos Pueblo sample. Males were responsible for rowing in those samples that practiced this form of transport. Consistent with expectations, males showed a progressive increase in humeral robusticity with increased dependence on water transport, while females showed no similar trend, except that Aleut females were the most robust. The author speculated that this latter result may be attributable to the other vigorous tasks performed by Aleut females, such as butchering whales, although factors such as overall body build related to climatic effects were considered possible contributors.

How to standardize long bone structural properties for body size and shape differences is a long-standing issue that has important implications for comparative studies (Ruff 1984, 2000b; Ruff et al. 1993; Trinkaus et al. 1999a; Polk et al. 2000; Shaw and Stock 2011). It is actually part of a more general issue regarding allometry, or size-shape relationships, within long bones, again first broached centuries ago (Galilei 1638; also see, e.g., Schultz 1953; McMahon 1973; Alexander et al. 1979). In earlier studies, bone length or powers of bone length were often used to size-standardize cross-sectional diaphyseal dimensions (Ruff et al. 1984, 1993; Bridges 1989; Ruff 1999; Pearson 2000). This was based in part on the strong allometric scaling relationships between bone length and cross-sectional size observed within human samples (Ruff 1984; Ruff et al. 1993). However, it was also realized that this procedure carries an implicit assumption that body proportions, i.e., bone or limb length relative to body size, are invariant. When this was manifestly not true, for example, in comparisons between Neandertals and modern humans, a correction factor was applied (Ruff et al. 1993). Even in comparisons between different modern (Holocene) samples, indiscriminant use of bone length alone to standardize properties can lead to misleading results (Ruff 2000a). For mechanically oriented studies, and based on a beam model of the diaphysis (Lovejoy et al. 1976; Ruff and Hayes 1983a; Gere and Goodno 2013), the best measure of body size is body mass, together with some measure of beam length for bending and torsional rigidity/strength parameters. For strength measures (i.e., section moduli), body mass * beam length is appropriate; for rigidity measures (i.e., second moments of area), body mass * beam length^2 should be used (see Ruff 2008 for description of properties). These factors apply to the upper as well as lower limb (Ruff 2000b). For most long bone diaphyseal locations, "beam length" here can be taken as bone length; however, for the proximal femur, body (maximum pelvic, or bi-iliac) breadth is a better measure of beam length (Ruff 2000b).

The importance of accounting for body shape variation in reconstructing mobility patterns was illustrated in a study of the "Ice Man," the late Neolithic body discovered in the Tyrolean Alps (Ruff et al. 2006b). The Ice Man's femoral strength relative to body size was about average for European Neolithic males, but his tibial relative strength was very high. In terms of cross-sectional shape, his femur was slightly rounder, i.e., ML reinforced, than average for Neolithic males, but his tibia had a high AP/ML bending strength ratio. Together these results imply relatively higher mechanical loads on his tibia, particularly AP bending loads, and relatively higher ML bending loads on his femur. This overall morphology is consistent with a combination of high mobility, increasing the AP loadings on his tibia, with his relatively "stocky" body shape, i.e., high body (bi-iliac) breadth to stature ratio, which increases ML bending of the more proximal lower limb (Ruff 1995). This illustrates that body shape must be factored into interpretations of mobility based on structural analyses. Similar conclusions were reached in a broader comparison involving a wide range of archaeological samples (Shaw and Stock 2011), and an even broader analysis of Pleistocene *Homo* specimens (Trinkaus and Ruff 2012). In fact, the relationship between long bone robusticity and body shape can be used to work backwards from cross-sectional geometry to reconstruction of body shape in more incomplete specimens (Trinkaus et al. 1999a, b).

Because climate has strong effects on body shape, in humans and other animals (Mayr 1963; Roberts 1978; Ruff 1994b), this is related to another potentially significant issue: how the effects of climate might modulate the relationship between long bone diaphyseal structure and activity patterns. Studies that have explicitly examined this issue (Pearson 2000; Stock 2006) have found, not surprisingly, that climate and various indices of relative long bone strength or rigidity are in fact correlated. However, these findings can largely be explained as indirect effects of climate on body shape, which then affects mechanical loadings (particularly of the lower limbs). In one study (Pearson 2000), diaphyseal breadths were divided by bone length to size-standardize them. As noted above, this systematically biases results for populations with different body proportions, i.e., it will systematically underestimate body mass in relatively stocky, short-limbed (e.g., arctic) populations, which have more mass per unit length compared to equatorial populations, and vice versa. Therefore, it was inevitable that "climate" would be strongly correlated with "robusticity" in this analysis, but this does not indicate any direct effect of climate on mechanical loadings per se. Incorporation of body mass is necessary in order to appropriately size-standardize structural properties. This was done in the other study (Stock 2006), which found some strong partial correlations between lower limb bone shape indices and degree of terrestrial mobility when controlling for average ("effective") temperature, especially among males. Significant partial correlations between temperature and lower limb bone shape and relative strength mainly occurred in the proximal femur. However, as noted above, the most appropriate measure of "beam" length in this region is body (bi-iliac) breadth (Ruff 2000b), and in this study bone lengths were used exclusively as beam lengths. Therefore, this result also likely simply reflects a climatic effect on relative body breadth, which shows very strong ecogeographic clines among humans (Roberts 1978; Ruff 1994b). There is, in fact, no plausible physiological mechanism that would directly link climatic variation with variation in long bone mechanical parameters. Thus, climate per se should not be a confounding factor in mobility reconstructions.

2.3 Recent Studies

The past decade has witnessed an explosion of interest in using long bone structural analyses to address issues concerning mobility and other activity patterns among past populations (Holt 2003; Stock and Pfeiffer 2004; Kimura 2006; Marchi et al. 2006, 2011; Sládek et al. 2006a, b; Wescott 2006; Wescott and Cunningham 2006; Carlson et al. 2007; Shackelford 2007; Maggiano et al. 2008; Marchi 2008; Sparacello and Marchi 2008; Nikita et al. 2011; Ogilvie and Hilton 2011; Sparacello et al. 2011; Stock et al. 2011). The geographic range represented in these studies has also spread beyond an earlier, largely North American focus to include many regions of the Old World. (A number of studies of modern and archaeological samples from Japan had also been carried out earlier (Kimura and Takahashi 1982, 1984; Nakatsukasa 1990)). With expansion into many different environments—both physical and cultural—has come a greater appreciation of the complexities involved in the relationship between behavior and cross-sectional morphology. In many respects—for example, sexual dimorphism in lower limb bone cross-sectional shape in relation to varying mobility levels—earlier observations have been largely confirmed and extended. The variety of ways in which "mobility" itself can be defined, and the different effects this can have on bone morphology and sexual dimorphism, have been addressed in a number of studies (e.g., Ogilvie 2000; Marchi et al. 2006, 2011; Sládek et al. 2006b, 2007; Carlson et al. 2007; Marchi 2008; Sparacello and Marchi 2008; Ogilvie and Hilton 2011; Stock et al. 2011). For example, in a series of investigations, Marchi and coworkers have shown that males in a Neolithic sample from Liguria, Italy, conform more closely in morphology to Upper Paleolithic and Mesolithic European samples, consistent with their highly seasonably mobile (transhumant) subsistence economy combined with a very rugged terrain (Marchi et al. 2006, 2011; Marchi 2008). Thus, the simple dichotomy between foraging and food production is actually much more complex, with actual degrees and kinds of mobility varying with local circumstances. Overall, however, there is a tendency for foragers to be more robust and more mobile than farmers.

Several studies have reported stronger associations between mobility levels and cross-sectional shape of lower limb bones (i.e., AP/ML or maximum/minimum bending rigidity) than with overall robusticity (average rigidity or strength) (Wescott 2006; Carlson et al. 2007; Maggiano et al. 2008; Sparacello and Marchi 2008), paralleling earlier findings (Ruff et al. 1984; Bridges 1989). This may be attributable in part to the difficulty of adequately standardizing for body size differences in robusticity analyses—a variety of techniques have been used, although the general approach advocated above (Ruff 2000b) has been employed in many of the most recent studies. Examining cross-sectional shape ratios avoids this problem, although the issue of possible body shape effects should still be considered, especially in analyses that include geographically disparate populations. Inclusion of upper limb bones in many of the more recent studies is useful in distinguishing between general activity levels and changes in mobility per se (since the upper limb should be much less affected by locomotor demands). Again, this approach was presaged in some earlier studies (Ruff and Larsen 1990, 2001; Ruff 1999). The addition of other less commonly studied long bones, such as the fibula, can also provide information on types of mechanical loadings of the lower limb, and thus mobility and terrain effects (Marchi and Shaw 2011; Marchi et al. 2011).

Recent experimental studies of humans and other animals have provided further context for interpreting bone shape differences in archaeological remains. Shaw and Stock (2009) compared midshaft tibial cross-sectional geometry in young adult male cross-country runners, field hockey players, and controls. Both groups of athletes had increased robusticity, or bone rigidity relative to body size, but only the runners had increased maximum/minimum bending rigidity, i.e., increased AP bending rigidity, compared to controls. Thus, field hockey players had robust, but more rounded cross sections, which the authors interpreted as a response to more varied mechanical loading (ML as well as AP) of the lower limb, due to frequent turning and change of direction, compared to the more straight-line movements of the runners. Similar findings were reported by Carlson and Judex (2007) in their study of the femora of young mice subjected to increased turning versus linear locomotion. Macdonald et al. (2009) found that a jumping activity (which would primarily load the tibia in AP bending) preferentially increased AP bending rigidity in the midshaft tibia in school children. Cowgill et al. (2010) reported age-related increases in AP/ML rigidity of the midshaft femur that corresponded to ontogenetic changes in lower limb loading, with relatively higher ML loadings in very young children learning to walk. Thus, there is experimental support for a relationship between AP/ML shape of the lower limb bones and the degree and kinds of mobility practiced. Preferential loading in the sagittal (AP) plane, which would be expected in long-distance travel in a relatively straight line, leads to relatively increased AP bending strength. Conversely, more irregular movements, such as those involved in negotiating steep and rough slopes, may produce relatively higher ML (as well as AP) loadings, and thus a rounder cross section (see Higgins 2014). Therefore, as with overall robusticity, terrain is a consideration in interpreting bone shape responses to increased mobility.

Results of some recent experimental and observational studies have also led to calls for caution in overly simplistic interpretations of bone structural responses to exercise (e.g., Lieberman et al. 2004; Cowgill 2010; Wallace et al. 2012). For example, genetic adaptation as well as developmental plasticity is likely involved in producing differences in bone form between and within populations and species. However, this in itself does not negate the effects of mechanical loadings applied during life (Ruff et al. 2006a). Population history as well environment should be considered when interpreting bone structural variation, but, as with all physical characteristics, within a functional and physiological framework (Ruff et al. 2013).

Finally, another very recent development in this field has been to derive bone cross-sectional properties from periosteal contours only, obtained using external molding or laser scanning (Stock and Shaw 2007; Sparacello and Pearson 2010; Marchi et al. 2011; Sparacello et al. 2011; Davies et al. 2012; Macintosh et al. 2013), a technique that has been used to help reconstruct mobility differences between populations (Marchi et al. 2011; Sparacello et al. 2011). In methodological tests, very high correlations between properties determined in this way and true cross-sectional properties were obtained in several of these studies. This follows from the fact that the most critical structural properties—second moments of area and section moduli—are highly dependent on the distribution of bone around a central axis or point; thus, the position of the outer contour of a section is much more important than that of the inner contour (see Ruff 2008). The use of only outer contours in analyses has several advantages, including more rapid acquisition of data (i.e., without the need for radiographing or CT scanning) and some simplification of reconstruction techniques (e.g., see Biknevicius and Ruff 1992; O'Neill and Ruff 2004; Sylvester et al. 2010). However, there are several important caveats that must be kept in mind when applying such methods: (1) While average errors in estimation of properties between samples may be relatively small, *individual* errors may be much larger (Sparacello and Pearson 2010; Macintosh et al. 2013). This has particular relevance for analyses that involve individual paleontological specimens, pathological specimens, or special subgroups of samples. (2) Error varies depending on the location of the sections, with rounder sections near midshaft showing smaller errors than those nearer bone ends (O'Neill and Ruff 2004; Macintosh et al. 2013); some of the above methodological tests examined only mid-diaphyseal regions. Thus, the approach is less applicable to regions such as the proximal tibia or femur. (3) Changes at the endosteal surface are critical in many kinds of analyses, including evaluation of nutritional effects (Garn et al. 1969), bone growth and development (Ruff et al. 1994), and aging (Ruff and Hayes 1983b). The endosteal surface also appears to be the most sensitive to mechanical influences after mid-adolescence (Ruff et al. 1991, 1994; Bass et al. 2002; Kontulainen et al. 2002), which may explain why earlier humans have relatively thick long bone cortices (Kennedy 1985; Ruff et al. 1993). (4) Related to this last point, not factoring in systematic differences in endosteal dimensions may lead to biased comparisons in broader studies. For example, chimpanzees and humans have systematically different relative cortical thicknesses of their femora and humeri; thus, consideration of external dimensions alone results in overlapping inter-limb proportions, while comparison of true cross-sectional properties yields complete separation between the species (Ruff 2009).

Therefore, as Sparacello and Pearson (2010: 620) noted in advocating the external approximation method: "Our conclusions apply best for researchers comparing estimates of population means drawn from fairly similar skeletal samples to make inferences about the behaviorally determined mechanical environment of past populations." Even in that situation, however, caution is advised: one of the most striking differences between the preagricultural and agricultural samples from the Georgia coast (Ruff et al. 1984) was the relatively inwardly "contracted" external and internal contours of the agricultural sample, which resulted in elevated relative cortical thickness combined with reduced rigidity (see Fig. 2.1). Reductions in cortical area in the agricultural group were thus very small compared to those in second moments of area. This had important implications regarding interpretations of dietary versus mechanical effects on morphology (diet might be expected to have more effect on amount of cortical bone, and behavior on distribution of bone), an observation that would have been missed had endosteal dimensions not been assessed. This also emphasizes the importance of considering non-mechanical as well as mechanical influences when interpreting bone structural variation (Ruff 1999; Ruff et al. 2006a).

Of course, even simpler linear external breadths or circumferences of long bone diaphyses have long been used to assess morphological and in some cases behavioral differences between populations (see references in Lovejoy et al. 1976; Ruff 1987). This may be most justifiable when applied in the form of "shape" indices, e.g., AP/ML breadth, in broader comparisons (Jungers and Minns 1979; Ruff 1987). However, in more fine-grained comparisons, linear dimensions, including shape ratios, do not accurately reflect variation in true cross-sectional properties (Stock and Shaw 2007).

2.4 Conclusions

The application of long bone structural analyses in archaeology and human paleontology had its beginnings in the 1970s. The development of more automated and noninvasive techniques in the next decade allowed larger-scale demographic studies and the first true population-level comparisons to be carried out. Methods for standardizing for body size and shape differences were developed, and the effects of terrain as well as behavior were factored into comparisons. Recent studies have emphasized the complexity of influences on bone structure, including local environmental variation, different forms of "mobility" and sex-related economic roles, and effects on the upper as well as lower limb. Despite this complexity, though, a number of general observations can be drawn: (1) The degree of sexual dimorphism in bone shape (AP/ML rigidity or strength) in the middle region of the lower limb (midshaft femur through midshaft tibia) parallels the degree of sexual division of labor, in particular, the relative mobility of males versus females. (2) Bone shape in general is a better indicator of mobility differences than relative bone size. This may be in part due to difficulties in adequately standardizing for body size differences, as well as the relative behavioral non-specificity of overall bone robusticity. (3) It is important to consider body shape (relative limb length, relative body breadth) when interpreting differences in structural properties. Climate per se does not have a direct influence on bone mechanical properties, but does affect them through its influence on body proportions.

Given the tremendous increase in interest in this field in recent years and the increasing availability of rapid data acquisition techniques, we look forward to many new advances in the coming decades. As noted elsewhere (Ruff et al. 2006a: 484): "While physical context and material culture give clues to past behavior, analysis of the skeletons themselves is the most direct way to reconstruct individual behavior, and to explore intra- and inter-populational differences in behavior." The continued application of engineering techniques to skeletal remains will yield insights into mobility patterns and other forms of past behavior that would not otherwise be achievable.

References

Alexander RM, Jayes AS, Maloiy GMO, Wathuta EM (1979) Allometry of the limb bones of mammals from shrews (*Sorex*) to elephants (*Loxodonta*). J Zool Lond 189:305–314

Amtmann VE (1971) Mechanical stress, functional adaptation, and the variation structure of the human femur diaphysis. Ergeb Anat Entwicklungsgesch 44:1–89

Bass SL, Saxon L, Daly RM, Turner CH, Robling AG, Seeman E, Stuckey S (2002) The effect of mechanical loading on the size and shape of bone in pre-, peri-, and postpubertal girls: a study in tennis players. J Bone Miner Res 17(12):2274–2280

Biknevicius AR, Ruff CB (1992) Use of biplanar radiographs for estimating cross-sectional geometric properties of mandibles. Anat Rec 232:157–163

Bridges PS (1989) Changes in activities with the shift to agriculture in the southeastern United States. Curr Anthropol 30:385–394

Brock SL, Ruff CB (1988) Diachronic patterns of change in structural properties of the femur in the prehistoric American Southwest. Am J Phys Anthropol 75:113–127

Carlson KJ, Judex S (2007) Increased non-linear locomotion alters diaphyseal bone shape. J Exp Biol 210(pt 17):3117–3125

Carlson DS, Armelagos GJ, VanGerven DP (1976) Patterns of age-related cortical bone loss (osteoporosis) within the femoral diaphysis. Hum Biol 48:295–314

Carlson KJ, Grine FE, Pearson OM (2007) Robusticity and sexual dimorphism in the postcranium of modern hunter-gatherers from Australia. Am J Phys Anthropol 134(1):9–23

Cowgill LW (2010) The ontogeny of Holocene and Late Pleistocene human postcranial strength. Am J Phys Anthropol 141(1):16–37

Cowgill LW, Warrener A, Pontzer H, Ocobock C (2010) Waddling and toddling: the biomechanical effects of an immature gait. Am J Phys Anthropol 143(1):52–61

Davies TG, Shaw C, Stock JT (2012) A test of a new method and software for the rapid estimation of cross-sectional geometric properties of long bone diaphyses from 3D laser surface scans. Archaeol Anthropol Sci 4:277–290

Dewey JR, Armelagos GJ, Bartley MH (1969) Femoral cortical involution in three Nubian archaeological populations. Hum Biol 41:13–28

Fresia A, Ruff CB, Larsen CS (1990) Temporal decline in bilateral asymmetry of the upper limb on the Georgia Coast. Anthropol Pap Am Mus Nat Hist 68:121–132

Galilei G (1638) Dialogues concerning two new sciences (trans: Crew H, De Salvio A, 1914). Macmillan, New York

Garn SM, Guzman MA, Wagner B (1969) Subperiosteal gain and endosteal loss in protein-calorie malnutrition. Am J Phys Anthropol 30(1):153–155

Gere JM, Goodno BJ (2013) Mechanics of materials, 8th edn. Cengage Learning, Stamford

Higgins RW (2014) The effects of terrain on long bone robusticity and cross-sectional shape in lower limb bones of bovids, neandertals, and upper paleolithic modern humans. In: Carlson KJ, Marchi D (eds) Reconstructing mobility: environmental, behavioral, and morphological determinants. Springer, New York

Holt BM (2003) Mobility in Upper Paleolithic and Mesolithic Europe: evidence from the lower limb. Am J Phys Anthropol 122(3):200–215

Jungers WL, Minns RJ (1979) Computed tomography and biomechanical analysis of fossil long bones. Am J Phys Anthropol 50:285–290

Kennedy GE (1985) Bone thickness in *Homo erectus*. J Hum Evol 14:699–708

Kimura T (1971) Cross-section of human lower limb bones viewed from strength of materials. J Anthropol Soc Nippon 79:323–336

Kimura T (2006) Robustness of the whole Jomon femur shaft assessed by cross-sectional geometry. Anthropol Sci 114(1):13–22

Kimura T, Takahashi H (1982) Mechanical properties of cross section of lower limb long bones in Jomon man. J Anthropol Soc Nippon 90:105–117

Kimura T, Takahashi H (1984) On the cross section of lower limb long bones in Minatogawa Man from Okinawa. J Anthopol Soc Nippon 98:69–78

Koch JC (1917) The laws of bone architecture. Am J Anat 21:177–298

Kontulainen S, Sievanen H, Kannus P, Pasanen M, Vuori I (2002) Effect of long-term impact-loading on mass, size, and estimated strength of humerus and radius of female racquet-sports players: a peripheral quantitative computed tomography study between young and old starters and controls. J Bone Miner Res 17(12):2281–2289

Larsen CS, and Ruff CB (2011) An external agency of considerable importance: the stresses of agriculture in the foraging-to-farming transition in eastern North America. In: Pinhasi R, Stock JT (eds). Human bioarchaeology of the transition to agriculture. Wiley-Blackwell, New York, pp 293–315

Lieberman DE, Polk JD, Demes B (2004) Predicting long bone loading from cross-sectional geometry. Am J Phys Anthropol 123(2):156–171

Lovejoy CO, Trinkaus E (1980) Strength and robusticity of the Neandertal tibia. Am J Phys Anthropol 53:465–470

Lovejoy CO, Burstein AH, Heiple KG (1976) The biomechanical analysis of bone strength: a method and its application to platycnemia. Am J Phys Anthropol 44:489–506

Macdonald HM, Cooper DM, McKay HA (2009) Anterior-posterior bending strength at the tibial shaft increases with physical activity in boys: evidence for non-uniform geometric adaptation. Osteoporos Int 20(1):61–70

Macintosh AA, Davies TG, Ryan TM, Shaw CN, Jay TS (2013) Periosteal versus true cross-sectional geometry: a comparison along humeral, femoral, and tibial diaphyses. Am J Phys Anthropol 150:442–452

Maggiano IS, Schultz M, Kierdorf H, Sosa TS, Maggiano CM, Tiesler Blos V (2008) Cross-sectional analysis of long bones, occupational activities and long-distance trade of the Classic Maya from Xcambo–archaeological and osteological evidence. Am J Phys Anthropol 136(4):470–477

Marchi D (2008) Relationships between lower limb cross-sectional geometry and mobility: the case of a Neolithic sample from Italy. Am J Phys Anthropol 137(2):188–200

Marchi D, Shaw CN (2011) Variation in fibular robusticity reflects variation in mobility patterns. J Hum Evol 61(5):609–616

Marchi D, Sparacello VS, Holt BM, Formicola V (2006) Biomechanical approach to the reconstruction of activity patterns in Neolithic Western Liguria, Italy. Am J Phys Anthropol 131(4):447–455

Marchi D, Sparacello V, Shaw C (2011) Mobility and lower limb robusticity of a pastoralist Neolithic population from north-western Italy. In: Pinhasi R, Stock JT (eds) Human bioarchaeology of the transition to agriculture. Wiley-Blackwell, New York, pp 317–346

Martin RB, Atkinson PJ (1977) Age and sex-related changes in the structure and strength of the human femoral shaft. J Biomech 10:223–231

Mayr E (1963) Animal species and evolution. Harvard University Press, Cambridge

McMahon T (1973) Size and shape in biology. Science (New York, NY 179(4079):1201–1204

Miller GJ, Piotrowski G (1977) Geometric properties of paired human femurs. In: Grood ES, Smith CR (eds) Advances in bioengineering. ASTM, New York, pp 73–74

Minns RJ, Bremble GR, Campbell J (1975) The geometric properties of the human tibia. J Biomech 8:253–255

Nagurka ML, Hayes WC (1980) An interactive graphics package for calculating cross-sectional properties of complex shapes. J Biomech 13:59–64

Nakatsukasa M (1990) Cross section of femoral shaft at the mid point: variation in modern Japanese and Jomon populations. J Anthopol Soc Nippon 98:1–12

Nikita E, Siew YY, Stock J, Mattingly D, Lahr MM (2011) Activity patterns in the Sahara Desert: an interpretation based on cross-sectional geometric properties. Am J Phys Anthropol 146(3):423–434

O'Neill MC, Ruff CB (2004) Estimating human long bone cross-sectional geometric properties: a comparison of noninvasive methods. J Hum Evol 47(4):221–235

Ogilvie MD (2000) A biological reconstruction of mobility patterns at the foraging to farming transition in the american southwest. Ph.D. Dissertation, University of New Mexico, Albuquerque

Ogilvie MD, Hilton CE (2011) Cross-sectional geometry in the humeri of foragers and farmers from the prehispanic American Southwest: exploring patterns in the sexual division of labor. Am J Phys Anthropol 144(1):11–21

Ohman JC (1993) Computer software for estimating cross-sectional geometric properties of long bones with concentric and eccentric elliptical models. J Hum Evol 25:217–227

Pearson OM (2000) Activity, climate, and postcranial robusticity: implications for modern human origins and scenarios of adaptive change. Curr Anthropol 41(4):569–607

Polk JD, Demes B, Jungers WL, Biknevicius AR, Heinrich RE, Runestad JA (2000) A comparison of primate, carnivoran and rodent limb bone cross-sectional properties: are primates really unique? J Hum Evol 39:297–325

Robbins DR, Rosenberg KR, Ruff CB (1989) Activity patterns in Late Middle Woodland, Delaware. Am J Phys Anthropol 78:290–291

Roberts DF (1978) Climate and human variability, 2nd edn. Cummings, Menlo Park

Ruff CB (1984) Allometry between length and cross-sectional dimensions of the femur and tibia in *Homo sapiens sapiens*. Am J Phys Anthropol 65:347–358

Ruff CB (1987) Sexual dimorphism in human lower limb bone structure: relationship to subsistence strategy and sexual division of labor. J Hum Evol 16:391–416

Ruff CB (1994a) Biomechanical analysis of Northern and Southern Plains femora: behavioral implications. In: Owsley DW, Jantz RL (eds) Skeletal biology in the Great Plains: a multidisciplinary view. Smithsonian Institution Press, Washington, DC, pp 235–245

Ruff CB (1994b) Morphological adaptation to climate in modern and fossil hominids. Yearb Phys Anthropol 37:65–107

Ruff CB (1995) Biomechanics of the hip and birth in early *Homo*. Am J Phys Anthropol 98:527–574

Ruff CB (1999) Skeletal structure and behavioral patterns of prehistoric Great Basin populations. In: Hemphill BE, Larsen CS (eds) Prehistoric lifeways in the Great Basin wetlands: bioarchaeological reconstruction and interpretation. University of Utah Press, Salt Lake City, pp 290–320

Ruff CB (2000a) Biomechanical analyses of archaeological human skeletal samples. In: Katzenburg MA, Saunders SR (eds) Biological anthropology of the human skeleton. Alan R. Liss, New York, pp 71–102

Ruff CB (2000b) Body size, body shape, and long bone strength in modern humans. J Hum Evol 38:269–290

Ruff CB (2005) Mechanical determinants of bone form: insights from skeletal remains. J Musculoskelet Neuronal Interact 5:202–212

Ruff CB (2008) Biomechanical analyses of archaeological human skeletal samples. In: Katzenburg MA, Saunders SR (eds) Biological anthropology of the human skeleton, 2nd edn. Wiley, New York, pp 183–206

Ruff CB (2009) Relative limb strength and locomotion in Homo habilis. Am J Phys Anthropol 138:90–100

Ruff CB, Hayes WC (1983a) Cross-sectional geometry of Pecos Pueblo femora and tibiae—a biomechanical investigation. I. Method and general patterns of variation. Am J Phys Anthropol 60:359–381

Ruff CB, Hayes WC (1983b) Cross-sectional geometry of Pecos Pueblo femora and tibiae—a biomechanical investigation. II. Sex, age, and side differences. Am J Phys Anthropol 60:383–400

Ruff CB, Hayes WC (1988) Sex differences in age-related remodeling of the femur and tibia. J Orthop Res 6:886–896

Ruff CB, Larsen CS (1990) Postcranial biomechanical adaptations to subsistence changes on the Georgia Coast. Anthropol Pap Am Mus Nat Hist 68:94–120

Ruff CB, Larsen CS (2001) Reconstructing behavior in Spanish Florida: the biomechanical evidence. In: Larsen CS (ed) Bioarchaeology of Spanish Florida: the impact of colonialism. University Press of Florida, Gainesville, pp 113–145

Ruff CB, Leo FP (1986) Use of computed tomography in skeletal structure research. Yearb Phys Anthropol 29:181–195

Ruff CB, Larsen CS, Hayes WC (1984) Structural changes in the femur with the transition to agriculture on the Georgia coast. Am J Phys Anthropol 64:125–136

Ruff CB, Scott WW, Liu AY-C (1991) Articular and diaphyseal remodeling of the proximal femur with changes in body mass in adults. Am J Phys Anthropol 86:397–413

Ruff CB, Trinkaus E, Walker A, Larsen CS (1993) Postcranial robusticity in *Homo*, I: temporal trends and mechanical interpretation. Am J Phys Anthropol 91:21–53

Ruff CB, Walker A, Trinkaus E (1994) Postcranial robusticity in *Homo*, III: ontogeny. Am J Phys Anthropol 93:35–54

Ruff CB, Holt BH, Trinkaus E (2006a) Who's afraid of the big bad Wolff? Wolff's Law and bone functional adaptation. Am J Phys Anthropol 129:484–498

Ruff CB, Holt BM, Sladék V, Berner M, Murphy WA Jr, Zur Nedden D, Seidler H, Recheis W (2006b) Body size, body proportions, and mobility in the Tyrolean "Iceman". J Hum Evol 51(1):91–101

Ruff CB, Garofalo E, Holmes MA (2013) Interpreting skeletal growth in the past from a functional and physiological perspective. Am J Phys Anthropol 150:29–37

Schultz AH (1953) The relative thickness of the long bones and the vertebrae in primates. Am J Phys Anthropol 11:277–311

Shackelford LL (2007) Regional variation in the postcranial robusticity of Late Upper Paleolithic humans. Am J Phys Anthropol 133(1):655–668

Shaw CN, Stock JT (2009) Intensity, repetitiveness, and directionality of habitual adolescent mobility patterns influence the tibial diaphysis morphology of athletes. Am J Phys Anthropol 140(1):149–159

Shaw CN, Stock JT (2011) The influence of body proportions on femoral and tibial midshaft shape in hunter-gatherers. Am J Phys Anthropol 144(1):22–29

Sládek V, Berner M, Sailer R (2006a) Mobility in Central European Late Eneolithic and Early Bronze Age: femoral cross-sectional geometry. Am J Phys Anthropol 130:320–332

Sládek V, Berner M, Sailer R (2006b) Mobility in Central European Late Eneolithic and Early Bronze Age: tibial cross-sectional geometry. J Archaeol Sci 33:470–482

Sládek V, Berner M, Sosna D, Sailer R (2007) Human manipulative behavior in the Central European Late Eneolithic and Early Bronze Age: humeral bilateral asymmetry. Am J Phys Anthropol 133(1):669–681

Sparacello V, Marchi D (2008) Mobility and subsistence economy: a diachronic comparison between two groups settled in the same geographical area (Liguria, Italy). Am J Phys Anthropol 136(4):485–495

Sparacello VS, Pearson OM (2010) The importance of accounting for the area of the medullary cavity in cross-sectional geometry: a test based on the femoral midshaft. Am J Phys Anthropol 143(4):612–624

Sparacello VS, Pearson OM, Coppa A, Marchi D (2011) Changes in skeletal robusticity in an iron age agropastoral group: the Samnites from the Alfedena necropolis (Abruzzo, Central Italy). Am J Phys Anthropol 144(1):119–130

Stock JT (2006) Hunter-gatherer postcranial robusticity relative to patterns of mobility, climatic adaptation, and selection for tissue economy. Am J Phys Anthropol 131:194–204

Stock J, Pfeiffer S (2001) Linking structural variability in long bone diaphyses to habitual behaviors: foragers from the southern African Later Stone Age and the Andaman Islands. Am J Phys Anthropol 115:337–348

Stock JT, Pfeiffer SK (2004) Long bone robusticity and subsistence behavior among Later Stone Age foragers of the forest and fynbos biomes of South Africa. J Archaeol Sci 31:999–1013

Stock JT, Shaw CN (2007) Which measures of diaphyseal robusticity are robust? A comparison of external methods of quantifying the strength of long bone diaphyses to cross-sectional geometric properties. Am J Phys Anthropol 134(3):412–423

Stock JT, O'Neill MC, Ruff CB, Zabecki M, Shackelford L, Rose J (2011) Body size, skeletal biomechanics, mobility and habitual activity from the Late Palaeolithic to the mid-Dynastic Nile Valley. In: Pinhasi R, Stock JT (eds) Human bioarchaeology of the transition to agriculture. Wiley-Blackwell, New York, pp 347–367

Stojanowski CM (2013) Mission cemeteries, mission peoples: historical and evolutionary dimensions of intracemetery bioarchaeology in spanish florida. University Press of Florida, Gainesville

Sumner DR (1984) Size, shape, and bone mineral content of the human femur in growth and aging. Thesis, University of Arizona, Arizona

Sumner DR, Mockbee B, Morse K (1985) Computed tomography and automated image analysis of prehistoric femora. Am J Phys Anthropol 68:225–232

Sylvester AD, Garofalo E, Ruff CB (2010) An R program for automating bone cross section reconstruction. Am J Phys Anthropol 142:665–669

Tate JR, Cann CE (1982) High-resolution computed tomography for the comparative study of fossil and extant bone. Am J Phys Anthropol 58:67–73

Trinkaus E, Ruff CB (1989) Diaphyseal cross-sectional morphology and biomechanics of the Fond-de-Forêt 1 femur and the Spy 2 femur and tibia. Bull Soc R Bel Anthropol Préhist 100:33–42

Trinkaus E, Ruff CB (2012) Femoral and tibial diaphyseal cross-sectional geometry in Pleistocene *Homo*. PaleoAnthropology 2012:13–62

Trinkaus E, Churchill SE, Ruff CB, Vandermeersch B (1999a) Long bone shaft robusticity and body proportions of the Saint-Césaire 1 Châtelperronian Neandertal. J Archaeol Sci 26:753–773

Trinkaus E, Stringer CB, Ruff CB, Hennessy RJ, Roberts MB, Parfitt SA (1999b) Diaphyseal cross-sectional geometry of the Boxgrove 1 Middle Pleistocene human tibia. J Hum Evol 37:1–25

Van Gerven DP (1973) Thickness and area measurements as parameters of skeletal involution of the humerus, femur, and tibia. J Gerontol 28:40–45

Van Gerven DP, Armelagos GJ (1970) Cortical involution in prehistoric Mississippian femora. J Gerontol 25:20–22

Van Gerven DP, Armelagos GL, Bartley MR (1969) Roentgenographic and direct measurement of femoral cortical involution in a prehistoric Mississippian population. Am J Phys Anthropol 31:23–38

Wallace IJ, Tommasini SM, Judex S, Garland T Jr, Demes B (2012) Genetic variations and physical activity as determinants of limb bone morphology: an experimental approach using a mouse model. Am J Phys Anthropol 148(1):24–35

Weiss E (2003) The effects of rowing on humeral strength. Am J Phys Anthropol 121:293–302

Wescott DJ (2006) Effect of mobility on femur midshaft external shape and robusticity. Am J Phys Anthropol 130:201–213

Wescott DJ, Cunningham DL (2006) Temporal changes in Arikara humeral and femoral cross-sectional geometry associated with horticultural intensification. J Archaeol Sci 33:1022–1036

Chapter 3
Bipedalism and Musculoskeletal Stress Markers: Variation and What It Reveals About Adaptation, Environmental Stress, and Reconstructing Activity Patterns

Elizabeth Weiss

Abstract Understanding the origins of bipedalism has been an important part of anthropological research. Many traits have been used to determine bipedalism, but musculoskeletal stress markers (MSMs) have not received much attention by paleoanthropologists. MSMs are observable locations on bone where muscles, tendons, or ligaments attach. Bioarchaeologists use MSMs to reconstruct activities. Using MSMs, anthropologists have addressed issues regarding differences in activity patterns (within and between populations) related to subsistence patterns, sex differences in specific activities, and the effects of terrain types. However, age and body weight are confounding factors of MSMs. Regardless of activity, it seems that older and larger individuals have greater MSM scores than younger and smaller individuals. Since these two factors are difficult to control for in the fossil record, paleoanthropologists may be reluctant to use MSMs to determine bipedalism. This chapter looks at the evidence of nonhuman primate research and modern human variation to deduce whether MSMs on fossils can be successfully analyzed to aid in understanding the evolution of bipedalism.

Keywords Musculoskeletal stress markers • Nonhuman primates • Bipedalism

3.1 Introduction

Examining mobility patterns of past populations has been a key component of anthropological research. Mobility is a general term that can be used to intend level of movements. In this chapter, mobility is intended as the daily ability to move one's

---

E. Weiss, Ph.D. (✉)
Department of Anthropology, SJSU, One Washington Square, San Jose, CA 95192-0113, USA
e-mail: Elizabeth.Weiss@sjsu.edu

K.J. Carlson and D. Marchi (eds.), *Reconstructing Mobility: Environmental, Behavioral, and Morphological Determinants*, DOI 10.1007/978-1-4899-7460-0_3,

entire body from one location to another. This kind of mobility is generally referred to as logistic mobility and can include both aquatic and terrestrial mobility (Stock and Pfeiffer 2001; Weiss 2003a). In humans, differently from all other mammals, mobility involves bipedal locomotion. Those studying human evolution have often focused on establishing bipedalism in early hominins and in trying to reconstruct the mode of bipedalism of our ancestors. When, how, and why bipedalism began is one of the most debated questions in the study of early human evolution (e.g., Lovejoy et al. 2009; Stanford 2012; Stern 2000; Ward 2002). Anthropologists have attempted to figure out which species of early hominins were habitual bipeds by examining a variety of traits that include, but are not limited to, femoral neck cross sections (e.g., Galik et al. 2004), lower limb morphology (e.g., Tardieu 1999; Latimer and Lovejoy 1989), upper limb morphology (e.g., Tallman 2012), vertebral morphology (e.g., Abitbol 1995), intermembral indices (e.g., McHenry and Berger 1998), and footprints (e.g., Raichlen et al. 2008). Many studies have focused on the investigation of the internal structure of bone (see Ruff and Larsen 2014). The examination of musculoskeletal stress markers (MSMs) is mentioned cursorily in various articles (see Davis 1964; Drapeau et al. 2005; Gebo and Schwartz 2006; Häusler and Berger 2001; Ward 2002; Ward et al. 2012), but anthropologists have not fully explored MSMs in the fossil record. In this chapter, I review the evidence of MSMs in the fossil record and in extant nonhuman primates. Also, meta-analyses of upper limb and lower limb patterns in MSMs of anatomically modern human populations will be presented to determine whether future fossil analyses on bipedalism would be enhanced through the use of MSM data. Thus, the aim of the paper is to determine whether MSMs can inform anthropologists about the origins and modes of bipedality in hominin evolution. The hypothesis, which will be tested using data from MSMs from Holocene human populations, is that lower limb MSMs (since they are used mainly in bipedal gait) should show less variation and correlate more strongly with one another than upper limb MSMs (which are used in many different activities).

3.2 MSM Research

MSMs, which are also known as entheseal changes, are observable locations on bone where muscles, tendons, or ligaments attach. MSMs appear as a result of resorptive and proliferative changes. The resorptive changes include stress lesions, which are pits in the cortex or vascular formation, and foramina, which are considered cysts. The proliferative changes involve raised and rough mounds and ridges, which are referred to as rugosity, and spurs, which are referred to as osteophytes.

There are two concepts that are cited as being responsible for MSMs: bone remodeling and microtrauma. In either case, in order for muscle insertion or origin sites to experience measureable morphological changes, muscle use is required. Activity is important in remodeling bones and maintaining bone strength because muscle usage places stress on bones necessary to activate bone forming cells (i.e., osteoblasts) (Chamay and Tschantz 1972; Ruff et al. 2006; Woo et al. 1981). Bone responds to stress through remodeling, which may change morphology, to prevent breakage. Hence, when a muscle is utilized, the origin and insertion site is stressed as the muscle pulls at the periosteum; this stress stimulates osteons and theoretically causes bone deposition at areas of greatest local stress. Muscle use should, therefore, result in bone hypertrophy, predominantly through the proliferative changes mentioned above, at the origin and insertion sites. Yet, some anthropologists have suggested that location-specific bone remodeling does not occur to a measurable degree and they suggest that bone remodeling as a result of every day stresses is not localized, but rather is systemic, and thus cannot be used to reconstruct specific activities by looking at specific insertion or origin sites from muscles typically functioning in these activities (Bertram and Swartz 1991). Nevertheless, many anthropologists still contend that MSM development is associated with activity via bone remodeling concepts (e.g., Cashmore and Zakrzewski 2013; Chapman 1997; Hawkey and Merbs 1995).

Conversely, some anthropologists theorize that MSM development is a result of microtrauma or repetitive strain without trauma (e.g., Steen and Lane 1998; Churchill and Morris 1998; Dutour 1986; Villotte et al. 2010a). The clinical literature on entheses has provided evidence that microtrauma results in osteophytes, cysts, and vascular formation; these traits are found in MSMs as well (Shaw and Benjamin 2007). Most of the research on sports injuries, however, involves fibrocartilaginous attachment sites and most of the bioarchaeological research on MSMs involves fibrous attachment sites. Muscles attach onto bone via a tendon (i.e., fibro cartilaginous attachment) or more directly onto the bone (i.e., fibrous attachment) (Shaw and Benjamin 2007). Changes in fibrocartilaginous sites are well-documented and include the formation of cysts, vascularization, and bone spurs (Villotte and Knüsel 2013). Fibrous sites are less well understood and appear to have changes even without microtrauma (Villotte and Knüsel 2013). Microtrauma MSM etiology gets around the bone remodeling dilemma mentioned above. MSMs then are pathologies that are the result of microtrauma that relate to repetitive stresses, such as long distance walking; they are not dependent on site-specific bone remodeling.

Regardless of their endorsed causal mechanism, both groups of anthropologists use MSMs to reconstruct activities. Using MSMs, anthropologists have addressed issues regarding sex differences in activity patterns (e.g., al-Oumaoui et al. 2004; Hawkey and Merbs 1995; Lovell and Dublenko 1999; Stefanović and Porčić 2013; Villotte et al. 2010b; Weiss 2007), within and between population differences related to subsistence patterns (e.g., Chapman 1997; Churchill and Morris 1998; Eshed et al. 2004; Havelková et al. 2011; Villotte et al. 2010b), population differences resulting from differences in specific activities (e.g., Lai and Lovell 1992; Lieverse et al. 2009; Molnar 2006; Peterson 1998; Steen and Lane 1998), and effects of terrain types (e.g., al-Oumaoui et al. 2004; Dutour 1986; Kutterer and Uerpmann 2010; Lukacs and Pal 2003).

Although many bioarchaeologists have reported success in using MSMs to reconstruct activity patterns, and thus these markers should be useful in resolving which species were bipedal, other researchers (e.g., Niinimäki 2011; Weiss 2003b) have also discovered confounding factors that hinder the usefulness of employing MSMs in activity pattern reconstructions. For example, there is a significant age effect on MSMs; older individuals have greater MSM scores than do younger individuals (e.g., al-Oumaoui et al. 2004; Alves Cardoso and Henderson 2010; Milella et al. 2012; Molnar 2006, 2010; Molnar et al. 2011; Niinimäki 2011; Villotte et al. 2010a; Weiss 2003b, 2004; Weiss et al. 2012; Wilczak 1998). Nagy (1998) and Molnar (2010) have suggested that the age correlation with MSMs could be the result of using muscles over a greater length of time, but Alves Cardoso and Henderson (2010), Cunha and Umbelino (1995), Milella et al. (2012), Mariotti et al. (2007), and Niinimäki (2011) have argued that age alone is a causative factor in the formation of MSMs. Age determination of skeletal remains is difficult even in the bioarchaeological record. Younger individuals, those who are unlikely to have pronounced MSMs, are more precisely aged than older individuals. For example, using either the Suchey-Brooks or Hamann-Todd pubic symphysis age techniques, the last phases are for 50 years or older. Thus, although age can be partially controlled for in anatomically modern human samples, early hominin age cannot be determined with the same precision.

Age is not the only confounding factor that affects MSMs. Body size has been shown to have confounds with MSMs (e.g., Godde and Taylor 2011; Lieverse et al. 2009; Myszka and Piontek 2011; Niinimäki 2011; Weiss 2003b, 2004, 2007). For many MSMs there is a positive correlation with body size. The body size effect is greater on the lower limb than on the upper limb (Weiss 2007), which may be especially troublesome for mobility studies. Since males and females differ in body size, determining sex differences in mobility may be difficult using MSMs. Plus, the fossil record seems to support great variation in body size, but whether these differences are species differences, sex differences, or individual differences is difficult to assess (Plavcan 2012). Some anthropologists (e.g., Lieberman et al. 1988; Gordon et al. 2008) have proposed that early hominin species may have even had a greater degree of sexual dimorphism than Holocene hominins; others have argued that sexual dimorphism is moderate in early hominins (e.g., Reno et al. 2010).

Regardless of confounding factors, asymmetry studies (e.g., Mays et al. 1999; Peterson 1998; Stirland 1993) and studies using specific MSMs that do not correlate with body size, such as fibrocartilaginous MSMs (e.g., Villotte et al. 2010a; Weiss 2012), have supported the role of activity in MSM formation. Thus, examining whether MSMs can be used to reconstruct mobility type may be fruitful.

3.3 MSM Research in the Fossil Record

Although there are few studies that employ MSMs in relation to determining bipedalism, there are many MSMs related to muscles utilized in bipedalism. Table 3.1 lists the muscles and related MSMs associated with bipedalism. Some studies have used evidence of muscle attachment sites to help in reconstructing bipedalism (e.g., Davis 1964; Marzke et al. 1988; Pickford et al. 2002; Rose 1984; Ward 2002).

Table 3.1 Common MSM locations of the lower limb associated with bipedalism

Bone	Muscle	Location
Femur	Common insertion of *adductor brevis, adductor longus, adductor magnus*	Linea aspera
	Insertion of *gluteus maximus*	Gluteal tuberosity
	Insertion of *gluteus minimus*	Anterior border of greater trochanter
	Insertion of *gluteus medius*	Lateral surface of greater trochanter
	Insertion of *iliopsoas*	Lesser trochanter
	Origin of lateral *gastrocnemius*	Lateral femoral condyle
	Origin of medial *gastrocnemius*	Medial femoral condyle
	Insertion of *quadriceps femoris*	Trochanteric crest
Tibia	Insertion of *popliteus*	Proximal posterior tibia
	Origin of *soleus*	Proximal posterior shaft of tibia, oblique line

Paleoanthropologists do not employ the same methodologies as bioarchaeologists. These studies often differ from bioarchaeological studies of MSMs in their emphasis on location and shape of the muscle markers rather than their degree of development. Paleoanthropologists do not typically publish research on the degree of muscle marker stress lesions, vascular formation, or osteophytes. For these reasons, I refer to these traits as muscle markers in this section. This difference makes comparisons between bioarchaeological and paleoanthropological research on muscle insertion and origin sites difficult. Paleoanthropologists who find muscle marker similarities in early hominins and modern humans use these similarities as evidence of early hominin bipedalism, but usually differences in muscle markers between early hominins and modern humans—or similarities between early hominins and extant apes—are not used as evidence of arboreality. Traits not directly referable to bipedalism are assumed to have little to do with activity patterns or mobility, rather they are often attributed to genetics (e.g., Drapeau et al. 2005; Ward et al. 2012).

Häusler and Berger (2001) discussed lower and upper limb muscle markers in australopithecines in comparison to muscle markers in apes and humans. They found that the ilium of *Australopithecus africanus*, the South African hominin dated between three and two million years ago, displayed evidence of a large anterior *gluteus medius* muscle and a broad attachment for the *gluteus minimus* muscle, both of which should be indicative of habitual bipedalism. *A. africanus*, they also found, lacked a chimpanzee-like extensive *latissimus dorsi* muscle, which is an adductor, medial rotator, and extensor of the humerus in the upper limb. Plus, the *latissimus dorsi* muscle has a broad extension in all of the great apes (Waterman 1929). The extensiveness of the *latissimus dorsi* muscle in chimpanzees relates to arboreal activities, such as climbing. Yet, other studies of early hominin MSMs display contradictory evidence.

Ward, in her 2002 review article, described similarities between *Australopithecus afarensis*, the early East African hominin that dates between 3.6 and 3 million years ago, and ape hip and thigh muscle attachment sites; she states that the attachment sites of early hominins and extant apes differ from the attachment sites found in modern humans. For example, the anterior fibers of the *gluteus minimus* muscle on the greater trochanter extend farther posterior in *A. afarensis* and apes than in humans. Furthermore, the ischial origin of the *adductor magnus* muscle is set at an angle to the rest of the tuberosity in *A. afarensis*, but not in humans. Ward concluded that *A. afarensis* appears to have been bipedal, a conclusion mainly drawn from traits other than the muscle markings, but that their bipedalism differed from modern human bipedalism, a conclusion supported with muscle marker data. Similarly, Ward and colleagues (2012) reported differences between modern human and *A. afarensis gluteal* muscles attachments, but they concluded that these differences were likely not reflective of mobility.

Drapeau et al. (2005) examined well-preserved muscle markers on *A. afarensis* fossil AL-438-1; this fossil skeleton from Hadar, Ethiopia is dated to three million years ago and consists of the humerus, radius, and ulna. They confirmed that on the upper limb the *flexor carpi ulnaris* and the *anconeus* muscles were more pronounced than those found on humans, but less pronounced than those found on chimpanzees. It also appears that there were no distinct muscle markers for the *abductor pollicis longus*, *extensor pollicis longus*, and the *extensor indicis* muscles, which suggests a less refined manipulatory capability of the hand than in humans. The same pattern is found in AL-288 (which is the smaller *A. afarensis* nicknamed Lucy). However, Drapeau and colleagues suggested that this pattern may not be reflective of arboreal climbing.

In 1964, Davis provided an assessment of the Olduvai Hominid (OH) 35 tibia and fibula that included muscle marker evidence; OH 35 is an approximately 2.5 million-year-old early *Homo* or robust australopithecine. OH 35's muscle markings of the *flexor digitorum longus*, the *tibialis posterior*, the *flexor hallucis longus*, and the *peronei* group muscles were compared with those of apes and modern humans. Davis used evidence of these muscle markers' extensiveness and texture to support that at 2.5 million years ago an early hominin was "adapting toward bipedalism." The *popliteus* muscle marker was also examined and Davis concluded that this specific muscle marker differed from modern humans. The *popliteus* muscle is usually associated with flexion of the knee and lateral rotation of the femur. Thus, Davis deduced that bipedalism utilized by the early hominin represented by OH 35 may have been considerably different from our own.

In 2006, Gebo and Schwartz examined fossilized foot bones dating between 2.2 and 2.36 million years ago from Omo, Ethiopia. These bones, which represent *Homo rudolfensis* or *Homo erectus*, display a large peroneal tubercle that differs from humans and suggests large peroneal muscles for foot eversion. Foot eversion is usually indicative of arboreality; for example, chimpanzee peroneal muscles involved in foot eversion are active only during arboreal locomotion (Stern and Susman 1983). But, Gebo and Schwartz (2006) contend that enlarged peroneal muscles are evidence of endurance running and support Bramble and Lieberman's (2004) hypothesis that places importance on running in early human evolution.

Mariotti and Belcastro (2011) compared Neanderthal and anatomically modern MSMs. Here the term MSM is used because the authors employed common bioarchaeological methods to collect the muscle marker data. They found that of the muscles they examined, which included the *gluteus maximus*, the *iliopsoas*, the *vastus medialis*, the *quadriceps femoris* (*tendon*), and the *soleus*, only the attachment regions of the *gluteus maximus* muscle appeared to be different between the two species. Neanderthal *gluteus maximus* MSMs were more pronounced than those of anatomically modern humans. The *gluteus maximus* differences between *H. sapiens* and Neanderthals seemed to arise at a young age. Additionally, it appears that the attachment sites of the *gluteus maximus* muscle of Neanderthals were fibrocartilaginous whereas in *H. sapiens* the attachment sites are fibrous. These factors along with different frequencies of the third trochanter, which is considered a nonmetric genetic trait in modern humans, led Mariotti and Belcastro (2011) to conclude that Neanderthal and *H. sapiens* differences in lower limb MSMs were attributable to genes rather than activity patterns.

Although the above-mentioned paleoanthropologists used muscle markers in assessing mobility, they most often did so without following a standardized methodology. Furthermore, the above-reviewed studies are examples where muscle markings have been inconsistently cited as related to activity patterns, especially when the muscle markers did not correlate to bipedal mobility. In the next sections of this chapter, I will address some of these issues, offering measures to be taken in order to advance the use of muscle markers in addressing questions related to mobility and bipedalism.

3.4 MSM Research in Nonhuman Primates

Although the literature on primate anatomy is vast, the literature on MSMs in primates is scant. Zihlman et al. (2011) used musculoskeletal data of gorillas and orangutans to provide evidence that orangutans are specialized outliers in the great ape family; for example, the orangutan *gluteus maximus* has a superior portion on the posterior ilium and a separate portion on the ischial tuberosity, but these two portions are continuous in gorillas. Zihlman and colleagues suggested that such muscular divisions provide orangutans with greater mobility. Research by Niskanen and Junno (2009) examined muscle insertion sizes, among other metrics, to determine whether muscle markers correlate with body size. When examining data on *Pan troglodytes* and *Gorilla gorilla*, they concluded that—when sex is accounted for—body size can be determined through muscle insertion sites. This research corroborates earlier research by Zumwalt and her colleagues (2000) who reported that the best indicator of MSM robusticity among nonhuman primates was body size rather than type of locomotion. Plus, Morimoto et al. (2011) examined femoral and pelvic musculoskeletal surface topography through virtual dissection and found that chimpanzees and humans—but not gorillas or orangutans—share similarities in the position of the *gluteus maximus* muscle insertion. They used this information to conclude that musculoskeletal surface topography and, therefore, MSMs may not be useful in reconstructing locomotion type. Suwa et al. (2012), however, found after using a broader comparative osteological approach coupled with reexamination of Morimoto et al. (2011) CT-images that the chimpanzee insertion of the *gluteus maximus* may appear to lie more posteromedially than it does in the other great apes, but the human and chimpanzee musculoskeletal surface topography of the femur and pelvis were, nevertheless, distinct.

Other nonhuman primate research has focused on bilateral asymmetry. Cashmore's (2009) examination of African ape data compared to modern human data suggests that MSMs may not be useful in determining handedness. Although this is unrelated to mobility, if MSMs do not correspond to asymmetry, then they are unlikely to correspond to other activities as well. One problem with nonhuman primate studies may be the lack of population-based asymmetry. The absence of population-based asymmetry has also been observed in musculature and behavioral studies (Carlson 2006). MSM asymmetry studies in humans have supported the link between activities and right dominance (e.g., Drapeau 2008; Peterson 1998; Milella et al. 2012).

Drapeau (2008) studied gorilla, chimpanzee, and human MSMs in relation to asymmetry in order to draw conclusions about mobility. Drapeau used the Hawkey and Merbs (1995) method to collect MSM scores; in the bioarchaeological literature, the Hawkey and Merbs (1995) method is used most frequently to collect MSM data. The Hawkey and Merbs method identifies three types of changes in MSMs: rugosity (which is general roughness and raised bone), stress lesions (which are pitting and porosity), and osteophytes (which are bony spurs). Using 18 upper limb and seven lower limb musculoskeletal markers, Drapeau tested whether human upper limb MSMs were less pronounced than their lower limb MSMs. She also noted that humans should have greater asymmetry in the upper limbs than in the lower limbs and African ape upper limb asymmetry should be lower than human upper limb asymmetry. Finally, lower limb asymmetry patterns should be unified across the three species.

Drapeau (2008) found that rugosity scores provided the most definitive conclusions; this is likely because stress lesions and osteophytes present more severe MSM expressions and, thus, show less variation. Humans have greater MSM scores in lower limbs compared to upper limbs, while the opposite is true for apes. Apes, however, should have similar upper limb and lower limb scores since both sets of limbs are used in locomotion. Whether apes are more terrestrial, like gorillas, or more arboreal, like orangutans, all four limbs are used in locomotion. All species have greater upper limb asymmetry compared to lower limb asymmetry, except humans who have greater lower limb asymmetry in osteophytes. Greater asymmetry in the human lower limb is unexpected and contrary to the hypothesis, but it could be an artifact of the small number of individuals with osteophytes. In both upper and lower limbs, humans display more asymmetry than either of the African apes. Another interesting aspect of this study is that chimpanzees and gorillas did not differ from one another in most aspects; thus, their similarity in MSMs may be reflective of a general similarity in locomotion. Drapeau's (2008) data provided some much-needed support for the use of MSMs in hominin mobility studies, and is the only study that explicitly examines nonhuman primate MSMs in a manner similar to those employed by bioarchaeologists to address whether paleoanthropologists could use these traits to draw conclusions about mobility.

3.5 MSM Research in Anatomically Modern Humans

Another area where MSMs may be useful in determining mobility types is in modern humans. Since all modern human populations are habitual bipeds, examining the patterns found in MSMs may help us establish the usefulness of these traits for reconstructing bipedalism. One expected trend is greater variation across populations in upper limb MSMs than in lower limb MSMs. Since upper limbs are used in a variety of activities that are culturally specific (e.g., atl-atl use, hide preparation, and mortar and pestle use), one might expect there to be much variation within and across populations. It is important to note that contrasting results have emerged on studies of identified skeletal collections about the link between specific activities and MSMs (e.g., Niinimäki 2011; Villotte et al. 2010a; Milella et al. 2012; Alves Cardoso and Henderson 2010). Given the difference in specific subsistence-related and other activities across populations, we expect different muscle insertion and origin sites, or complexes of sites, to show higher MSM scores. Conversely, the lower limb is mainly used in walking and running, and the same complex of muscles is activated in all populations, albeit at different degrees of intensity. However, the lower limb pattern could perhaps be complicated by the presence of activities, such as balancing on boats and horseback riding.

3.5.1 Meta-analyses

I decided to conduct two types of meta-analyses. The first meta-analysis looks at upper limb versus lower limb MSM scores within populations; the second meta-analysis examined the highest ranking MSM scores in published data to see whether the highest ranked scores for the upper limb vary more than for the lower limb. In both cases, the hypothesis is that the myriad of activities conducted with the upper limb should result in greater variability in MSM scores in upper limbs compared to lower limbs, which are used mainly for bipedal locomotion. The main limitations to these meta-analyses include that inter-observer error rates may be high due to the many different researchers involved, there is no standardized method of data collection, and the sample sizes are small.

The first meta-analysis conducted for this chapter involved 12 samples from five studies (Table 3.2) that separated upper and lower limb data by sex. I examined whether upper limb or lower limb scores showed more variation by looking at the averages for the highest and lowest means, or the highest and lowest frequencies for each MSM since these values were available in each study. Then, I calculated the score range for the lower limbs and the upper limbs. For example, Steen and Lane (1998) reported male Golvin Bay Alaskan Eskimos' with upper limb MSM scores ranging from a mean of 1.3 for the *subclavius* muscle to a mean of 3.0 in the *teres major*; for the lower limb they reported mean ranges from 1.2 (*tibialis anterior* and *flexor digitorum* muscles) to 3.1 (*gluteus maximus*). Thus, there was more variation

Table 3.2 Populations examined for greater upper limb compared to lower limb variation

Publication	Location	Culture	Male sample size	Female sample size
Steen and Lane (1998)	Alaska	Eskimo, Hunter-gathering	111	127
Havelková et al. (2011)	Great Moravia, Central Europe	Farming	35	45
	Great Moravia, Central Europe	Elite, Urban	68	49
Churchill and Morris (1998)	South Africa	Khoisan Foragers, Fynbos	22	18
		Khoisan Foragers, Forest	12	5
		Khoisan Foragers, Savanna	8	5
al-Oumaoui et al. (2004)	Iberian Pennisula	La Carada, Agriculture	39	39
		Argar, Hunter-gatherer and Agriculture	41	32
		La Torrecilla, Agriculture	47	44
		Villanueva, Agriculture	30	26
		S. Baudelio, Herding	16	11
Milella et al. (2012)	Italy	Contemporary population	274	210

in lower limb MSMs among male Golvin Bay Eskimos than in their upper limb MSMs. I discovered that there was no difference in the frequency of variation between upper limb MSM scores and lower limb MSM scores. Among males, five populations had greater lower limb MSM variation, five populations had greater upper limb MSM variation, and two populations showed no differences. Among females, five populations had greater lower limb MSM variation and seven populations had greater upper limb MSM variation. Differences between these distributions were nonsignificant ($X^2=2.33$, d$f=2$, $P=0.311$).

Due to the limitations of a small meta-analysis, I examined the same question using a different technique. To determine whether there is more variation in the upper limb than in the lower limb, I looked at which MSMs were ranked highest across 45 populations with upper limb data and 25 populations with lower limb data. The populations included preagricultural, agricultural, and industrial samples; additionally, samples from the New World (Chapman 1997; Hawkey and Merbs 1995; Steen and Lane 1998; Weiss 2003b, 2007) and the Old World (al-Oumaoui et al. 2004; Havelková et al. 2011; Lukacs and Pal 2003; Peterson 1998; Molnar 2006; Eshed et al. 2004) were represented. More studies could be included because the studies did not require data on both upper and lower limbs. Unexpectedly, upper limbs displayed less variation than lower limbs in which MSMs were ranked highest. The results are displayed in Figs. 3.1 and 3.2. In the upper limb, the *pectoralis major* muscle received the highest rank in 35.5 % of the populations; this same pattern was found in males (41 %) and females (32 %) separately. Over a third of the populations ranked the *pectoralis major* muscle higher than any other upper limb site, which is surprising considering cultural variation. In the lower limb, no

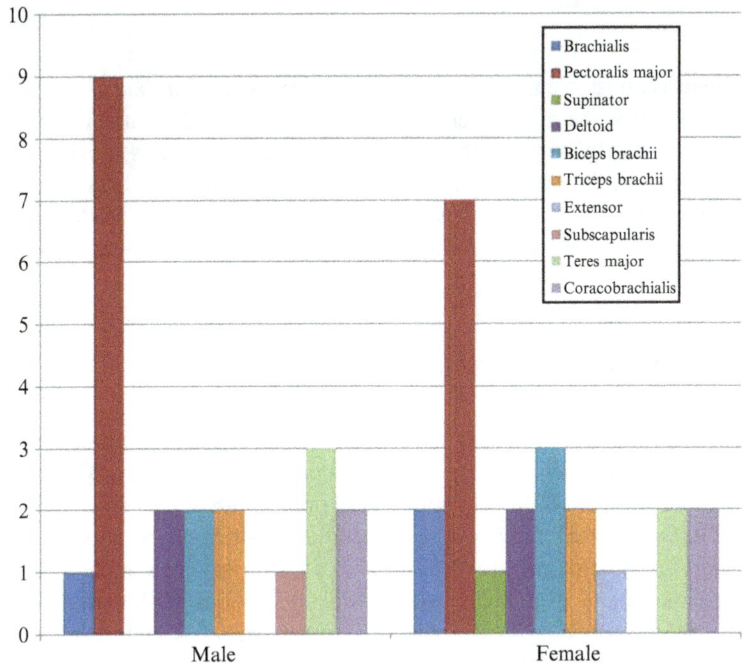

Fig. 3.1 Upper limb MSM data from 45 populations. Bars display all the MSMs that were ranked highest in at least one population. *Y*-axis displays the number of times the specific MSM is ranked highest

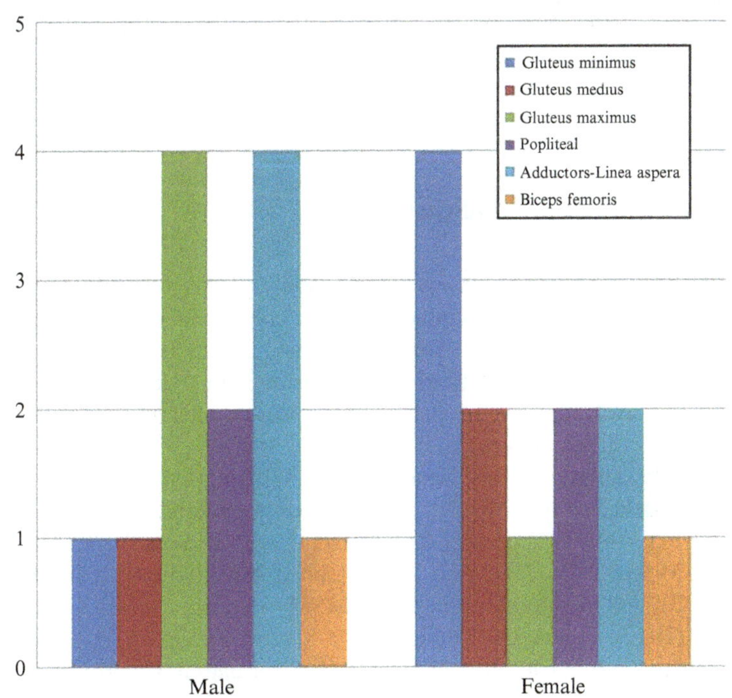

Fig. 3.2 Lower limb MSM data from 25 populations. Bars display all the MSMs that were ranked highest in at least one population. *Y*-axis displays the number of times the specific MSM is ranked highest

single site was most common; the *gluteus minimus*, *gluteus maximus*, and the *adductor* muscles at the linea aspera were all the highest ranking in 5 out of 25 different populations. In females, the *gluteus minimus* muscle was most frequently highest ranked (42 %) and in males the *gluteus maximus* muscle was most frequently highest ranked (33 %). Thus, it appears that lower limb MSMs are more varied and this variation can be found across populations and between sexes.

3.5.2 Single Study of Bipedalism and MSMs

Meta-analyses support my recent study looking at MSMs of a single California hunter-gatherer population. I hypothesized that in bipeds, lower limb MSMs should correlate more strongly with one another than upper limb MSMs since bipedalism engages the entire lower limb, whereas specific activities of the upper limb may require more specific muscle use (Weiss 2011). I tested this in a sample of 105 individuals (52 males and 53 females) using fibrocartilaginous MSMs of upper limbs (four humeral, two ulnar, and two radial) and lower limbs (six femoral and one tibial). I found correlations in upper limb MSMs only. However, upper limbs did also express more asymmetry than lower limbs. The main problem occurred with the lack of variation in lower limb fibrocartilaginous MSMs. For example, the *quadriceps femoris* MSM scores were the same for each individual. With no visible variation in some of the fibrocartilaginous MSM scores, statistical tests could not be performed with them; after all, a variable must vary to be able to perform either parametric or nonparametric tests. The results presented in this section suggest that MSMs may not be useful in mobility reconstructions.

3.6 Discussion and Conclusion

MSMs have been used to reconstruct activity patterns in bioarchaeology for nearly three decades; mobility reconstruction has been a component of this research (e.g., al-Oumaoui et al. 2004; Dutour 1986; Churchill and Morris 1998; Havelková et al. 2011; Lukacs and Pal 2003; Mariotti and Belcastro 2011). However, MSMs have not been extensively utilized by paleoanthropologists attempting to understand mobility in even early hominins.

Perhaps paleoanthropologists' resistance to employing MSMs is a result of the contradictory evidence that MSMs relate to activity patterns. For example, most researchers (e.g., Alves Cardoso and Henderson 2010; Chapman 1997; Havelková et al. 2011; Molnar 2006; Niinimäki 2011; Robb 1998; Villotte et al. 2010a; Weiss 2003b, 2007) find relationships between MSMs and age. Alves Cardoso and Henderson (2010), Cunha and Umbelino (1995), and Niinimäki (2011) state that age in itself is likely a causative factor of MSM formation. Controls for age are difficult in bioarchaeological samples, but they may be nearly impossible in even early hominin samples.

Another problem with MSMs is their correlation with body size (Lieverse et al. 2009; Niinimäki 2011; Weiss 2003b, 2004, 2007; Zumwalt et al. 2000). But, the type of muscle attachment examined may be important in determining whether MSMs can be useful in mobility research. Much of the MSM previous research has focused on fibrous sites, but fibrocartilaginous sites may be more suitable for activity reconstructions of any type. Fibrocartilaginous muscle attachment sites occur at secondary ossification sites of long bones and along the carpals, tarsals, and vertebrae (Benjamin et al. 2004). They can be thought of as organs that involve four distinct tissue zones and their changes in response to activities are better understood than fibrous muscle site changes (Benjamin et al. 2004). Recent research by Weiss (2012), and clinical research summarized by Villotte and Knüsel (2013), indicate that fibrocartilaginous MSMs change with age, but that they may not be affected by body size. Fibrocartilaginous MSMs should have no correlation to body size due to their anatomy (Villotte et al. 2010a). Body size correlations with MSMs, it seems, should be present only where muscles attach on extensive areas of bone, which is not the case at fibrocartilaginous sites, but is the case at fibrous sites. The lack of need to control for body size would be useful in hominin research due to the variation in body size found in the fossil record. Some fibrocartilaginous MSM scores, however, show less variation than fibrous sites, which puts the utility of fibrocartilaginous sites into question as well (Weiss 2012). Upper limb fibrocartilaginous MSMs, for instance, display more variation than lower limb fibrocartilaginous MSMs, but the opposite is true for fibrous MSMs (Weiss 2011, 2012).

Review of the literature and meta-analyses provided here offer a mixed message. In some aspects, ape data compared to human data support the use of MSMs for bipedalism studies (e.g., Drapeau 2008), but research on asymmetry by Cashmore (2009) suggests that MSMs do not help us distinguish between ape and human patterns of upper limb use. Further, Zumwalt et al. (2000) found that body size—and not locomotor type—was the best predictor of MSMs development in nonhuman primates.

In regards to anatomically modern human samples, variation was not greater in the upper limb compared to the lower limb. It appears that even if various populations engaged in a multitude of different activities, their upper limb MSMs were similar. This was true when sexes were separated or combined. And, lower limbs displayed as much or more variation than upper limbs; for example, males and females had different highest ranking lower limb muscle markers. The sex difference, however, may be at least partially the result of sex differences in pelvic morphology due to the obstetrics of giving birth to a large-brained baby. The wider pelvis of females may result in greater muscular stresses, and during pregnancy carrying frontal load may result in excessive stress on their lower back and upper thigh muscles. During pregnancy, the *gluteal* muscles are strained and often result in lower back pain (Sivrioglu et al. 2013). But, even within a population, correlations are found within both upper limb and lower limb MSM scores; this is contradictory to what is expected. When engaging in bipedal behavior, the whole lower limb is utilized and, although not all muscles are used simultaneously, the pattern of use should be repeated with each step. Thus, this should result in greater correlations in lower limb MSMs than in upper limb MSMs since upper limbs engage in a variety of activities that do not necessarily employ the entire limb. Hence, we may expect greater correlations between scores of MSMs on the femur and tibia than between the humerus and the radius or ulna. With such mixed results, it seems necessary to further explore MSM etiologies before supporting their use in mobility level and type of research.

References

Abitbol MM (1995) Lateral view of *Australopithecus afarensis*: primitive aspects of bipedal positional behavior in the earliest hominids. J Hum Evol 28:211–229

al-Oumaoui I, Jiménez-Brobeil S, du Souich R (2004) Markers of activity patterns in some populations of the Iberian Peninsula. Int J Osteoarchaeol 14:343–359

Alves Cardoso F, Henderson CY (2010) Enthesopathy formation in the humerus: data from known age-at-death and known occupation skeletal occupations. Am J Phys Anthropol 141:550–560

Benjamin M, Moriggl B, Brenner E, Emery P, McGonagle D, Redman S (2004) The "enthesis organ" concept: why enthesopathies may not present as focal insertional disorders. Arthritis Rheum 50:3306–3313

Bertram JE, Swartz SM (1991) The 'law of bone transformation': a case of crying Wolff? Biol Rev 66:245–273

Bramble DM, Lieberman DE (2004) Endurance running and the evolution of Homo. Nature 432:345–352

Carlson KJ (2006) Muscle architecture of the common chimpanzee (*Pan troglodytes*): perspectives for investigating chimpanzee behavior. Primates 47:218–229

Cashmore LA (2009) Can hominin 'handedness' be accurately assessed? Ann Hum Biol 36:624–641

Cashmore LA, Zakrzewski SR (2013) Assessment of musculoskeletal stress marker development in the hand. Int J Osteoarchaeol 23:334–347

Chamay A, Tschantz P (1972) Mechanical influences in bone remodeling. Experimental research on Wolff's law. J Biomech 5:173–180

Chapman NE (1997) Evidence for Spanish influence on activity induced musculoskeletal stress markers at Pecos Pueblo. Int J Osteoarchaeol 7:497–506

Churchill SE, Morris AG (1998) Muscle marking morphology and labour intensity in prehistoric Khoisan foragers. Int J Osteoarchaeol 8:390–411

Cunha E, Umbelino C (1995) What can bones tell about labour and occupation: the analysis of skeletal markers of occupational stress in the Identified skeletal collection of the Anthropological Museum of the University of Coimbra (preliminary results). Antropol Port 13:49–68

Davis PR (1964) Hominid fossils from Bed I, Olduvai Gorge, Tanganyika. Nature 201:967–988

Drapeau M (2008) Enthesis bilateral asymmetry in humans and African apes. Homo 59:93–109

Drapeau M, Ward CV, Kimbel WH, Johanson DC, Rak Y (2005) Associated cranial and forelimb remains attributed to *Australopithecus afarensis* from Hadar, Ethiopia. J Hum Evol 48:593–642

Dutour O (1986) Entheses (lesions of muscular insertions) as indicators of the activities of Neolithic Saharan populations. Am J Phys Anthropol 71:221–224

Eshed V, Gopher A, Gage TB, Hershkovitz I (2004) Has the transition to agriculture reshaped the demographic structure of prehistoric populations? New evidence from the Levant. Am J Phys Anthropol 123:303–315

Galik K, Senut B, Pickford M, Gommery D, Treil J, Kuperavage AJ, Eckhardt RB (2004) External and internal morphology of the BAR 1002'00 *Orrorin tugenensis* femur. Science 305:1450–1453

Gebo DL, Schwartz GT (2006) Foot bones from Omo: implications for hominid evolution. Am J Phys Anthropol 129:499–511

Godde K, Taylor RW (2011) Musculoskeletal stress marker (MSM) differences in the modern American upper limb and pectoral girdle in relation to activity level and body mass index (BMI). Forensic Sci Int 210:237–242

Gordon AD, Green DG, Richmond BG (2008) Strong postcranial size dimorphism in *Australopithecus afarensis*: results from two new resampling methods from multivariate data sets with missing data. Am J Phys Anthropol 135:311–328

Häusler M, Berger L (2001) Stw 441/465: a new fragmentary ilium of a small-bodied *Australopithecus africanus* from Sterkfontein, South Africa. J Hum Evol 40:411–417

Havelková P, Villotte S, Velemínský P, Poláček L, Dobisíková M (2011) Entheses and activity patterns in the Early Medieval Great Moravian population: evidence of division of labour. Int J Osteoarchaeol 21:487–504

Hawkey DE, Merbs CF (1995) Activity-induced musculoskeletal stress markers (MSM) and subsistence strategy changes among ancient Hudson Bay Eskimos. Int J Osteoarchaeol 5:324–338

Kutterer A, Uerpmann M (2010) Human remains from a post-Hellenistic grave-chamber in Dibba al-Hisn (Sharjah, UAE). Arab Arch Epig 21:170–198

Lai P, Lovell N (1992) Skeletal markers of occupational stress in the fur trade: a case study from a Hudson's Bay Company fur trade post. Int J Osteoarchaeol 2:221–234

Latimer B, Lovejoy CO (1989) The calcaneus of *Australopithecus afarensis* and its implications for the evolution of bipedalism. Am J Phys Anthropol 78:369–386

Lieberman DE, Pilbeam DR, Wood BA (1988) A probabilistic approach to the problem of sexual dimorphism in *Homo habilis*: a comparison of KNM-ER 1470 and KNM-ER 1813. J Hum Evol 17:503–511

Lieverse AR, Bazaliiskii VI, Goriunova OI, Weber AW (2009) Upper limb musculoskeletal stress markers among middle Holocene foragers of the Siberia's Cis-Baikal region. Am J Phys Anthropol 138:458–472

Lovejoy CO, Suwa G, Spurlock L, Asfaw B, White TD (2009) The pelvis and femur of *Ardipithecus ramidus*: the Emergence of upright walking. Science 326:71–76

Lovell NC, Dublenko AA (1999) Further aspects of fur trade life depicted in the skeleton. Int J Osteoarchaeol 9:248–256

Lukacs JR, Pal JN (2003) Skeletal variation among Mesolithic people of the Ganga Plains: new evidence of habitual activity and adaptation to climate. Asian Persp 42:329–351

Mariotti V, Belcastro MG (2011) Lower limb entheseal morphology in the Neanderthal Krapina population (Croatia, 130 000 BP). J Hum Evol 60:694–702

Mariotti V, Facchini F, Belcastro MG (2007) The study of entheses: proposal of a standardised scoring method for twenty-three entheses of the postcranial skeleton. Coll Antropol 31:191–313

Marzke MW, Longhill JM, Rasmussen SA (1988) Gluteus maximus muscle function and the origin of hominid bipedalism. Am J Phys Anthropol 77:519–528

Mays S, Steele J, Ford M (1999) Directional asymmetry in the human clavicle. Int J Osteoarchaeol 9:18–28

McHenry HM, Berger LR (1998) Body proportions of *Australopithecus afarensis* and *A. africanus* and the origin of the genus *Homo*. J Hum Evol 35:1–22

Milella M, Belcastro MG, Zollikofer CPE, Mariotti V (2012) The effects of age, sex, and physical activity on entheseal morphology in a contemporary Italian skeletal collection. Am J Phys Anthropol 148:379–388

Molnar P (2006) Tracing prehistoric activities: musculoskeletal stress marker analysis of a stone-age population on the island of Gotland in the Baltic Sea. Am J Phys Anthropol 129:12–23

Molnar P (2010) Patterns of physical activity and material culture on Gotland, Sweden during the Middle Neolithic. Int J Osteoarchaeol 20:1–14

Molnar P, Ahlstrom TP, Leden I (2011) Osteoarthritis and activity—an analysis of the relationship between eburnation, musculoskeletal stress markers (MSM) and age in two Neolithic hunter-gatherer populations from Gotland, Sweden. Int J Osteoarchaeol 21:283–291

Morimoto N, De León MSP, Nishimura T, Zollikofer CP (2011) Femoral morphology and femoropelvic musculoskeletal anatomy of humans and great apes: a comparative virtopsy study. Anat Rec 294:1433–1445

Myszka A, Piontek J (2011) Shape and size of the body vs. musculoskeletal stress. J Biol Clin Anthropol 68(2):139–152

Nagy BLB (1998) Age, activity, and musculoskeletal stress markers. Am J Phys Anthropol 26S:168–169

Niinimäki S (2011) What do muscle marker ruggedness scores actually tell us? Int J Osteoarchaeol 21:292–299

Niskanen M, Junno JA (2009) Estimation of African apes' body size from postcranial dimensions. Primates 50:211–220

Peterson J (1998) The Natufian hunting conundrum: spears, atlatls, or bows. Musculoskeletal and armature evidence. Int J Osteoarchaeol 8:378–389

Pickford M, Senut B, Gommery D, Treil J (2002) Bipedalism in *Orrorin tugenensis* revealed by its femora. Hum Paleontol Prehist 1:191–203

Plavcan JM (2012) Body size, size variation, and sexual size dimorphism in early homo. Curr Anthropol 53:S409–S423

Raichlen DA, Pontzer H, Sockol MD (2008) The Laetoli footprints and early hominin locomotor kinematics. J Hum Evol 54:112–117

Reno PL, McCollum MA, Meindl RS, Lovejoy CO (2010) An enlarged postcranial sample confirms Australopithecus afarensis dimorphism was similar to modern humans. Philos Trans R Soc Lond B Biol Sci 365:3355–3363

Robb JE (1998) The interpretation of skeletal muscle sites: a statistical approach. Int J Osteoarchaeol 8:363–377

Rose MD (1984) A hominine hip bone, KNM-ER 3228, from East Lake Turkana, Kenya. Am J Phys Anthropol 63:371–378

Ruff C, Holt B, Trinkaus E (2006) Who's afraid of the big bad Wolff?:"Wolff's law" and bone functional adaptation. Am J Phys Anthropol 129:484–498

Ruff CB, Larsen CS (2014) Long bone structural analyses and the reconstruction of past mobility—a historical review. In: Carlson KJ, Marchi D (eds) Reconstructing mobility: environmental, behavioral, and morphological determinants. Springer, New York

Shaw HM, Benjamin M (2007) Structure–function relationships of entheses in relation to mechanical load and exercise. Scand J Med Sci Sports 17:303–315

Sivrioglu AK, Ozyurek S, Mutlu H, Sonmez G (2013) Piriformis syndrome occurring after pregnancy. BMJ Case Rep. doi:10.1136/bcr-2013-008946

Stanford C (2012) Chimpanzee and the behavior of *Ardipithecus ramidus*. Annu Rev Anthropol 41:139–149

Steen SL, Lane RW (1998) Evaluation of habitual activities among two Alaskan Eskimo populations based on musculoskeletal stress markers. Int J Osteoarchaeol 8:341–353

Stefanović S, Porčić M (2013) Between-group differences in the patterning of musculo-skeletal stress markers: avoiding confounding factors by focusing on qualitative aspects of physical activity. Int J Osteoarchaeol 23:94–105

Stern JT (2000) Climbing to the top: a personal memoir of Australopithecus afarensis. Evol Anthropol 9:113–133

Stern JT Jr, Susman RL (1983) The locomotor anatomy of Australopithecus afarensis. Am J Phys Anthropol 60:279–317

Stirland A (1993) Asymmetry and activity-related change in the male humerus. Int J Osteoarchaeol 3:105–113

Stock J, Pfeiffer S (2001) Linking structural variability in long bone diaphyses to habitual behaviors: foragers from southern African Later Stone Age and the Andaman Islands. Am J Phys Anthropol 115:337–348

Suwa G, Lovejoy CO, Asfaw B, White T (2012) Proximal femoral musculoskeletal morphology of chimpanzees and its evolutionary significance: a critique of Morimoto et al. (2011). Anat Rec 295:2039–2044

Tallman M (2012) Morphology of the distal radius in extant hominoids and fossil hominins: implications for the evolution of bipedalism. Anat Rec 295:454–464

Tardieu C (1999) Ontogeny and phylogeny of femoro-tibial characters in humans and hominid fossils: functional influence and genetic determinism. Am J Phys Anthropol 110:365–377

Villotte S, Knüsel CJ (2013) Understanding entheseal changes: definition and life course changes. Int J Osteoarchaeol 23:135–146

Villotte S, Castex D, Couallier V, Dutour O, Knüsel CJ, Henry-Gambier D (2010a) Entheses as occupational stress markers: evidence from the upper limb. Am J Phys Anthropol 142:224–234

Villotte S, Churchill SE, Dutour OJ, Henry-Gambier D (2010b) Subsistence activities and the sexual division of labor in the European Upper Paleolithic and Mesolithic: evidence from upper limb enthesopathies. J Hum Evol 59:35–43

Ward CV (2002) Interpreting the posture and locomotion of *Australopithecus afarensis*: where do we stand? Yearb Phys Anthropol 45:185–215

Ward CV, Kimbel WH, Harmon EH, Johanson DC (2012) New postcranial fossils of *Australopithecus afarensis* from Hadar, Ethiopia (1990–2007). J Hum Evol 63:1–51

Waterman HC (1929) Studies on the evolution of the pelvis of man and other primates. Bull Am Mus Nat Hist 58:585–642

Weiss E (2003a) The effects of rowing on humeral strength. Am J Phys Anthropol 121:293–302

Weiss E (2003b) Understanding muscle markers: aggregation and construct validity. Am J Phys Anthropol 121:230–240

Weiss E (2004) Understanding muscle markers: lower limbs. Am J Phys Anthropol 125:232–238

Weiss E (2007) Muscle markers revisited: activity pattern reconstructed with controls in a central California Amerind Population. Am J Phys Anthropol 133:931–940

Weiss E (2011) Can mobility patterns be revealed through enthesopathies? Am J Phys Anthropol 52S:308

Weiss E (2012) Examining activity patterns and biological confounding factors: differences between fibrocartilaginous and fibrous MSMs. Int J Osteoarchaeol. doi:10.1002/oa.2290

Weiss E, Corona L, Schultz B (2012) Sex differences in MSMs: problems with activity pattern reconstructions. Int J Osteoarchaeol 22:70–80

Wilczak CA (1998) Consideration of sexual dimorphism, age, and asymmetry in quantitative measurements of muscle insertion sites. Int J Osteoarchaeol 8:311–325

Woo SL, Kuei SC, Amiel D, Gomez MA, Hayes WC, White FC, Akeson WH (1981) The effect of prolonged physical training on the properties of long bone: a study of Wolff's law. J Bone Joint Surg Am 63A:780–786

Zihlman AL, McFarland RK, Underwood CE (2011) Functional anatomy and adaptation of Male Gorillas (*Gorilla gorilla gorilla*) with comparison to male orangutans (*Pongo pygmaeus*). Anat Rec 294:1842–1855

Zumwalt A, Ruff C, Wilczak C (2000) Primate muscle insertions: what does size tell you? Am J Phys Anthropol 30s:331

Chapter 4
Does the Distribution and Variation in Cortical Bone Along Lower Limb Diaphyses Reflect Selection for Locomotor Economy?

Colin N. Shaw, Jay T. Stock, Thomas G. Davies, and Timothy M. Ryan

Abstract It has been hypothesised that limb tapering reflects an energetic trade-off between bone strength and weight, and selection for tissue economy, resulting in lighter distal limb segments. If adaptive mechanisms constrain the response of osseous tissue to mechanical loading one might expect a higher level of constraint, and therefore less variability, in more distal aspects of the limb. High-resolution CT was used to quantify the distribution and variation in strength (Z_p), cortical area (CA) and shape (I_{max}/I_{min}) at 5 % intervals along the femoral and tibial diaphysis for

C.N. Shaw, Ph.D. (✉)
PAVE Research Group, Department of Archaeology & Anthropology, University of Cambridge, Downing Street, Cambridge CB2 3ER, UK

McDonald Institute for Archaeological Research, Department of Archaeology and Anthropology, University of Cambridge, Downing Street, Cambridge CB2 3ER, UK
e-mail: cns30@cam.ac.uk; cshaw111@mac.com

J.T. Stock, Ph.D.
PAVE Research Group, Department of Archaeology & Anthropology, University of Cambridge, Downing Street, Cambridge CB2 3ER, UK
e-mail: jts34@cam.ac.uk

T.G. Davies, Ph.D.
PAVE Research Group, Department of Archaeology & Anthropology, University of Cambridge, Downing Street, Cambridge CB2 3ER, UK

Churchill College, Storey's Way, Cambridge, UK
e-mail: td274@cam.ac.uk

T.M. Ryan, Ph.D.
Department of Anthropology, Pennsylvania State University, University Park, State College, PA 16868, USA

Center for Quantitative X-Ray Imaging, EMS Energy Institute, Pennsylvania State University, University Park, State College, PA 16868, USA
e-mail: tmr21@psu.edu

K.J. Carlson and D. Marchi (eds.), *Reconstructing Mobility: Environmental, Behavioral, and Morphological Determinants*, DOI 10.1007/978-1-4899-7460-0_4,

a skeletal sample of mid- to late Holocene Native American agriculturalists and foragers ($M=21$, $F=19$). Z_p and CA are highest in the proximal femur, decrease at a fairly consistent rate (Z_p increases in the distal femur) and reach their lowest values at the distal tibia. By contrast, inherent morphological variability (coefficient of variation) for both Z_p and CA are relatively constant along both the femur and tibia. The distribution and variation in I_{max}/I_{min} is greater than that of CA or Z_p. These findings support earlier studies that have identified tapering in human limbs, yet, because morphological plasticity appears to be generally consistent across the diaphyses of the femur and tibia, morphological constraint (canalisation) does not seem to be the overriding mechanism dictating the tapering of limb bone structure.

Keywords *Homo sapiens* • Femur • Tibia • Cortical bone • Geometric property variation and distribution • High-resolution computed tomography

4.1 Introduction

The relationship between habitual behaviour patterns and long bone diaphyseal morphology is the basis upon which inferences of prehistoric activity patterns, and more often mobility patterns, are often based (cf., Holt 2003; Marchi 2008; Ruff 2008b, 2009; Stock 2006; Stock and Pfeiffer 2001). For the purposes of this study, the term 'mobility' refers to distance covered during bipedal locomotor travel (e.g. throughout a lifetime). However, suggesting that any measure of 'mobility' merely reflects distance travelled is inadequate. Such a definition of mobility ignores factors such as load magnitude or intensity (influenced by, among other factors, body weight, load carried, speed of travel, substrate angle, terrain undulation) and frequency (how often load is applied: constantly, hourly, daily, weekly, etc.). Factors such as load magnitude and frequency influence the stress imposed upon the bone, the strain to which the bone must ultimately adapt, and therefore bone morphology. Thus, as well as a consideration of distance travelled, these additional factors are also understood to be integral to a definition of mobility, as much as is reasonable. While it has been argued that the relationship between activity/mobility and skeletal structure, termed 'Bone Functional Adaptation' (Ruff et al. 2006), is not necessarily straightforward (Pearson and Lieberman 2004), the correspondence between variation in diaphyseal strength and shape, and the performance of habitual activities has been repeatedly demonstrated experimentally (cf., Hseih et al. 2001; Judex et al. 1997; Lanyon 1992; Rubin et al. 1990).

Variation in robusticity along a limb has been attributed to selective pressure for tissue economy and lighter distal limb segments, reflecting an energetic trade-off between bone strength and weight (Alexander 1998; Lieberman and Crompton 1998; Skedros et al. 2003). If adaptive mechanisms constrain the response of osseous tissue to mechanical loading, limiting the maximisation of strength while optimising biomechanical efficiency, one might expect a higher level of 'constraint', and therefore less variability, in more distal aspects of the limb. Theoretically, the morphology of distal limb segments should face relatively greater selective pressure for structural optimisation. The more distal aspects must be strong enough to prevent fracture, yet light enough to minimise the mass of the segment and the energetic requirements of movement. Proximal aspects of a limb, by contrast, can afford to maintain a greater amount of variation that is not directly related to functional constraints on the bone (Stock 2006). If genetically imposed constraints (i.e. canalisation) differ throughout the skeleton, distinct morphological patterns among postcranial elements should be expected (Hallgrimsson et al. 2002; Lieberman et al. 2003). It has been predicted that variability in bone robusticity would decrease towards the distal end of the limb (Stock 2006). If this is the case, there may be adaptive mechanisms that constrain the ability of the bones of different limb segments to respond to mechanical loading. Until recently, it has been difficult to collect the morphological data necessary to adequately test these ideas.

Due to recent improvements in imaging technology and computing power, the quantification of morphological variation can now be performed for increasingly greater areas of the skeleton. The quantification of larger sections of long bone diaphyses has advanced osteometric analyses and descriptions of morphological variation along the entire limb. Previous studies have detected profound changes in hominin postcranial morphology over the last five to six million years; however, until now the opportunity has not existed to adequately assess differences in more than five or six locations on each bone diaphysis (c.f., Ruff and Hayes 1983b; Trinkaus and Ruff 2012).

The diaphyseal structure of long bones is often quantified by applying beam theory (see Ruff 2008a) to estimate cross sectional properties, including diaphyseal strength (Z_p), cortical area (CA) and diaphyseal 'shape' (e.g. I_{max}/I_{min}). The application of beam theory to the study of mechanical characteristics of human long bones provides a method to estimate the mechanical performance of a bone diaphysis under various types of loading (Holt 2003; Ruff et al. 1993; Shackelford 2007; Shaw and Stock 2009a, b; Stock and Pfeiffer 2004; Westcott 2014). High-resolution micro-CT scanning provides a means to further consider variation in the distribution of cortical bone throughout an entire diaphysis using quite small voxel sizes (i.e. at a large number of cross section locations). With a slice thickness (z) of 0.12 mm, ~3,000 CT slices are needed to construct a 3D rendering of an adult femoral or tibial diaphysis, and for each of these 2D slices geometric properties can be calculated. Previous studies have also considered such variation in the distribution of properties, but the resolution and extent of the analyses have been constrained by the need to manually digitise every cross section (cf., Ruff and Hayes 1983a, b; Trinkaus and Ruff 2012). To determine whether this abundance of data along bone diaphyses reflects more subtle mechanical variation that could lead to improvements in our ability to interpret habitual mobility from morphology, we must first gain a better understanding of the variation that exists throughout the lower limb.

The purpose of this study is to investigate the pattern of proximal-distal variation in cortical bone along the length of the femoral and tibial diaphyses, using samples of pre-industrial village agriculturalists and foragers from the late Holocene of North America. With this dataset, two hypotheses are tested:

1. The suggestion has been made that, for reasons of locomotor efficiency, the human lower limb has been tapered through selective mechanisms (cf., Lieberman et al. 2003). The first hypothesis posits that diaphyseal strength (Z_p) and cortical area (CA) are relatively tapered, becoming less robust moving from the proximal femur down the lower limb to the distal tibia, while the distribution of cross-sectional shape (I_{max}/I_{min}) will be independent of such patterning.
2. The second hypothesis posits that one of the factors controlling the tapering of robusticity along the lower limb diaphyses is greater levels of canalisation (more constraint, less plasticity) in more distal aspects of the lower limb (Hallgrimsson et al. 2002; Lieberman et al. 2003). This greater level of constraint will be reflected in lower levels of inherent variation in the morphology of distal segments, compared to more proximal segments.

These two hypotheses, the first concerning the distribution and magnitude of bone cross-sectional properties, and the second concerning relative levels of variation between different section locations, are of great significance to our understanding of the roles for mobility in influencing the structure of limb bones. Theoretically limb tapering and greater canalisation in the distal limb would suggest that the most mechanically relevant tissue for signatures of mobility may be located proximally. However, direct skeletal analyses have suggested that correlation between robusticity and mobility increases moving from proximal to distal aspects of the lower limb (Stock 2006). Thus, while enhancing our understanding of the mechanisms acting in the functional adaptation of limb bone morphology, testing these hypotheses is also central to refining interpretations of mobility that are drawn from skeletal morphology.

4.2 Materials and Methods

Matched femoral and tibial remains of 40 individuals, from the Norris Farms #36 and Black Earth archaeological skeletal collections, were used in this study. The 21 males ranged in age from 21.5 to 37.5 years (mean age at death: 31.4 years), while the age range for the 19 females was 18–47 years (mean age at death: 31.0 years). Age at death was estimated for the Norris Farms individuals using Transition Analysis (Boldsen et al. 2002; Milner and Smith 1990), while the multivariate methods used to determine age at death and sex for the Black Earth individuals are described in the original Carrier Mills Site Report (Jefferies and Butler 1982, pp. 1035–1039). The Norris Farms #36 site is a late Prehistoric cemetery site from the central Illinois River Valley dating to approximately AD 1300 with graves containing one or more individuals associated with the Oneota cultural tradition of village agriculturalists (Santure et al. 1990). Archaeologists from the Illinois State Museum excavated the site, and the remains were generally well preserved as a result of deep burial in thick loess deposits. By contrast, Black Earth is a multi-component site in the Carrier Mills Archaeological District in southern Illinois that dates to the Middle Archaic. The Black Earth modern human hunter-gatherers relied upon white-tailed deer and to a lesser extent a wide variety of aquatic avifauna, while also gathering an array of nuts and seeds (Brietburg 1980; Lopinot and Lynch 1979). All individuals included in these analyses were free from disease or pathology. Inclusion criteria dictated that for each individual the femoral and tibial remains were in a state of preservation that allowed for the measurement of maximum limb segment length, and the accurate measurement of diaphyseal cross-sectional images.

All remains were scanned at the Center for Quantitative Imaging (CQI), Pennsylvania State University. The Norris Farms collection was scanned using an OMNI-X HD-600 HRCT scanner using the PANTAK High-Energy X-ray sub-system, while the Black Earth collection was scanned using the X-TEK microfocus High-Resolution X-ray computed tomography sub-system (Varian Medical Systems, Lincolnshire, IL). Each specimen was mounted inside an acrylic tube (0.6 cm wall thickness) and was secured using radio-translucent low-density polyethylene (LDPE) foam disks to position each bone in anatomically accurate vertical position throughout scanning. Serial cross-sectional scans (taken perpendicular to the long axis of the long bone) were collected beginning at the most distal aspect of the bone and proceeded proximally to transect the entire element. All HRCT scans were collected using source energy settings of 180 kV/0.3 mA, 2,400 views, two samples per view, and a Feldkamp reconstruction algorithm. For each scan, approximately 89 slices were collected during each rotation. The images were reconstructed as 16-bit TIFF grayscale images with a $1{,}024 \times 1{,}024$ pixel matrix and an isotropic voxel size of 0.117 mm for Norris Farms and 0.113 mm for Black Earth. The resulting datasets contained between 3,000 and 4,100 slices representing the entire femur or tibia.

Long bone robusticity can be quantified by applying the engineering principle of 'beam theory' to cross-sectional geometric properties. The application of beam theory to the study of the mechanical characteristics of human long bones provides a method to estimate the mechanical performance of a bone diaphysis under various types of loading (Holt 2003; Ruff et al. 1993; Shackelford 2007; Shaw and Stock 2009a, b; Stock and Pfeiffer 2004). In this study, beam theory was used to estimate the mechanical performance of the diaphysis for each raw cross-sectional CT image through the calculation of specific geometric properties including maximum and minimum principle (second) moments of area (I_{max}, I_{min}), measures of maximum and minimum bending rigidity; section modulus (Z_p), a measure of torsional and twice average bending strength and cortical area (CA), a measure of axial compressive and tensile strength. From these measurements diaphyseal 'shape' (I_{max}/I_{min}) was calculated, an indicator of the distribution of bone about the neutral axis.

Adjusting absolute measures of bone strength to body size and shape of the individual expresses these properties relative to a common baseline of mechanical bone function (Ruff 2000; Ruff et al. 1993; Trinkaus 1997). This process enables the comparison of the mechanical capability of bones of different limb proportions. To control for variation in body size, polar second moment of area (J) is standardised by the product of body mass and moment arm length^2, while cortical area (CA) is standardised by body mass (Ruff 2008a; Sparacello and Marchi 2008; Shaw and Stock 2009a, b). In the present study, standardisation for body mass was unnecessary as it would have altered measures of J and CA from the femur and tibia equally, and therefore would not have influenced comparisons along the lower limb. Additionally, standardisation for limb length was also not employed. The aim of this study was to quantify variation in the lower limb, thus standardisation for limb segment length would have erroneously attenuated variation in the longer bone (femur) while simultaneously accentuating that in the shorter bone (tibia).

Analyses were performed on datasets derived from whole diaphysis CT scans. To properly quantify cross-sectional geometric properties of the diaphyseal cortical bone for the Norris Farms remains, a dual threshold technique was used to separate the cortical bone compartment from the non-cortical regions of the diaphysis (i.e. the medullary cavity) (Buie et al. 2007). This was a necessary step that was taken to remove the loess deposits that had infiltrated and filled the medullary cavities of many Norris Farms individuals, and would have adversely influenced the later calculation of cross-sectional properties. This method uses two threshold inputs and a series of dilation and erosion procedures to define the endosteal and periosteal surfaces of the cortical bone. The result of this algorithm is a binary image mask of the cortical bone with a two to five pixel buffer on all sides. This binary mask, in which '1' represents cortical bone and '0' represents all non-cortical bone regions, was then multiplied by the original dataset to produce a new dataset. In the 'new' dataset the cortical bone compartment was fully segmented from any loess that had been contained in the medullary cavity. A relatively conservative approach was taken with the masking operation to ensure that no cortical bone was removed during this step. Specialised code running in Image Processing Language (IPL) was used to generate the dual threshold mask, and the final masking step was performed in Avizo Fire 6.2 (VSG). This processing step ensured that no loess or other extraneous material was included in the cross-sectional slices to be quantified in later steps.

The number of cross-sectional CT images ('slices') required to translate through an entire long bone at 0.117 and 0.113 mm resolution was over 3,000 'slices' in some cases. For the femur and tibia, the proximal and distal 20 % were excluded from each bone. Following extraction of the diaphysis, the cross-sectional properties for the diaphysis were calculated using the BoneJ plugin (Doube et al. 2010, http://bonej.org/) within ImageJ (http://rsbweb.nih.gov/ij/). For each individual, cross-sectional properties were calculated along the diaphysis between 20 and 80 % of whole bone length at 5 % increments (i.e. 20, 25, 30 … . 80 %). At each 5 % section location, the coefficient of variation (CV) was calculated for CA, Z_p and I_{max}/I_{min} to provide a metric indicator of the inherent morphological variation of each property along the lower limb (see Simpson et al. 1960). Comparisons of diaphyseal properties were conducted separately for male and female individuals within the Norris Farms and Black Earth groups, and also as a pooled sample which combined both populations, yet still separated individuals by sex prior to analysis.

4.3 Results

4.3.1 Distribution of Diaphyseal Cortical Area (CA), Strength (Z_p) and Shape (I_{max}/I_{min}) Along the Femur and Tibia

Comparisons of diaphyseal Z_p (Table 4.1, Fig. 4.1) reveal that, for both males and females, diaphyseal strength is at its highest at the proximal aspect of the femur (80 % of bone length), decreases consistently until the proximal-midshaft (60–50 %), where it then increases steadily and peaks at the distal aspect of the femur (20 %). The strength of the proximal tibia (80 %) is comparable to the distal femur and decreases relatively consistently to the weakest point along the whole leg at the distal tibia (30–20 %). Distribution of diaphyseal cortical area (CA; Table 4.1, Fig. 4.2) also follows a consistent pattern for both the male and female sub-samples.

Table 4.1 Mean values for cortical area (CA), section modulus (Z_p) and cross-sectional shape (I_{max}/I_{min}) at each 5 % section along the femoral and tibial diaphysis for males and females, Black Earth and Norris Farms combined

	Z_p		CA		I_{max}/I_{min}	
% Bone length	Male	Female	Male	Female	Male	Female
Femur						
80	3,516.72	2,635.53	432.19	329.09	1.427	1.609
75	3,071.96	2,381.18	406.85	339.10	1.873	2.050
70	2,914.02	2,285.90	414.46	346.50	1.981	2.066
65	2,887.38	2,258.45	414.21	345.40	1.583	1.651
60	2,837.63	2,203.50	414.04	346.18	1.257	1.332
55	2,684.91	2,102.33	407.46	338.57	1.248	1.245
50	2,604.81	2,011.94	396.38	328.50	1.344	1.267
45	2,604.00	1,906.77	386.61	317.08	1.415	1.303
40	2,649.11	1,979.19	376.26	306.52	1.404	1.306
35	2,822.25	2,080.96	365.99	290.24	1.351	1.226
30	3,029.86	2,223.43	353.02	278.98	1.244	1.192
25	3,259.30	2,297.96	350.44	271.01	1.218	1.229
20	3,476.58	2,509.55	349.16	267.14	1.320	1.364
Tibia						
80	3,391.24	2,165.38	332.34	235.17	2.506	2.361
75	2,841.84	1,873.96	326.03	232.88	2.731	2.521
70	2,533.55	1,723.21	324.15	235.31	2.928	2.564
65	2,362.00	1,637.10	330.74	243.79	2.994	2.544
60	2,219.43	1,543.97	334.48	247.32	2.960	2.453
55	2,086.61	1,453.47	332.44	245.76	2.866	2.387
50	1,983.27	1,399.38	330.27	248.27	2.741	2.290
45	1,851.34	1,329.61	322.32	243.25	2.587	2.167
40	1,752.91	1,247.04	308.89	228.63	2.392	2.028
35	1,672.99	1,180.33	289.34	211.66	2.165	1.853
30	1,620.44	1,145.96	263.75	193.68	1.874	1.634
25	1,625.87	1,136.88	242.31	174.62	1.554	1.410
20	1,680.38	1,139.34	222.70	155.52	1.303	1.237

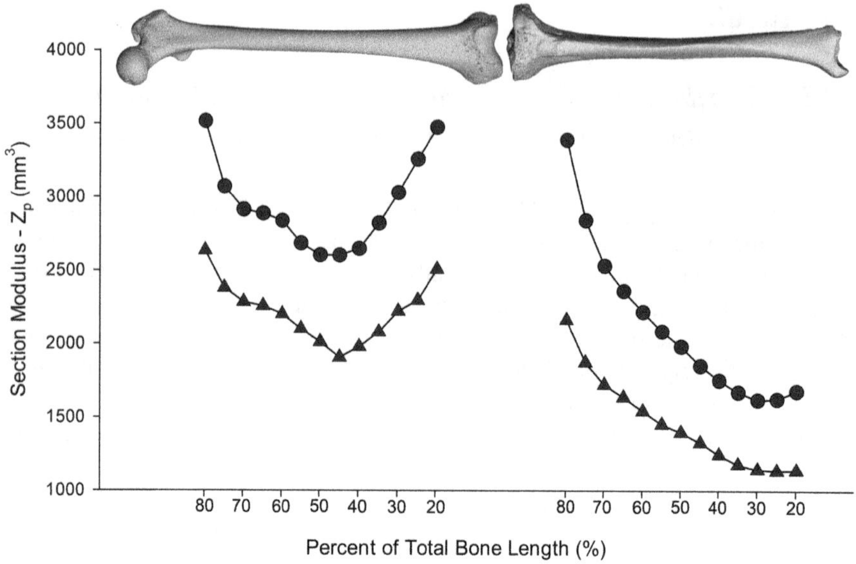

Fig. 4.1 Diaphyseal strength (Z_p, Y-axis) raw data plotted at 5 % intervals from the proximal (1 on X-axis) to distal (13 on X-axis) femoral diaphysis and proximal (14 on X-axis) to distal (26 on X-axis) tibial diaphysis. Males: *circles*, females: *triangles*

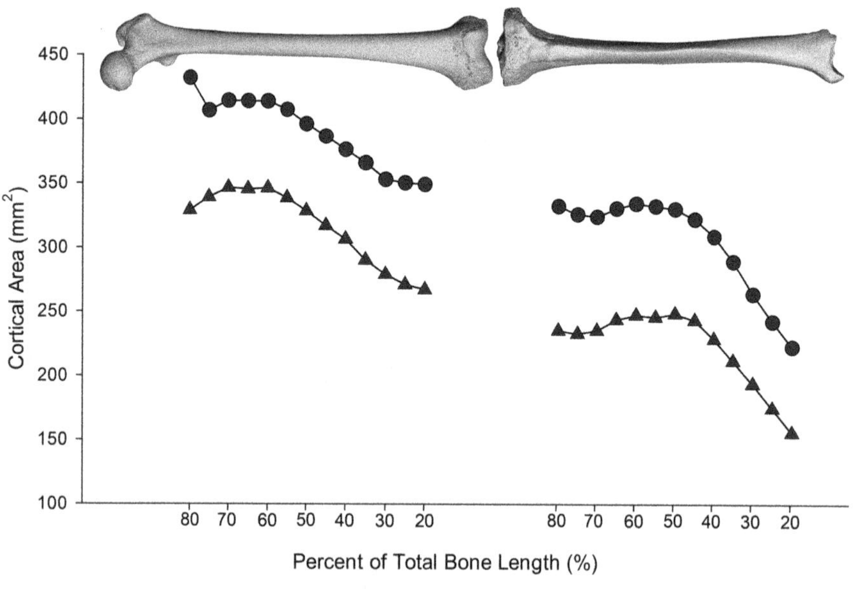

Fig. 4.2 Cross-sectional cortical area (CA, Y-axis) raw data plotted at 5 % intervals from the proximal (1 on X-axis) to distal (13 on X-axis) femoral diaphysis and proximal (14 on X-axis) to distal (26 on X-axis) tibial diaphysis. Males: *circles*, females: *triangles*

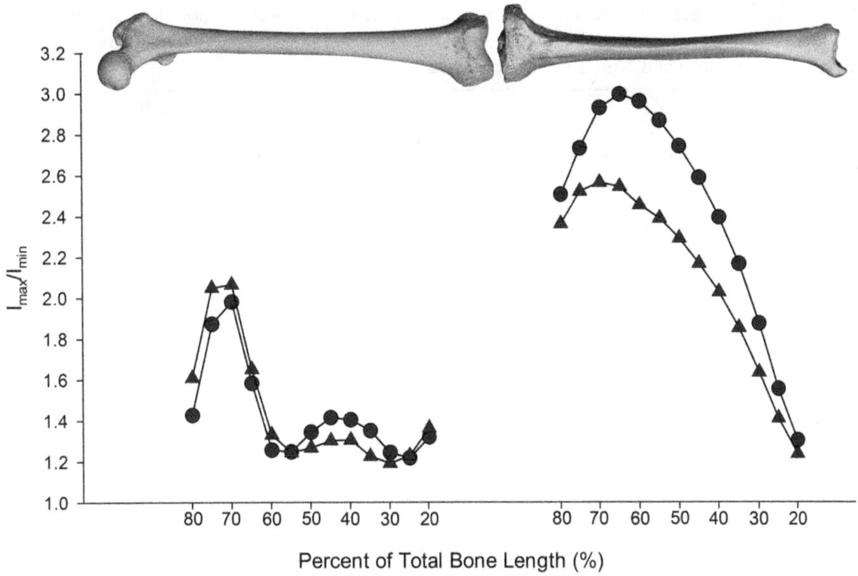

Fig. 4.3 Diaphysis cross-sectional shape (I_{max}/I_{min}, Y-axis) raw data plotted at 5 % intervals from the proximal (1 on X-axis) to distal (13 on X-axis) femoral diaphysis and proximal (14 on X-axis) to distal (26 on X-axis) tibial diaphysis. Males: *circles*, females: *triangles*

From the high point at the proximal femur (80–65 %), the amount of cortical bone decreases fairly consistently, levelling off around the proximal tibia until the tibial mid-shaft (80–50 %), then decreases markedly, reaching a low point at the distal tibia (20 %).

Cross sectional diaphyseal shape (I_{max}/I_{min}; Table 4.1, Fig. 4.3) along the lower limb also follows a similar pattern in both the male and female sub-samples. In the femur, diaphyseal shape is most oval at more proximal aspects (75–70 %), which decreases until the 60 % section at which point the diaphysis becomes comparably circular, and maintains this shape to the end of the distal femur (20 %). In the proximal tibia I_{max}/I_{min} is much higher than in the femur, reflective of a more asymmetric diaphyseal shape. Tibial I_{max}/I_{min} peaks at the mid-proximal sections (65–60 %), and is higher than at any other point along the lower limb, then decreases relatively consistently to the distal tibia (20 %), where shape is comparable to the mid-distal portion of the femoral diaphysis.

4.3.2 Variance (CV) in CA, Z_p and I_{max}/I_{min} Along the Lower Limb

Analyses comparing the level of inherent variance (coefficient of variation: CV) for Z_p, CA and I_{max}/I_{min}, at each 5 % section along both the femoral and tibial diaphysis, reveal patterns markedly different from those described above for the raw data of

Table 4.2 Coefficient of variation (CV) for cortical area (CA), section modulus (Z_p) and cross-sectional shape (I_{max}/I_{min}) at each 5 % section along the femoral and tibial diaphysis for males and females, Black Earth and Norris Farms combined

	Z_p		CA		I_{max}/I_{min}	
% Bone length	Male	Female	Male	Female	Male	Female
Femur						
80	24.91	23.72	25.41	22.55	14.37	16.41
75	19.10	19.93	14.84	17.19	14.31	13.66
70	19.89	19.42	15.17	16.35	13.33	15.83
65	20.34	19.81	15.49	15.58	13.83	18.05
60	21.43	19.98	15.77	15.71	14.48	12.24
55	22.08	20.67	15.30	15.31	10.10	11.08
50	22.54	20.75	15.24	16.32	11.68	14.36
45	21.96	23.70	15.27	16.30	13.39	16.35
40	20.76	21.27	16.19	16.91	15.24	14.78
35	20.79	21.84	16.84	18.25	14.21	12.07
30	20.06	22.91	16.31	18.21	12.54	12.50
25	19.12	25.51	15.83	18.62	11.33	14.64
20	19.61	20.72	18.63	18.53	15.83	17.60
Tibia						
80	23.41	23.22	20.83	19.73	19.27	18.09
75	22.15	19.45	18.77	17.35	19.70	19.08
70	20.36	18.09	16.64	17.15	22.23	18.44
65	20.89	17.69	16.46	16.52	23.08	18.83
60	21.50	18.78	16.62	16.86	21.96	18.14
55	21.66	19.61	16.26	17.25	21.77	18.43
50	22.59	20.79	15.72	17.81	22.33	19.17
45	22.03	20.59	16.24	17.85	22.84	19.84
40	22.57	20.29	16.46	17.78	22.74	20.66
35	21.70	18.27	15.93	16.28	22.17	18.62
30	20.59	16.94	15.50	15.65	21.24	17.69
25	21.28	18.06	16.31	16.42	17.25	15.32
20	21.12	19.84	18.04	16.70	13.66	9.94

these variables. Patterns of variance displayed by males and females are generally similar, with the exception of Z_p in the mid-distal femur (see below). Overall, variance in strength and cortical area moving from the proximal femoral diaphysis to the distal tibial diaphysis occupies a relatively narrow band (CV ~18–24 for Z_p, CV ~15–25 for CA; Table 4.2, Figs. 4.4 and 4.5). Coefficients of variation for diaphyseal shape, by contrast, indicate a larger range of variance throughout the lower limb bones (CV as low as ~11 in the femur and as high as ~23 in the tibia; Table 4.2, Fig. 4.6). Lower CV values are assumed to reflect greater morphological constraint while relatively higher CV values reflect greater morphological plasticity. Variance in Z_p throughout the lower limb diaphyses is somewhat patterned, with males displaying lower CV values (higher constraint) near the proximal femur (75 % bone length), followed by a consistent increase in variance (greater plasticity) that peaks

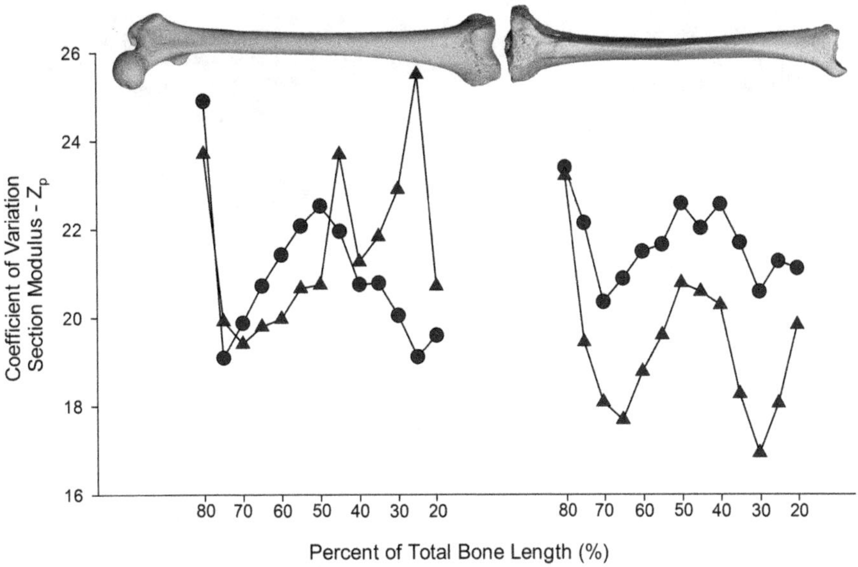

Fig. 4.4 Variation (CV) in diaphyseal strength (Z_p, *Y*-axis) plotted at 5 % intervals from the proximal (1 on *X*-axis) to distal (13 on *X*-axis) femoral diaphysis and proximal (14 on *X*-axis) to distal (26 on *X*-axis) tibial diaphysis. Males: *circles*, females: *triangles*

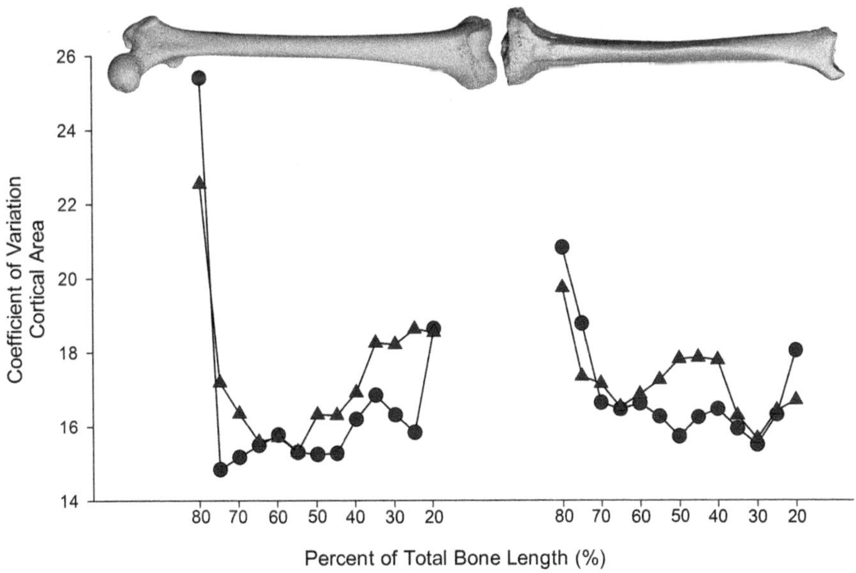

Fig. 4.5 Variation (CV) in cross-sectional cortical area (CA, *Y*-axis) plotted at 5 % intervals from the proximal (1 on *X*-axis) to distal (13 on *X*-axis) femoral diaphysis and proximal (14 on *X*-axis) to distal (26 on *X*-axis) tibial diaphysis. Males: *circles*, females: *triangles*

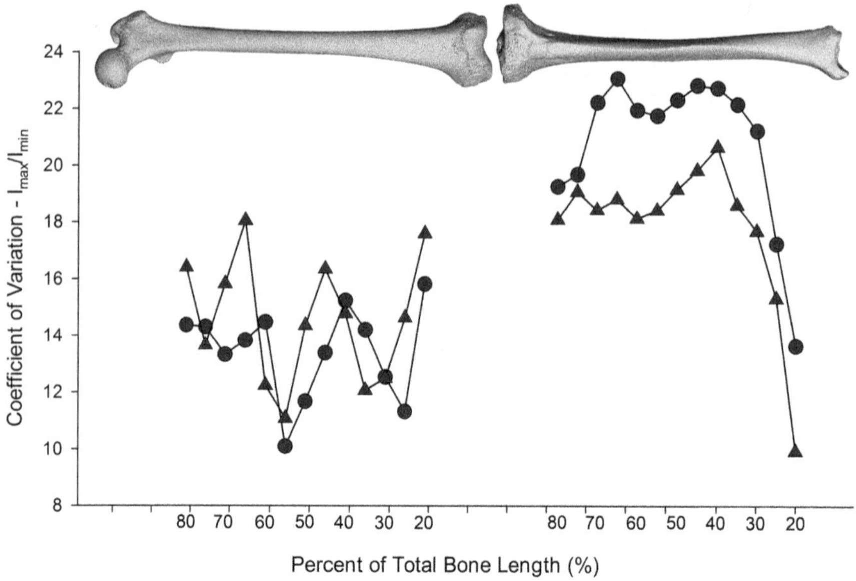

Fig. 4.6 Variation (CV) in diaphysis cross-sectional shape (I_{max}/I_{min}, Y-axis) plotted at 5 % intervals from the proximal (1 on X-axis) to distal (13 on X-axis) femoral diaphysis and proximal (14 on X-axis) to distal (26 on X-axis) tibial diaphysis. Males: *circles*, females: *triangles*

at the femoral midshaft (55 %) then decreases consistently to a low point at the distal-most sections of the femoral diaphysis (25, 20 %) (Table 4.2, Fig. 4.4). Females display a similar pattern of variation in strength in the proximal half of the femoral diaphysis compared to males, yet differ markedly in the mid to distal portion of the femur, displaying less consistent patterning and higher CV values. Variance in tibial strength is similar for males and females and when plotted effectively forms a 'W' (Fig. 4.4), reflective of greater variance at the proximal tibia (75 %), the midshaft (40–50 %) and distal tibia (20 %), and lower variance between those sections.

Variance in cortical area along the femoral and tibial diaphysis (Table 4.2, Fig. 4.5) was similar for males and females: peaked at the proximal femur (80 %) (likely because of the impact of the lesser trochanter on the 80 % section in some if not all individuals), dropped much lower at the 75 % section and remained low (CV ~16) until midshaft when variance increases slightly towards a small peak at the distal femur (20 %) and proximal tibia (80 %). For the vast majority of the tibial diaphysis (70–20 %), variance in cortical area remains relatively constant (~17), and is similar to the level of variance found throughout much of the femoral diaphysis. In contrast, variance in I_{max}/I_{min} (Table 4.2, Fig. 4.6) throughout the limb is much less consistent than variance for cortical area and displays a much wider range. Variance for I_{max}/I_{min} is similar for both males and females, and displays peaks reflective of higher variance at the mid-proximal femoral diaphysis. This initial peak is followed by a distinct drop towards the midshaft, followed by a secondary peak and drop at the mid-distal femur and a final peak at the distal-most aspect of the femoral diaphysis. When plotted, variance in tibial I_{max}/I_{min} is comparable to an inverted 'U' (Fig. 4.6), with less variance found at the most proximal (80 %) and distal (25, 20 %) aspects of the tibial diaphysis.

4.4 Discussion

The results for both the male and female sub-samples reveal structural variability along the leg, yet a general trend of decreasing cortical bone area and strength moving from the proximal aspects of the femoral diaphysis to the distal aspects of the tibial diaphysis. The strongest (Z_p) region of the leg appears to be the proximal and distal femur, while the proximal femur contains more cortical bone (CA) than all other aspects of the femoral or tibial diaphyses. Other than the most proximal aspects of the femoral diaphysis that are fairly asymmetric, most of the femoral shaft is virtually circular in cross section. This contrasts with the shape of nearly all of the tibial diaphysis, which is highly asymmetric and anteroposteriorly (AP) strengthened (Shaw and Stock 2009b). The exception is the generally circular distal portion of the tibia. These results support the idea of distal limb gracility (Stock 2006), and suggest that diaphyseal robusticity decreases in a somewhat linear pattern from more proximal to distal segments, while cross-sectional diaphyseal shape is variable across the limb (peak AP elongation at tibia 65 %).

If adaptive mechanisms constrain the response of osseous tissue to mechanical loading, limiting the maximisation of strength while optimising biomechanical efficiency, then one might expect a higher level of constraint, and therefore less variability (lower CV values), in the more distal aspects of the limb. The second question posed in this study asked whether sections of the tibial and femoral diaphyses differ in their structural variance ('plasticity'). The results indicate that the distribution of 'allowable' structural variance along the diaphyses of the femur and tibia does not generally conform to this expectation. Unlike the obvious pattern of decreasing diaphyseal strength and cortical bone area moving from more proximal to more distal aspects of the lower limb diaphysis, a corresponding trend of decreasing structural variability is not apparent. Overall, variance in strength and cortical area along both the femoral and tibial diaphysis is fairly consistent, and falls within a relatively narrow band for both Z_p and CA (Table 4.2, Figs. 4.4 and 4.5). Variation in diaphyseal shape (I_{max}/I_{min}) is less straightforward with greater variation at the mid-proximal and mid-distal femoral diaphysis, increased constraint around the femoral midshaft, and quite high levels of variation along the vast majority of the tibial diaphysis, other than at the most proximal and distal segments, which appear as constrained as the femoral midshaft. This variation may be indicative of greater levels of plasticity in these regions.

One of the earliest quantitative descriptions of the shape and strength of human femoral and tibial diaphyses was performed by Ruff and Hayes (1983a) who measured cross-sectional properties at five equally spaced locations along the diaphysis.

Similar to the results presented here, the authors concluded that in the tibia a gradual change in 'shape' to a more circular distal cross section reflected a decrease in AP bending, and consequently a (relative) increase in the importance of torsional loading in the distal tibia. The results of the present study also support the conclusion of Ruff and Hayes (1983a) that the human femur is adapted for relatively greater mediolateral (ML) bending loads in the proximal shaft, and relatively greater AP bending loads in the mid-distal portion (proximal) to the ML flaring of the diaphysis at the distal end (Ruff and Hayes 1983a).

Experimental and comparative research has provided support to the idea that canalisation influences limb structure in both humans and other animals. Lieberman and colleagues (2003) assessed rates of periosteal modelling and haversian remodelling in the hind limb midshafts of sheep in response to moderate levels of exercise. Exercising juveniles displayed higher rates of periosteal modelling in proximal skeletal elements, and higher rates of haversian remodelling in distal elements. The authors concluded that this varied response to comparable loading stimuli reflected a morphological compromise that optimised limb strength while maintaining locomotor efficiency. An analysis of diaphyseal robusticity that involved 12 catarrhine species also found that distal limb segments displayed greater variability than proximal limb segments (Buck et al. 2009). Similarly, analyses of the upper and lower limb morphology of protohistoric hunter-gatherers revealed a correspondence between patterns of activity and distal limb diaphyseal rigidity, and a stronger correspondence in the proximal limbs with ecogeographic influences (Stock 2006). In partial contrast, Drapeau and Streeter (2006) found that in the human skeleton the tibial diaphysis did not display less resistance to bending and torsion, and did not show a greater level of haversian remodelling, compared to the diaphysis of the femur.

Perhaps counter to theoretical expectation, with the inclusion of both soft and hard tissue, the relationship between limb segment inertial properties and locomotor economy appears inconsistent among primate taxa. Isler and colleagues (2006) found large inter-individual variation in segment inertial properties among both apes and humans, and, as a result, extensive overlap among genera. Differences in the distribution of mass between the limb segments resulted in the centre of mass of both the arms and legs being located more distally in apes compared with humans. The authors concluded that hominoid limbs are not optimised for efficiency in quadrupedal walking, but rather reflect a compromise between various locomotor modes in contrast to the limbs of cursorial mammals and cercopithecoid primates. Recent experimental work has also shown that variation in the distribution of mass along the limbs does not necessarily hinder locomotor economy. Studies involving infant baboons with relatively heavy distal limbs describe 'trade-off' mechanisms that allow infants to maintain similar total power outputs (and likely similar energy costs) compared to mammals with more proximally concentrated limb mass (Raichlen 2006).

In theory the distribution of muscle mass along the entire human leg should also conform to the expectation that, to optimise economy, the more distal aspects of the leg should be lighter than the more proximal aspects of the leg. Unfortunately, measurements of cross-sectional muscle area are generally restricted to the midshaft (cf., Schantz et al. 1983). Conversely, measures of muscle cross-sectional area taken at 1 cm intervals along the lower leg have demonstrated that the majority of muscle tissue is located between 50 and 75 % of tibial shaft length and peaks at ~66 % (Fukunaga et al. 1992, p. 930, Fig. 4.3). While gross observations indicate that the human lower limb is tapered distally, and that this may influence locomotor economy, factors that determine muscle performance, including muscle fibre type, overall muscle volume, muscle architecture (orientation of muscle fibres) and the bi-articular or uni-articular nature of the muscle (the number of joints that the muscle crosses from origin to insertion) (Enoka 2008), may override or compensate for the effect of tissue distribution on the economy of bipedal travel.

The findings of Lieberman and colleagues (2003), Stock (2006) and Buck and colleagues (2009) support the idea that adaptive mechanisms may constrain the ability of osseous tissue from different limb segments to respond to mechanical loading. Nevertheless, the work of Isler and colleagues (2006), Raichlen (2006) and Enoka (2008) suggest that: (a) attempts to determine the relationship between locomotor economy and morphology would benefit from a consideration of soft tissue function and distribution, and (b) that alterations to gait structure may allow an individual to overcome economically 'disadvantageous' skeletal attributes. Nevertheless, the question still remains whether, for reasons of economy, constraints exist in the range of 'allowable variation' in skeletal structure along the human lower limb, and if so, how these constraints may have influenced morphology.

The results obtained here are somewhat inconclusive for predicting the most appropriate locations along the limb for detecting signatures of mobility. The observation of relatively constant (though variable) CVs across much of the femoral and tibial diaphyses perhaps suggests that much of the limb may retain information relevant to the interpretation of past mobility. The assessment of larger samples will help to address this issue. At present, these results should not be seen to override previous interpretations that signatures of mobility may be best located in mid-distal limb segments (e.g. Stock 2006).

4.5 Conclusion

The present study offered two hypotheses: (1) the robusticity of the lower limb diaphyses are tapered and become less robust moving distally from the proximal femur down the leg to the distal tibia; (2) a greater level of constraint within the limb as one moves distally, should be reflected in lower levels of inherent or allowable variation in the morphology of more distal segments, compared to more proximal segments.

Overall the results of this study support the first hypothesis; both diaphyseal strength (Z_p) and cortical area (CA) were highest in the proximal aspects of the femur, and decreased at a relatively consistent rate until their lowest points at the distal aspects of the tibia. The second hypothesis, however, was not supported: variability, quantified using coefficient of variation (CV), for both diaphyseal strength and cortical area remained relatively unchanged along diaphyses of both lower limb bones. We interpret this finding to indicate that the level of constraint differs little along most of the femoral and tibial diaphysis. Additional testing with larger samples, from a range of geographically distributed populations will be necessary to take the next step forward to more effectively test the hypotheses posed here.

Acknowledgements Grant Sponsorship: National Science Foundation BCS-0617097 and OISE-1158603 (TMR), Cushing Anthropological Research Fund (CNS).

References

Alexander M (1998) Symmorphosis and safety factors. In: Weibel E, Taylor C, Bolis L (eds) Principles of animal design: the optimization and symmorphosis debate. Cambridge University, New York

Boldsen J, Milner G, Konigsberg L, Wood J (2002) Transition analysis: a new method for estimating age from skeletons. In: Hoppa R, Vaupel J (eds) Paleodemography: age distributions from skeletal samples. Cambridge University, Cambridge, pp 73–106

Brietburg E (1980) Vertebrate remains from 11SA87a. Center for Archaeological Investigations, Southern Illinois University, Carbondale

Buck L, Stock JT, Foley R (2009) Levels of intraspecific variation within the Catarrhine skeleton. Int J Primatol 31(5):779–795

Buie H, Campbell G, Klinick R, MacNeil J, Boyd S (2007) Automatic segmentation of cortical and trabecular compartments based on a dual threshold technique for in vivo micro-CT bone analysis. Bone 41(4):505–515

Doube M, Klosowski M, Agranda-Carreras I, Cordelieres F, Dougherty R, Jackson J, Schmid B, Hutchinson J, Shefelbine S (2010) BoneJ: free and extensible bone image analysis in Image J. Bone 47:1076–1079

Drapeau M, Streeter M (2006) Modeling and remodeling responses to normal loading in the human lower limb. Am J Phys Anthropol 129:403–409

Enoka R (2008) Neuromechanics of human movement. Human Kinetics, Champaign

Fukunaga T, Roy R, Shellock F, Hodgson J, Day M, Lee P (1992) Physiological cross-sectional area of human leg muscles based on magentic resonance imaging. J Orthop Res 10:926–934

Hallgrimsson B, Willmore K, Hall B (2002) Canalization, developmental stability, and morphological integration in primate limbs. Am J Phys Anthropol 45:131–158

Holt BM (2003) Mobility in upper paleolithic and mesolithic Europe: evidence from the lower limb. Am J Phys Anthropol 122(3):200–215

Hseih Y, Robling A, Abmbrosius W, Burr B, Turner C (2001) Mechanical loading of diaphyseal bone in *vivo*: the strain threshold for an osteogenic response varies with location. J Bone Miner Res 16(12):2291–2297

Isler K, Payne R, Gunther M, Thorpe S, Li Y, Savage R, Crompton A (2006) Inertial properties of hominoid limb segments. J Anat 209(2):201–218

Jefferies R, Butler B (1982) In: The Carrier Mills archaeological project, human adaptation in the Saline Valley, Illinois, vol II. Southern Illinois University at Carbondale Center for Archaeological Investigations

Judex S, Gross T, Zernicke R (1997) Strain gradients correlate with sites of exercise-induced bone-forming surfaces in the adult skeleton. J Bone Miner Res 12:1737–1745

Lanyon LE (1992) Control of bone architecture by functional load bearing. J Bone Miner Res 7:S369–S375

Lieberman DE, Crompton A (1998) Responses to stress: constraints on symmorphosis. In: Weibel E, Taylor C, Bolis L (eds) Principles of animal design: the optimization and symmorphosis debate. Cambridge University, Cambridge, pp 78–86

Lieberman DE, Pearson OJ, Polk JD, Demes B, Crompton A (2003) Optimization of bone growth and remodeling in response to loading in tapered mammalian limbs. J Exp Biol 206:3125–3138

Lopinot N, Lynch M (1979) An empirical examination of Crab Orchard Settlement-subsistence in Southern Illinois. Center for Archaeological Investigations, Southern Illinois University, Carbondale

Marchi D (2008) Relationships between lower limb cross-sectional geometry and mobility: the case of a Neolithic sample from Italy. Am J Phys Anthropol 137(2):188–200

Milner G, Smith V (1990). Oneota human skeletal remains. In: Santure S, Harn A, Esarey D (eds) Archaeological Investigations at the Morton Village and Norris Farms 36 Cemetery. Reports of Investigations 45 Springfield, Illinois State Museum, pp 111–148

Pearson OM, Lieberman DE (2004) The aging of Wolff's "law": ontogeny and responses to mechanical loading in cortical bone. Yearbook of Phys Anthropol 47:63–99

Raichlen D (2006) Effect of limb mass distribution on mechanical power outputs during quadrupedalism. J Exp Biol 209:633–644

Rubin C, McLeod K, Basin S (1990) Functional strains and cortical bone adaptation: epigenetic assurance of skeletal integrity. J Biomech 23(Suppl 1):43–54

Ruff C (2008a) Biomechanical analyses of archaeological human skeletons. In: Katzenberg M, Saunders A (eds) Biological anthropology of the human skeleton. Wiley, Hoboken, pp 183–206

Ruff C (2008b) Femoral/humeral strength in early African *Homo erectus*. J Hum Evol 54:383–390

Ruff C (2009) Relative limb strength and locomotion in Homo habilis. Am J Phys Anthropol 138:90–100

Ruff CB (2000) Body size, body shape and long bone strength in modern humans. J Hum Evol 38:269–290

Ruff CB, Hayes WC (1983a) Cross-sectional geometry of Pecos Pueblo femora and tibiae—a biomechanical investigation: I. Method and general patterns of variation. Am J Phys Anthropol 60:359–381

Ruff CB, Hayes WC (1983b) Cross-sectional geometry of Pecos Pueblo femora and tibiae—a biomechanical investigation: II. Sex, age, and side differences. Am J Phys Anthropol 60:383–400

Ruff CB, Holt BM, Trinkaus E (2006) Who's afraid of the big bad Wolff? "Wolff's Law" and bone functional adaptation. Am J Phys Anthropol 129(4):484–498

Ruff CB, Trinkaus E, Walker A, Larsen CS (1993) Postcranial robusticity in *Homo*. I: temporal trends and mechanical interpretation. Am J Phys Anthropol 91:21–53

Santure S, Harn A, Esarey D (1990) Archaeological investigations at the Morton Village and Norris Farms 36 Cemetery. Illinois State Museum, Springfield

Schantz P, Randall-Fox E, Hutchison W, Tyden A, Astrand P (1983) Muscle fibre type distribution, muscle cross-sectional area and maximal voluntary strength in humans. Acta Physiol Scand 117(2):219–226

Shackelford L (2007) Regional variation in the postcranial robusticity of late upper Palaeolithic humans. Am J Phys Anthropol 133(1):655–668

Shaw C, Stock J (2009a) Habitual throwing and swimming correspond with upper limb diaphyseal strength and shape in modern human athletes. Am J Phys Anthropol 140(1):160–172

Shaw C, Stock J (2009b) Intensity, repetitiveness, and directionality of habitual adolescent mobility patterns influence the tibial diaphysis morphology of athletes. Am J Phys Anthropol 140:149–159

Simpson G, Roe A, Lewontin R (1960) Measures of dispersion and variability. Quantitative zoology. Harcourt Brace, New York

Skedros J, Sybrowsky C, Parry T, Bloebaum R (2003) Regional differences in cortical bone organisation and micro-damage prevalence in Rocky Mountain mule deer. Anat Rec 274A:837–850

Sparacello V, Marchi D (2008) Mobility and subsistence economy: a diachronic comparison between two groups settled in the same geographic area (Liguria, Italy). Am J Phys Anthropol 136(4):485–495

Stock J (2006) Hunter-gatherer postcranial robusticity relative to patterns of mobility, climatic adaptation, and selection for tissue economy. Am J Phys Anthropol 131:194–204

Stock J, Pfeiffer S (2001) Linking structural variability in long bone diaphyses to habitual behaviors: foragers from the southern African Later Stone Age and the Andaman Islands. Am J Phys Anthropol 115:337–348

Stock JT, Pfeiffer S (2004) Long bone robusticity and subsistence behavior among Later Stone Age foragers of the forest and fynbos biomes of South Africa. J Archaeol Sci 31:999–1013

Trinkaus E (1997) Appendicular robusticity and the paleobiology of modern human emergence. Proc Natl Acad Science 94:13367–13373

Trinkaus E, Ruff C (2012) Femoral and tibial diaphyseal cross-sectional geometry in Pleistocene *Homo*. Paleoanthropology 2012:13–62

Wescott DJ (2014) The relationship between femur shape and terrestrial mobility patterns. In: Carlson KJ, Marchi D (eds) Reconstructing mobility: environmental, behavioral, and morphological determinants. Springer, New York

Chapter 5
Human Variation in the Periosteal Geometry of the Lower Limb: Signatures of Behaviour Among Human Holocene Populations

Thomas G. Davies and Jay T. Stock

Abstract Behavioural influences upon human diaphyseal morphology are well documented; however, relatively little is known about the relationship between eco-geographic variation in body shape and the variation in diaphyseal strength parameters throughout long bone diaphyses. Diaphyseal rigidity of the femur and tibia was assessed among a range of human Holocene populations, using 3D laser surface scans to extract cross-sectional biomechanical data from periosteal contours.

Within particular regions of the lower limb, most notably the mid-diaphyseal region of the tibial diaphysis, signatures of high mobility may override the confounding effects of different body size and shape signatures. Populations display specific distributions of cross-sectional properties that should be matched to independent indicators of underlying mobility levels. Variation in location of minimal cross-sectional rigidity is also considered and merits further investigation. This whole-limb approach may enable further isolation of behavioural signatures in long bone geometry.

Keywords Cross-sectional geometry • Morphology • Femur • Tibia • Mobility

---

T.G. Davies (✉)
PAVE Research Group, Department of Archaeology & Anthropology, University of Cambridge, Pembroke Street, Cambridge CB2 3DZ, UK

Churchill College, Storey's Way, Cambridge CB3 0DS, UK
e-mail: td274@cam.ac.uk

J.T. Stock
PAVE Research Group, Department of Archaeology & Anthropology, University of Cambridge, Pembroke Street, Cambridge CB2 3DZ, UK
e-mail: jts34@cam.ac.uk

K.J. Carlson and D. Marchi (eds.), *Reconstructing Mobility: Environmental, Behavioral, and Morphological Determinants*, DOI 10.1007/978-1-4899-7460-0_5,

Abbreviations

CSG	Cross-sectional geometry
3D	Three dimensional
I_x	Second moment of area, *x*-axis
I_y	Second moment of area, *y*-axis
J_{solid}	Polar second moment of area
TA	Total subperiosteal area

5.1 Introduction

Behavioural influences on bone diaphyseal morphology are well documented. Studies of cross-sectional geometry (CSG) of human long bones over the past 40 years have demonstrated: (a) the presence of bilateral asymmetry in response to unilateral behaviours (Jones et al. 1977; Roy et al. 1994; Shaw and Stock 2009a), (b) the influence of mobility (terrestrial or marine) or subsistence economy on lower and upper limb robusticity (Ruff and Hayes 1983; Ruff et al. 1993; Stock and Pfeiffer 2001; Weiss 2003; Stock 2006; Shaw and Stock 2009b) and (c) the role of ruggedness of terrain in modulating bone adaptation (Ruff 1995a; Marchi et al. 2006; Ruff et al. 2006b; Sparacello and Marchi 2008; Higgins and Ruff 2011). Signals of mobility encompass both the overall size of bone cross sections as well as the ratios of properties, either within a section (cross-sectional shape ratios) or between skeletal elements (e.g. humero-femoral or humero-tibial indices, Ruff 2003; Shaw and Stock 2013). In this chapter, we consider mobility in terms of the types and extent of movement employed by an individual in their environment throughout their life, for example, in relation to marine vs. terrestrial movement, or degrees of mobility associated with subsistence behaviours. It is not possible to differentiate more subtle variations of mobility (logistical, residential, seasonal) from the skeletal properties that are examined, and so we use the term to indicate aggregate mobility of different varieties over longer time frames.

Some studies have questioned the extent to which mobility can be inferred from limb bone geometry (Ohman and Lovejoy 2001; Wescott 2006), but the weight of evidence supports the argument that at least a portion of limb diaphyseal variability can be attributed to the plastic influences of functional loading on limb bone diaphyses (Ruff et al. 1993, 2006a). Experimental evidence has provided sufficient support to allow the application of the model of bone functional adaptation. Traditional cross-sectional geometric properties provide the best available estimates of in vivo mechanical competence (Ruff et al. 2006a for review).

One area for debate, however, is the extent to which populations with high levels of mobility display high limb bone rigidity. For example, long distance runners do not necessarily display high overall lower limb bone rigidity (Frost 1997), although they do appear to be characterized by unique diaphyseal cross-sectional shapes (Shaw and Stock 2009b). Furthermore, some results obtained in studies of highly mobile forager populations, for example Australian Aborigines for whom high mobility is documented ethnographically, have failed to show significantly more robust limb bone morphology than control groups (Carlson et al. 2007). Collier (1989) found relatively gracile (for length) bones among Australian Aborigines, whilst similarly Australian Aborigines did not stand out in Pearson's diaphyseal robusticity quantification relative to other recent human populations (Pearson 2000). Whilst a body of literature clearly supporting limb bone functional adaptation has accumulated (Ruff et al. 1984, 1993, 2006a; Trinkaus et al. 1994; Stock and Pfeiffer 2001; Lieberman et al. 2004; Shaw and Stock 2009a, b), detection of high mobility from skeletal data can in some contexts be challenging. There is merit, therefore, in expanding the type of skeletal features that are quantified, and assessing diaphyseal morphology in novel alternative ways.

Body mass and limb length are important determinants of forces imposed upon the limbs (Ruff 1987, 2000; Ruff et al. 1993). Variation in body shape, combining both influences in a way, also affects long bone diaphyseal morphology, particularly in selected anatomical locations such as the proximal femur (Ruff 1995b; Weaver 2003; Ruff et al. 2006b; Stock 2006; Shaw and Stock 2011). However, relatively little is known about the relationship between ecogeographic variation in body shape and variation in diaphyseal strength parameters throughout the limb. Stock (2006) tested the extent to which climatic vs. behavioural influences most affect different locations in limb diaphyses, finding support for an influence of body shape on cross-sectional properties at the subtrochanteric femur location. Pearson (2000), Churchill (1994, 1996) (for the upper limb), and Davies (2013) considered issues regarding body shape and behaviour across a range of either human or hominin groups.

Studies of mobility signatures in the limbs focus primarily on bone midshafts, but they also have considered differences at other locations in some cases (e.g. Ruff and Hayes 1983). Examination of multiple cross sections throughout a bone diaphysis is usually applied via CT scanning, for example in the context of assessing variability introduced into calculated cross-sectional properties when erroneously locating bone midshafts (Sladek et al. 2010). Sladek et al. (2010) further highlighted the degree to which there is considerable inter-individual variation not only in the magnitude of properties but also in the patterning of variation throughout bone diaphyses.

Recent technological advances increase the potential number of locations throughout diaphyses at which it is practical to assess cross-sectional properties (Sladek et al. 2010; Davies et al. 2012; Shaw et al. 2014). This chapter highlights a number of results concerning the influence of mobility on limb architecture among very diverse forager populations. In drawing conclusions regarding mobility, early studies treated different regions of the limbs fairly equally with regard to insights that may be obtained. A number of studies have now argued that proximal regions of lower limb bones, particularly the subtrochanteric femur section, are influenced by variation in body shape (Ruff et al. 2006b; Stock 2006; Shaw and Stock 2011). In this context, tibial diaphyses may provide better (in some instances) insights into mobility (Stock 2006; Marchi 2008). The extent to which morphological signals of mobility may be resilient to highly contrasting body size and shape backgrounds remains an open question.

The aim of this chapter is to quantify variability in the distribution of bone rigidity from proximally to distally in the femur and tibia in past populations involved in different subsistence strategies using a 3D laser scan method (Davies et al. 2012). Once quantified, patterns exhibited by these groups will be statistically compared. One important contrast among forager populations is the contrast between robusticity observed among populations practicing predominantly either marine or terrestrial mobility (Stock and Pfeiffer 2001; Stock 2006). Following from these previous studies, highly terrestrially mobile groups would be expected to display greater rigidity throughout the lower limb than those practicing marine transportation, which may be associated with reduced mobility terrestrially. The method applied here quantifies only the periosteal contour of a cross section: the effect of the medullary cavity is not assessed, and calculated cross-sectional properties are therefore only estimates of "true" values (Shaw and Stock 2011; Davies et al. 2012; Macintosh et al. 2013). However, such an approach makes it practical to assess cross-sectional properties calculated at multiple locations throughout the limb bones, as well as to identify the locations throughout the diaphyses at which each minimal value is observed. Ruff (1995b) argued that the distal location of minimum mediolateral (ML) breadth of the femur in early *Homo* is related to increased ML bending loads associated with a long femoral neck. In relation to mobility, Stock (2006) argued that selection for tissue economy and structural optimization of the distal limb segment may contribute to the evidence for greater correspondence between mobility and tibial midshaft properties than observed for the femur. There are also several lines of evidence that bone adaptation results from a combination of both systemic and local influences (see Gosman et al. 2011 for review). Given localized response to loading and the aforementioned pressure for bone optimization in more distal limb segments, it is expected that localized adaptation in the tibial midshaft region of high mobility individuals would be associated with a more distal shift in the cross-sectional location of minimum rigidity.

In this analysis, we specifically address three questions:

1. Do midshaft locations in lower limb bones retain signatures of terrestrial vs. marine mobility in each shaft relative to other locations within the diaphyses?
2. Does diaphyseal rigidity of highly mobile Australian Aborigine individuals exceed that of populations thought to have had reduced terrestrial mobility (either due to predominance of marine mobility or increased sedentism)?
3. Does variation in the diaphyseal location of minimum bending rigidity in the tibia reflect mobility level?

5.2 Materials and Methods

Diaphyseal rigidity of the femur and tibia was assessed for a series of Holocene hunter-gatherers and an agricultural population, incorporating contrasts in mobility (e.g. terrestrial vs. marine) (Table 5.1). These populations also incorporate variation

Table 5.1 Sample sizes and contextual information on the populations included

	Femur		Tibia		
	Males	Females	Males	Females	Subsistence category
Andaman Is.	14	13	14	13	Marine hunter-gatherer
Australian Aborigines	15	8	8	6	Terrestrial hunter-gatherer
Santa Cruz Is.	16	13	12	12	Marine hunter-gatherer
Kerma	13	14	11	14	Terrestrial agriculturalist
Yahgan	9	10	7	5	Marine hunter-gatherer
Sadlermiut	16	13	16	13	Marine hunter-gatherer

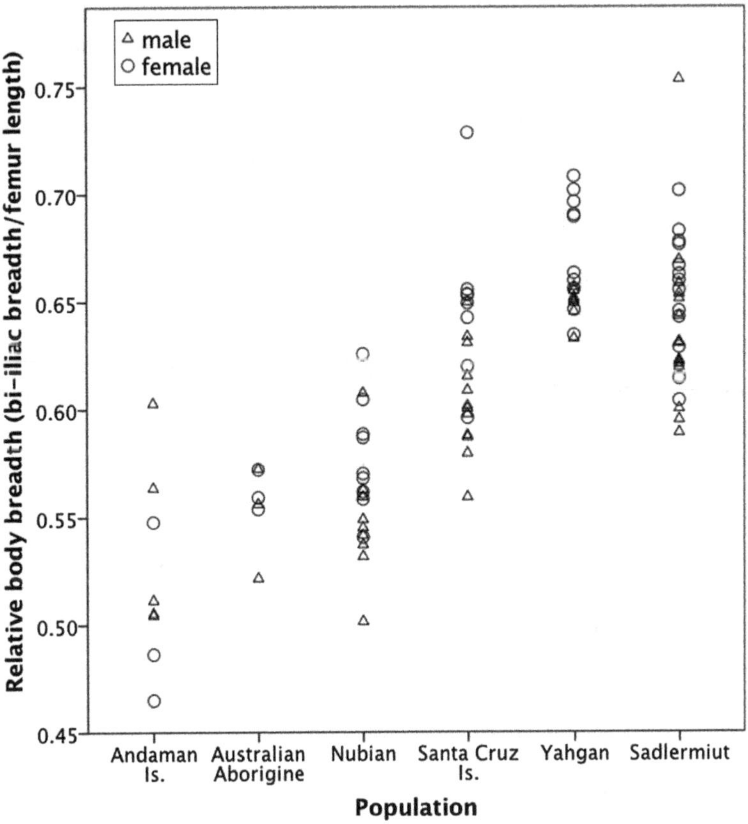

Fig. 5.1 Relative body breadth of individuals from the six populations studied: Andaman Islanders, Australian Aborigines, Nubian, Santa Cruz Islanders, Yahgan and Sadlermiut

in body size and shape, most notably in relative body breadth (Fig. 5.1). Summary details on each of the populations included are provided below. Further information on their skeletal proportions, as well as description of ethnographic and archaeological evidence for their mobility patterns, has been provided elsewhere (see Davies 2013).

Andaman Islanders. The Andaman Islanders represent proto-historic individuals from the Andaman Islands in the Indian Ocean dating to the period immediately following European settlement in 1858 (Stock and Pfeiffer 2001). The Andaman Islanders represent a small-bodied, tropically adapted population, and can be characterized as marine or coastal foragers, with subsistence strategies likely incorporating a considerable proportion of marine resources, and marine locomotion as a means of travel.

Australian Aborigines. Australian Aborigines are characterized as highly mobile hunter-gatherers, although with considerable regional variation (Carlson et al. 2007). The skeletal remains studied here consist of pooled isolated skeletons from across Australia and likely encompass substantial underlying variation in mobility. The sample differs from that assessed previously for cross-sectional geometric properties (Carlson et al. 2007).

Kerma, Nubian. The 12th dynasty site of Kerma in the Upper Nile region dates to 2100–1500 BC and is attributed to the hierarchical Nubian Kerma civilization at a time of Egyptian state development and agricultural intensification (Stock et al. 2011). This agriculturalist population from the site of Kerma has been studied in a diachronic sequence of Egyptian populations, and has been seen to display relatively gracile limbs interpreted as supporting reduced mobility with the rise of agriculture and Egyptian state development (Stock et al. 2011).

Santa Cruz Islanders. The archaeological Santa Cruz Islander skeletons studied date to 5000–4000BC (Walker 1986). They likely reflect individuals associated with the Chumash culture, which practiced marine hunting and gathering subsistence and exploitation of terrestrial forageable resources. Early period subsistence adaptation appears to have combined terrestrial and marine resources with reliance upon plants and shellfish, as well as fish and sea mammals to varying extents (Walker 1986; Glassow 1993; Colten and Arnold 1998; Perry 2005).

Sadlermiut. The Sadlermiut from Native Point inhabited Southampton Island and the archaeological skeletons represent individuals dating primarily to the last 500 years (Coltrain 2009). A subset of the skeletons is likely constituted by individuals that died following the introduction of diseases brought to Southampton Island via whalers (Merbs 1983). The Sadlermiut display cold-adapted physique, with relative short stature, high body mass, wide-bodied phenotype, and can be characterized as marine foragers, with subsistence strategies incorporating marine resources, marine transportation and terrestrial behaviours (e.g. sledding, terrestrial hunting).

Yahgan. The Yahgan of Tierra del Fuego also display a cold-adapted physique, and represent proto-historic foragers inhabiting the shores of the Beagle channel (Bridges 1948). They were reliant upon marine travel and have been characterized as having low to moderate terrestrial mobility, in the context of high terrain relief (Stock 2006).

Femora and tibiae of individuals were 3D laser scanned using a NextEngine 3D Desktop laser scanner (www.nextengine.com) and ScanStudio HDPro software (version 1.3.2). Three-dimensional virtual models of whole bones were created from one or two 360° rotations (8–10 individual scan faces per rotation) emphasizing diaphyses, plus additional individual scans of the articular surfaces. Scan processing (trimming, alignment and fusion) was conducted in ScanStudio HDPro. Finished models were imported into RapidWorks 64 software (version 2.3.5), where the bone was orientated onto a 3D coordinate system by realigning x-, y- and z-axes onto anatomical planes following definitions provided by Ruff (2002). Finally, each bone model was exported as an ascii (.asc) file. The ascii file (a list of all 3D coordinates for points that constitute the virtual model) was then used as input in custom-designed software—AsciiSection (www.pave.bioanth.cam.ac.uk), which automatically calculates solid cross-sectional properties (Stock and Shaw 2007; Davies et al. 2012; Macintosh et al. 2013) at 1 % intervals of bone length throughout the diaphysis using equations for polygons (see Davies et al. 2012 for further details of the method).

Solid cross-sectional properties (i.e. from the periosteal contour only) calculated by this method have been shown to be consistent with those calculated using periosteal moulding (Davies et al. 2012). The cross-sectional properties discussed here include total area of the cross section (*TA*), which is an indicator of compressive strength, the polar moment of area (J_{solid}), an indicator of twice average bending or torsional rigidity, and the second moments of area I_x and I_y, which indicate antero-posterior (AP) and ML bending rigidity, respectively. Additionally, the location along the diaphysis at which cross-sectional properties reach a rigidity minimum was recorded. Total area of section was standardized to body mass and polar moment of area to the product of body mass and the second power of bone length (following Ruff 2000, 2008). Body mass estimates were calculated from femoral head diameters (Ruff et al. 1997). Relative body breadth was calculated as bi-iliac breadth divided by maximum femur length.

One-Way ANOVA statistical tests of population means were used to compare *TA* and J_{solid} among the six groups. Hochberg's GT2 or Games-Howell pairwise post-hoc tests were used (selection depending upon homogeneity of variance test). Tests were conducted at 10 % intervals along femoral and tibial diaphyses in order to provide insights into differences throughout the whole lower limb. For the assessment of minimum rigidity location within the tibia, Kruskal-Wallis non-parametric tests of median differences were conducted. Statistical significance was achieved for $p < 0.05$. All analyses were carried out in SPSS version 21.

5.3 Results

5.3.1 *Whole-Limb Variation in the Distribution of Cross-sectional Properties*

Figures 5.2 and 5.3 plot mean profiles for standardized *TA* and standardized J_{solid} throughout the diaphyses of the femur and tibia (means calculated at every 1 % of bone length), respectively. One-Way ANOVA tests were conducted at a subset of

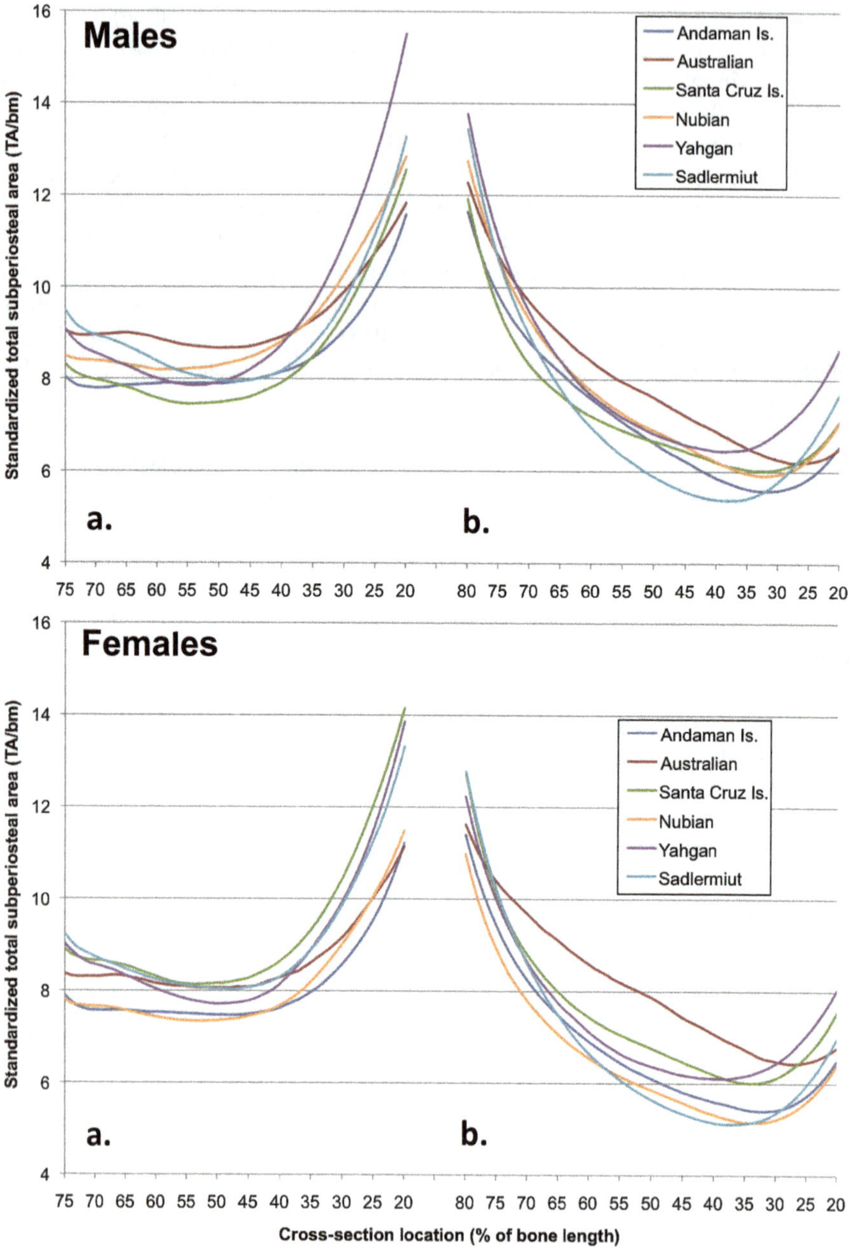

Fig. 5.2 Population mean standardized total area (*TA*) from proximal to distal locations throughout the femur and tibia. Each line connects mean values calculated at every 1 % of bone length (from distal end). (a) femur, (b) tibia

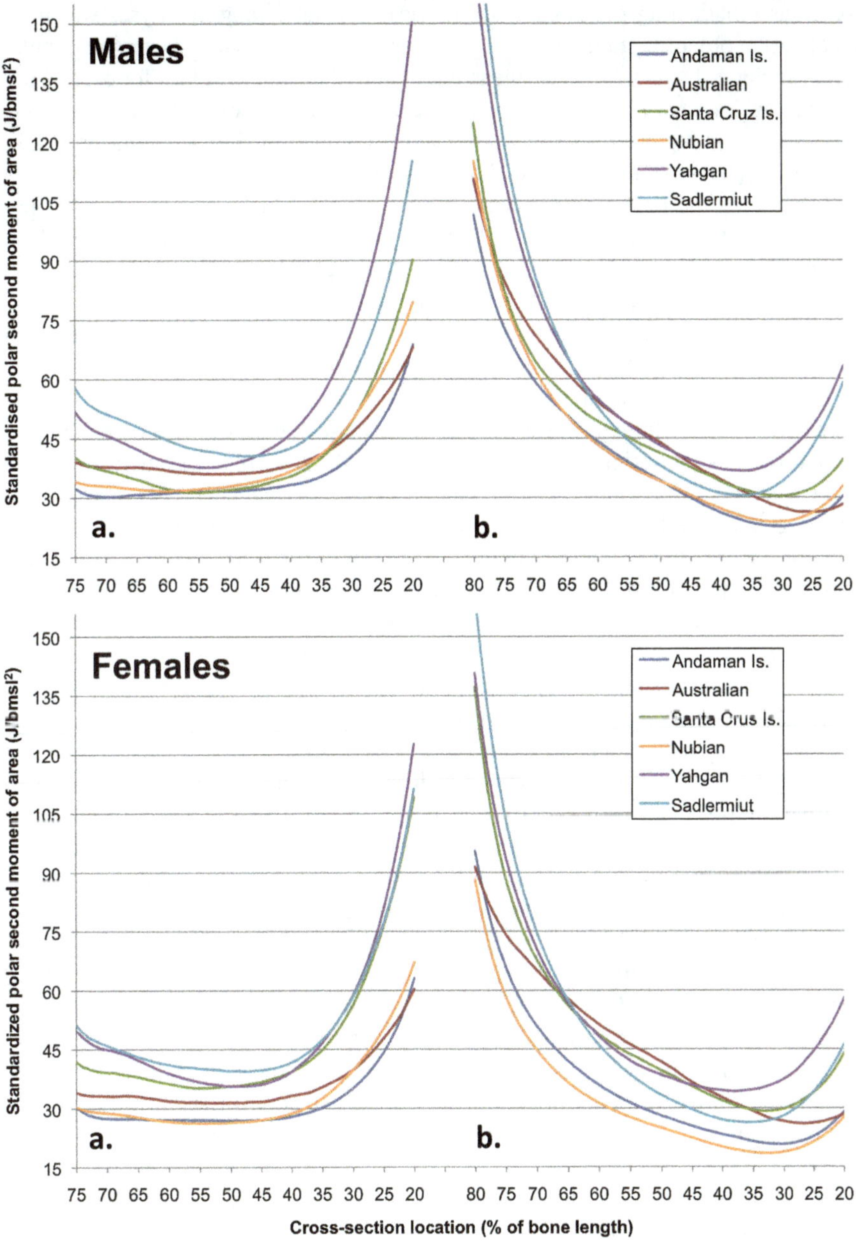

Fig. 5.3 Population mean standardized polar moment of area (J_{solid}) from proximal to distal locations throughout the femur and tibia. Each line connects mean values calculated at every 1 % of bone length (from distal end). (a) femur, (b) tibia

locations (conducting statistical tests at 10 % intervals along diaphyses was most practical) in order to establish significance for pairwise comparisons among populations. Tables 5.2 and 5.3 display mean and standard deviation values for each population, for *TA*, and J_{solid}, respectively. Results of pairwise comparisons among populations are reported for *TA* throughout the femur (Fig. 5.4) and tibia (Fig. 5.5), and for J_{solid} throughout the femur (Fig. 5.6) and tibia (Fig. 5.7).

Comparing the Kerma Nubians and Australian Aborigines, two terrestrial populations expected to have practiced differing levels of mobility, the Australians show significantly greater torsional rigidity (J_{solid}) than the Nubians only at the tibial midshaft among males, and at the tibial 30 % section among females (Fig. 5.7). For standardized *TA*, the Australians and Nubians do not show any significant difference throughout the limb either among males or females.

Comparing terrestrial and marine groups for *TA*, there are no significant differences involving the female Australians throughout the femur, whereas among males, the Australians show significantly higher *TA* than the Andaman Islanders and Santa Cruz Islanders throughout the proximal half of the femur (Fig. 5.4), and significantly greater rigidity than Santa Cruz Islanders and Sadlermiut at a number of locations in the mid-diaphyseal region of the tibia (Fig. 5.5). For J_{solid}, Australian males are not significantly different from any other population at the femur midshaft, whilst among females Sadlermiut display significantly greater rigidity than the Australians (Fig. 5.6). Australian males show significantly greater rigidity than the Andaman Islanders at the tibial midshaft (Fig. 5.7). Santa Cruz Islanders, Yahgan and Sadlermiut, all of which are coastal, marine hunter-gatherers or foragers, show comparable standardized mean J_{solid} to Australian Aborigines among both males and females at a large number of locations throughout the limb, however, both Australian males and females show lower rigidity than Yahgan and Sadlermiut at locations towards both the proximal and distal ends of each diaphysis (Figs. 5.6 and 5.7). The Nubians show lower rigidity throughout the lower limb than other populations, with significant differences particularly among females relative to the Santa Cruz Islanders, Yahgan, and Sadlermiut.

Among the populations involved in marine transportation, comparisons tend to separate the Andaman Islanders from the other three groups (Santa Cruz Islanders, Yahgan and Sadlermiut). Andaman Islanders exhibit lower values of both *TA* and J_{solid} with significant differences most visible towards the ends of each bone's diaphysis. There are no differences in *TA* among the marine populations at either femoral or tibial midshafts, however, the Sadlermiut display lower values than the Andaman Islanders in the mid-diaphyseal region of the tibia (Fig. 5.5). Significant differences were observed for J_{solid}, however, both at midshaft locations and elsewhere in the diaphysis (see Figs. 5.6 and 5.7).

5.3.2 Location of Minimum Rigidity Within the Tibia

One novel opportunity provided by analysing a high number of cross sections is the ability to estimate the location of minimum rigidity along the shaft. We assessed variation in the observed tibial location (% of bone length) of minimum rigidity for

Table 5.2 Standardized total subperiosteal area (*TA*) means and standard deviations at 10 % intervals throughout the femur and tibia (75 % instead of 80 % used in the femur to avoid including the lesser trochanter)

		Femur				Tibia			
		Males		Females		Males		Females	
Location	Population	Mean	SD	Mean	SD	Mean	SD	Mean	SD
75 %/80 %	Andaman Is.	8.01	0.48	7.89	0.85	11.55	0.88	11.40	1.35
	Australian	9.08	0.93	8.37	0.88	12.29	1.11	11.63	2.16
	Santa Cruz Is.	8.31	0.59	8.90	0.52	11.92	1.52	12.73	1.18
	Kerma	8.58	0.44	7.81	0.31	13.01	1.63	10.99	1.15
	Yahgan	9.09	0.91	9.01	0.76	13.78	1.64	12.23	0.56
	Sadlermiut	9.49	1.18	9.23	0.70	13.44	1.80	12.77	1.11
70 %	Andaman Is.	7.77	0.47	7.57	0.63	8.75	0.75	8.25	0.97
	Australian	8.97	0.80	8.31	1.02	9.70	0.78	9.69	2.12
	Santa Cruz Is.	7.98	0.56	8.65	0.45	8.32	0.85	8.81	0.84
	Kerma	8.46	0.46	7.66	0.37	9.51	0.91	7.81	0.59
	Yahgan	8.55	0.95	8.57	0.69	9.47	1.00	8.64	0.44
	Sadlermiut	8.95	0.94	8.74	0.66	8.95	1.17	8.60	0.88
60 %	Andaman Is.	7.88	0.43	7.53	0.57	7.54	0.67	6.91	0.85
	Australian	8.86	0.73	8.16	1.24	8.41	0.57	8.59	1.70
	Santa Cruz Is.	7.58	0.49	8.31	0.47	7.21	0.71	7.44	0.75
	Kerma	8.24	0.52	7.44	0.36	7.93	0.75	6.55	0.45
	Yahgan	8.00	0.91	8.04	0.57	7.67	0.85	7.12	0.57
	Sadlermiut	8.38	0.75	8.25	0.70	6.97	0.85	6.65	0.69
50 %	Andaman Is.	7.89	0.44	7.48	0.62	6.59	0.57	6.10	0.75
	Australian	8.65	0.75	8.07	1.20	7.62	0.58	7.88	1.48
	Santa Cruz Is.	7.49	0.52	8.17	0.56	6.67	0.66	6.77	0.70
	Kerma	8.36	0.55	7.36	0.48	7.03	0.70	5.87	0.48
	Yahgan	7.92	0.91	7.72	0.57	6.83	0.67	6.37	0.48
	Sadlermiut	7.97	0.80	8.04	0.77	5.89	0.68	5.65	0.59
40 %	Andaman Is.	8.13	0.51	7.63	0.66	5.81	0.36	5.61	0.79
	Australian	8.91	0.86	8.30	1.21	6.88	0.48	7.08	1.29
	Santa Cruz Is.	7.92	0.74	8.65	0.64	6.21	0.61	6.20	0.61
	Kerma	8.91	0.63	7.69	0.70	6.34	0.61	5.33	0.46
	Yahgan	8.75	0.89	8.15	0.54	6.46	0.64	6.11	0.39
	Sadlermiut	8.18	0.93	8.31	0.80	5.39	0.58	5.15	0.55
30 %	Andaman Is.	9.03	0.66	8.58	1.03	5.55	0.31	5.42	0.63
	Australian	9.93	1.05	9.17	1.33	6.27	0.48	6.49	0.86
	Santa Cruz Is.	9.44	0.90	10.45	1.00	6.05	0.69	6.12	0.65
	Kerma	10.44	0.91	9.02	1.08	6.05	0.56	5.22	0.50
	Yahgan	11.01	1.15	9.95	0.76	6.87	0.75	6.44	0.33
	Sadlermiut	9.72	1.34	9.88	0.99	5.78	0.66	5.36	0.56
20 %	Andaman Is.	11.56	0.99	11.22	1.85	6.55	0.47	6.49	0.79
	Australian	11.87	1.48	11.14	1.63	6.50	0.75	6.78	0.83
	Santa Cruz Is.	12.56	1.18	14.16	1.38	7.09	0.94	7.54	0.84
	Kerma	13.08	1.40	11.49	1.52	7.18	0.76	6.40	0.76
	Yahgan	15.52	1.48	13.86	1.39	8.65	1.02	8.04	0.31
	Sadlermiut	13.28	2.15	13.31	1.25	7.68	1.09	6.97	0.76

Table 5.3 Standardized polar second moment of area (J_{solid}) means and standard deviations at 10 % intervals throughout the femur and tibia (75 % instead of 80 % used in the femur to avoid including the lesser trochanter)

		Femur				Tibia			
		Males		Females		Males		Females	
Location	Population	Mean	SD	Mean	SD	Mean	SD	Mean	SD
75 %/80 %	Andaman Is.	32.38	4.09	30.30	6.72	101.45	12.77	95.51	18.06
	Australian	39.24	10.29	34.07	6.61	110.65	26.88	91.48	28.26
	Santa Cruz Is.	40.30	5.97	41.90	6.06	124.77	25.50	137.29	30.30
	Kerma	34.86	3.76	30.40	3.86	118.93	27.29	88.16	20.32
	Yahgan	51.84	9.81	49.71	7.72	171.80	41.88	140.76	17.81
	Sadlermiut	57.91	14.09	51.31	7.81	192.89	52.74	161.67	19.45
70 %	Andaman Is.	30.26	3.73	27.48	4.47	59.02	6.43	50.79	10.93
	Australian	37.89	8.58	33.15	6.54	70.88	15.00	65.07	23.80
	Santa Cruz Is.	36.97	5.36	39.15	4.50	64.22	9.33	67.59	13.77
	Kerma	33.56	4.10	28.85	3.91	63.90	8.90	44.65	7.39
	Yahgan	45.95	9.15	44.99	6.36	82.27	15.79	70.35	6.02
	Sadlermiut	51.32	10.07	45.97	6.07	85.66	17.86	74.54	9.65
60 %	Andaman Is.	31.26	3.78	27.15	3.99	44.20	4.38	35.85	9.05
	Australian	36.99	7.31	31.87	7.87	54.22	10.89	51.37	16.81
	Santa Cruz Is.	32.26	4.43	36.30	4.41	49.27	7.02	48.65	9.52
	Kerma	32.14	4.11	26.75	3.34	44.72	5.27	31.32	4.54
	Yahgan	39.26	8.52	38.90	4.96	54.86	10.35	48.18	3.81
	Sadlermiut	44.68	7.54	41.12	6.31	52.85	9.92	45.96	7.00
50 %	Andaman Is.	31.69	3.59	26.98	4.34	34.20	3.46	28.08	6.91
	Australian	36.14	6.96	31.49	7.56	43.87	9.91	41.74	12.45
	Santa Cruz Is.	31.98	4.64	35.74	4.79	41.23	5.89	39.50	7.50
	Kerma	33.50	3.84	26.46	3.74	35.01	4.29	25.02	4.07
	Yahgan	38.50	8.51	35.64	4.79	43.17	7.11	38.40	3.17
	Sadlermiut	40.91	7.80	39.49	6.68	38.08	6.56	33.26	5.46
40 %	Andaman Is.	33.30	4.18	27.92	4.45	26.06	2.15	23.31	6.10
	Australian	38.22	8.43	33.23	7.86	34.40	7.94	32.67	9.32
	Santa Cruz Is.	35.47	6.61	39.35	6.01	33.94	4.78	31.89	5.52
	Kerma	37.57	3.77	28.75	5.10	27.81	3.50	20.20	3.29
	Yahgan	46.19	8.98	39.30	4.70	37.21	6.54	34.49	2.43
	Sadlermiut	42.58	8.26	41.68	7.08	31.01	4.88	27.04	4.81
30 %	Andaman Is.	40.55	5.43	35.16	7.53	22.64	1.95	20.80	4.96
	Australian	46.48	11.57	39.82	8.65	27.25	6.52	26.65	5.77
	Santa Cruz Is.	49.48	9.57	56.48	12.47	30.32	5.12	29.84	5.94
	Kerma	51.00	6.91	39.81	9.68	24.62	3.97	18.91	3.45
	Yahgan	72.19	14.58	58.87	8.42	40.59	7.41	37.41	4.15
	Sadlermiut	59.77	14.51	58.48	9.33	34.12	6.71	28.12	4.85
20 %	Andaman Is.	68.78	12.04	63.10	18.45	30.40	3.69	29.14	7.62
	Australian	67.97	20.56	60.38	13.92	28.32	8.73	28.56	6.32
	Santa Cruz Is.	90.14	18.65	109.11	25.29	39.73	8.37	44.07	10.22
	Kerma	82.29	14.88	67.20	19.26	33.98	7.83	27.90	6.83
	Yahgan	150.22	26.44	122.75	22.15	63.30	13.81	58.00	9.17
	Sadlermiut	115.29	33.56	111.28	15.40	59.08	16.43	46.27	8.39

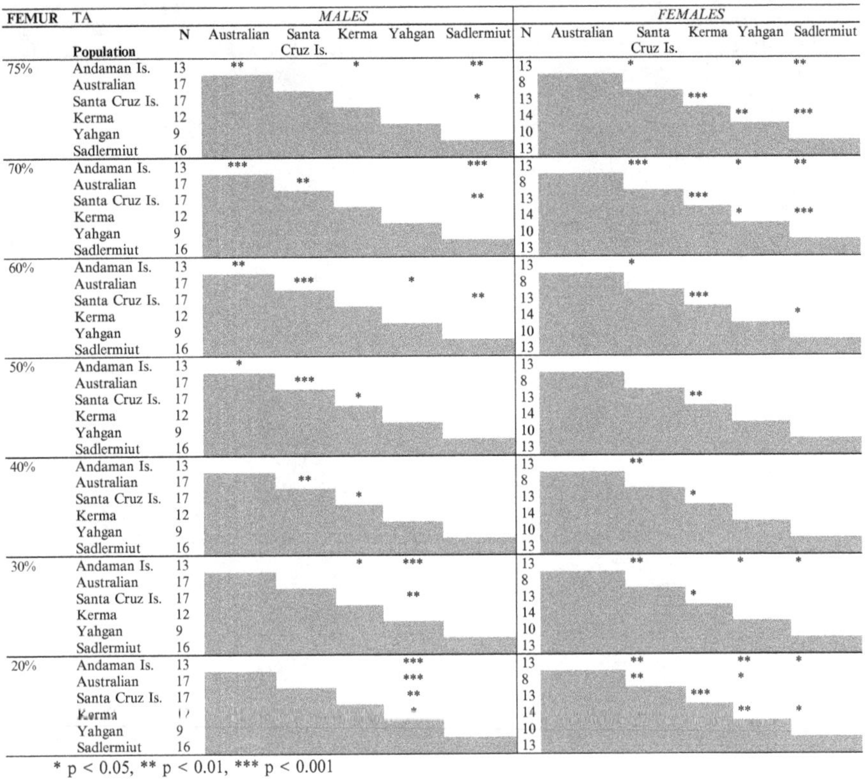

Fig. 5.4 Results of One-Way ANOVA pairwise post hoc tests of standardized total subperiosteal area (*TA*) population means at 10 % intervals throughout the femur (75 % instead of 80 % used in the femur to avoid including the lesser trochanter)

the cross-sectional properties examined above (Table 5.4). Figure 5.8 displays box-plots of minima for each property. General patterns in relative ranks expressed among populations are consistent across the different cross-sectional properties, although absolute locations vary somewhat between the properties. Three samples in particular have a highly consistent pattern in location, with median values focused around 30 % of bone length: the Andaman Islanders, Santa Cruz Islanders and Nubian agriculturalists. Australian Aborigines consistently display minimal values more distally along the tibial diaphysis than other groups. In contrast, the Yahgan and Sadlermiut display the opposite pattern, reaching a minimum at more proximal locations along the shaft. Comparing observed positional differences, Kruskal-Wallis non-parametric tests of multiple independent samples are all highly significant ($p<0.001$) among males, with post-hoc pairwise comparisons (Fig. 5.9) indicating that significant differences are most frequently observed between the Yahgan and Sadlermiut relative to the Andaman Islanders, Australian Aborigines and Santa Cruz Islanders. Among females, whilst three of the tests are significant (*TA*, I_x, and J_{solid}), thus rejecting the hypothesis of the same median expressed in all groups, no pairwise comparisons are statistically significant ($p<0.05$).

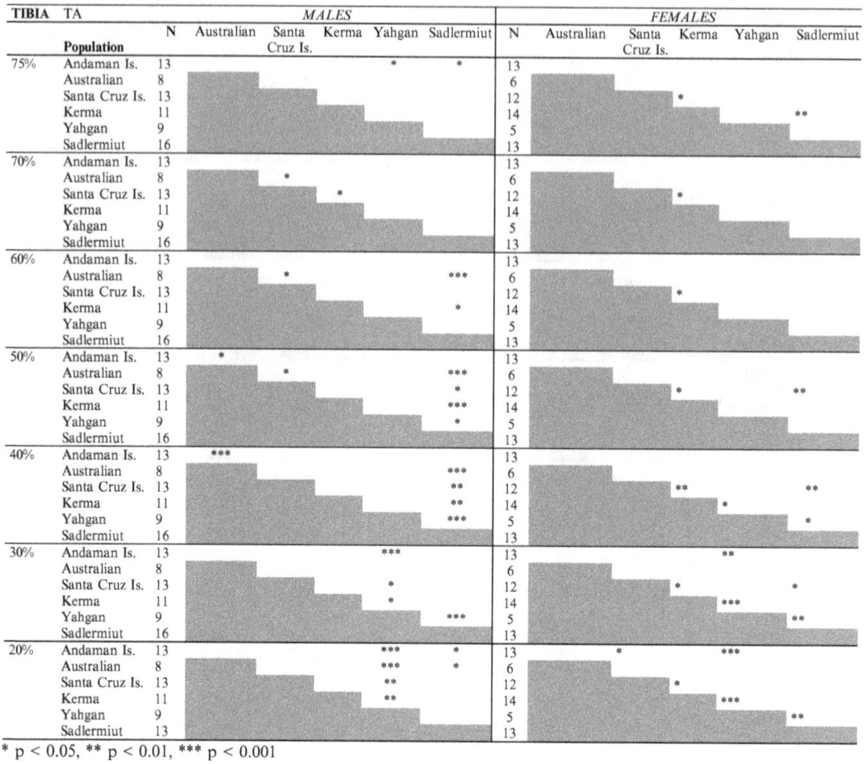

Fig. 5.5 Results of One-Way ANOVA pairwise post-hoc tests of standardized total subperiosteal area (*TA*) population means at 10 % intervals throughout the tibia

5.4 Discussion

5.4.1 Distribution of Cross-sectional Properties Along the Diaphyses

The large extent of phenotypic variation in body size and shape observed across modern human populations presents challenges to the interpretation of skeletal results used for assessing underlying mobility differences (Stock 2006; Ruff et al. 2006a, b). However, the results observed here suggest that there are a number of different underlying processes to which the bone morphology is adapted, and specifically that the effects may be localized to particular bones, or regions within each bone diaphysis. The pattern of differences in femoral and tibial midshafts observed among the populations were in general highly consistent with differences observed at other mid-diaphyseal locations (40 and 60 %). However, towards the proximal and distal ends of each diaphysis, the observed pairwise differences reflect quite

FEMUR Jsolid		MALES						FEMALES					
	Population	N	Australian	Santa Cruz Is.	Kerma	Yahgan	Sadlermiut	N	Australian	Santa Cruz Is.	Kerma	Yahgan	Sadlermiut
75%	Andaman Is.	13		**		**	***	13		***		***	***
	Australian	17					**	8				***	***
	Santa Cruz Is.	17					**	13			***		**
	Kerma	12				**	***	14				***	***
	Yahgan	9						10					
	Sadlermiut	16						13					
70%	Andaman Is.	13	*	**		**	***	13		***		***	***
	Australian	17					**	8				***	***
	Santa Cruz Is.	17					***	13			***		*
	Kerma	12				*	***	14				***	***
	Yahgan	9						10					
	Sadlermiut	16						13					
60%	Andaman Is.	13					***	13		***		***	***
	Australian	17					**	8					
	Santa Cruz Is.	17					***	13			***		
	Kerma	12					***	14				***	***
	Yahgan	9						10					
	Sadlermiut	16						13					
50%	Andaman Is.	13					**	13		**		**	***
	Australian	17						8					*
	Santa Cruz Is.	17					**	13			***		
	Kerma	12					**	14				**	***
	Yahgan	9						10					
	Sadlermiut	16						13					
40%	Andaman Is.	13				***	*	13		***		***	***
	Australian	17						8					*
	Santa Cruz Is.	17				**		13			***		
	Kerma	12						14				***	***
	Yahgan	9						10					
	Sadlermiut	16						13					
30%	Andaman Is.	13		*	**	**	**	13		***		***	***
	Australian	17				**		8		**		**	***
	Santa Cruz Is.	17				*		13			***		
	Kerma	12				*		14				***	***
	Yahgan	9						10					
	Sadlermiut	16						13					
20%	Andaman Is.	13		**		***	***	13		***		***	***
	Australian	17		*		***	***	8		***		***	***
	Santa Cruz Is.	17				***		13			***		
	Kerma	12				***	*	14				***	***
	Yahgan	9						10					
	Sadlermiut	16						13					

* $p < 0.05$, ** $p < 0.01$, *** $p < 0.001$

Fig. 5.6 Results of One-Way ANOVA pairwise post-hoc tests of standardized polar second moment of area (J_{solid}) population means at 10 % intervals throughout the femur (75 % instead of 80 % used in the femur to avoid including the lesser trochanter)

different patterns, driven primarily by increased cross-sectional properties in the Yahgan and Sadlermiut populations. These two populations share a similar physique, characterized by higher body mass and wider relative body breadths than the other populations included (Fig. 5.1), and thus it is possible that these regions of the bone diaphyses reflect body shape differences more so than an influence of mobility.

Throughout the mid-diaphyseal portion of the tibial diaphysis, Australian Aborigines display greater total subperiosteal area than that observed for other populations (Fig. 5.2), whereas they do not show marked differences in *TA* elsewhere throughout the lower limb relative to the other populations. This result, observed both among males and females, is consistent with localized bone remodelling of the distal limb segment in response to high mobility levels. This observation, of a possible influence of mobility most apparent in the midshaft tibia, fits with the argument that the distal limb is under strongest selective pressures for limb economy (Stock 2006). However, whilst high values of standardized *TA* were observed among

TIBIA	Jsolid		*MALES*						*FEMALES*				
	Population	N	Australian	Santa Cruz Is.	Kerma	Yahgan	Sadlermiut	N	Australian	Santa Cruz Is.	Kerma	Yahgan	Sadlermiut
75%	Andaman Is.	13				***	***	13		***		**	***
	Australian	8				**	***	6		**		**	***
	Santa Cruz Is.	13				*	***	12			***		
	Kerma	11				*	***	14				***	***
	Yahgan	9						5					
	Sadlermiut	16						13					
70%	Andaman Is.	13				*	***	13		*		**	***
	Australian	8						6					
	Santa Cruz Is.	13					**	12			**		
	Kerma	11					**	14				***	***
	Yahgan	9						5					
	Sadlermiut	16						13					
60%	Andaman Is.	13					*	13		*		*	*
	Australian	8						6					
	Santa Cruz Is.	13						12			***		
	Kerma	11						14				***	***
	Yahgan	9						5					
	Sadlermiut	16						13					
50%	Andaman Is.	13	*			*		13		**		**	
	Australian	8			*			6					
	Santa Cruz Is.	13						12			***		
	Kerma	11						14				***	**
	Yahgan	9						5					
	Sadlermiut	16						13					
40%	Andaman Is.	13		***		**	*	13		*		***	
	Australian	8						6					
	Santa Cruz Is.	13			*			12			***		
	Kerma	11				*		14				***	**
	Yahgan	9						5					**
	Sadlermiut	16						13					
30%	Andaman Is.	13		**		***	***	13		***		***	**
	Australian	8				*		6			*	**	
	Santa Cruz Is.	13				*		12			***		
	Kerma	11				***	**	14				***	***
	Yahgan	9						5					**
	Sadlermiut	16						13					
20%	Andaman Is.	13		*		***	***	13		***		***	***
	Australian	8				***	***	6		**		***	***
	Santa Cruz Is.	13				**	**	12			***	*	
	Kerma	11				**	***	14				***	***
	Yahgan	9						5					
	Sadlermiut	16						13					

* $p < 0.05$, ** $p < 0.01$, *** $p < 0.001$

Fig. 5.7 Results of One-Way ANOVA pairwise post-hoc tests of standardized polar second moment of area (J_{solid}) population means at 10 % intervals throughout the tibia

the Australians, for J_{solid} the results do not fit as well with an explanation based upon mobility. This lack of correspondence between ethnographic evidence for mobility and bone rigidity in a highly mobile forager population is consistent with other studies of Australian Aborigine individuals (Pearson 2000; Carlson et al. 2007), and may be seen as a challenge to studies of CSG in relation to mobility. However, Shaw and Stock (2009b) found very high I_x/I_y ratios in runners, and it may be that such a "mobility index", rather than overall cross section size, provides the best means of differentiating mobility. Additionally, assessing results for individual second moments of area (I_x, I_y, I_{max}, I_{min}) throughout the limb bones may offer further insights. Requirements for low mass in the distal limb may further limit bone deposition in distal limb segments in order to maximize energetic efficiency (Stock 2006). Further studies of a greater range of highly mobile groups need to be considered in order to ascertain the extent to which high levels of postcranial robusticity can be seen as characterizing other terrestrially mobile hunter-gatherers.

Concerning observed differences between populations practicing either predominantly marine or terrestrial mobility, population-specific patterns emerge in the deposition of diaphyseal bone, perhaps to an extent reflecting relative magnitudes of terrestrial mobility, but not to the point of supporting a generic classification of subsistence economy as either "terrestrial" or "marine". For example, the Nubian terrestrial agriculturalists display very similar mean limb rigidity profiles to the Andaman Islander marine foragers, whilst the Sadlermiut and Yahgan, also predominantly marine foragers, display some of the highest observed values for both

Table 5.4 Descriptive statistics for the location (% diaphyseal length from distal end) of minimum tibial cross-sectional properties

Property	Sex	Population	*N*	Median	Q1	Q3	Mean	St. Dev.
TA[a]	Male	Andaman Is.	13	32.0	30.0	34.0	32.2	2.0
		Australian	8	26.5	24.0	28.0	26.0	2.1
		Santa Cruz Is.	13	32.0	31.0	34.0	32.5	2.7
		Nubian	12	32.5	29.5	35.0	32.3	3.5
		Yahgan	9	38.0	35.0	39.0	37.6	2.7
		Sadlermiut	16	38.0	36.0	39.5	37.7	2.5
	Female	Andaman Is.	13	32.0	30.0	34.0	32.3	3.3
		Australian	6	27.5	25.0	32.0	28.8	5.0
		Santa Cruz Is.	12	33.0	32.5	35.0	33.6	1.9
		Nubian	13	34.0	33.0	35.0	34.0	2.2
		Yahgan	6	39.0	35.0	44.0	38.3	6.3
		Sadlermiut	13	37.0	35.0	38.0	36.7	2.2
I_x[b]	Male	Andaman Is.	13	29.0	28.0	29.0	28.7	1.3
		Australian	8	22.5	20.5	24.0	22.6	2.4
		Santa Cruz Is.	13	27.0	25.0	31.0	27.9	2.7
		Nubian	12	29.5	28.0	31.5	29.7	2.8
		Yahgan	9	35.0	34.0	36.0	35.3	2.9
		Sadlermiut	16	34.0	33.0	36.0	34.1	2.5
	Female	Andaman Is.	13	29.0	28.0	33.0	29.8	2.7
		Australian	6	25.0	22.0	27.0	25.7	4.1
		Santa Cruz Is.	12	31.0	30.0	32.0	30.9	2.9
		Nubian	13	31.0	30.0	32.0	31.5	2.2
		Yahgan	6	35.5	30.0	40.0	34.7	5.4
		Sadlermiut	13	34.0	32.0	35.0	34.1	2.3
I_y[c]	Male	Andaman Is.	13	33.0	30.0	35.0	33.1	2.9
		Australian	8	26.5	25.0	29.0	26.9	2.3
		Santa Cruz Is.	13	35.0	33.0	39.0	37.5	6.9
		Nubian	12	32.5	30.5	35.5	33.1	3.8
		Yahgan	9	40.0	35.0	45.0	41.3	7.3
		Sadlermiut	16	41.0	39.0	42.0	40.6	3.2
	Female	Andaman Is.	13	32.0	30.0	38.0	33.8	5.4
		Australian	6	27.0	26.0	38.0	30.2	6.1
		Santa Cruz Is.	12	37.5	34.5	38.0	36.3	3.2
		Nubian	13	36.0	34.0	38.0	35.9	3.2
		Yahgan	6	39.5	37.0	44.0	40.5	7.1
		Sadlermiut	13	39.0	38.0	41.0	38.8	2.9

(continued)

Table 5.4 (continued)

Property	Sex	Population	*N*	Median	Q1	Q3	Mean	St. Dev.
J_{solid}[d]	Male	Andaman Is.	13	30.0	30.0	32.0	30.3	1.7
		Australian	8	25.0	24.0	27.5	25.3	2.4
		Santa Cruz Is.	13	31.0	28.0	32.0	30.5	2.7
		Nubian	12	31.0	28.5	34.5	31.6	3.7
		Yahgan	9	37.0	34.0	39.0	37.0	2.9
		Sadlermiut	16	36.0	35.0	38.5	36.3	2.8
	Female	Andaman Is.	13	30.0	29.0	32.0	31.1	2.9
		Australian	6	27.5	25.0	29.0	27.7	3.3
		Santa Cruz Is.	12	33.0	32.0	35.0	33.2	2.0
		Nubian	13	33.0	31.0	34.0	33.1	2.4
		Yahgan	6	36.5	33.0	42.0	36.3	5.5
		Sadlermiut	13	35.0	33.0	37.0	35.6	2.4

[a]*TA* total subperiosteal area
[b]I_x second moment of area, *x*-axis
[c]I_y second moment of area, *y*-axis
[d]J_{solid} polar second moment of area. All properties estimated from periosteal contour only

femoral and tibial torsional rigidity. The results here appear to corroborate to an extent the previous finding of a lack of relatively high limb bone rigidity in the Australian Aborigines (Carlson et al. 2007), although a signature does appear for *TA* in the mid-distal region of the tibial diaphysis in the present study. Comparisons of tibial cross-sectional properties appear to reflect differences associated with mobility levels more than comparisons of cross-sectional properties throughout the femur, in line with previous studies (Stock 2006). However, further assessment of the distribution of bone throughout cross sections along a limb in a wider array of "high mobility" groups is required to confirm the degree to which femoral vs. tibial sections should be placed at the forefront of studies of past populations' mobility.

Variation in the distribution of properties along the limb may alternatively reflect other underlying factors, unrelated to behaviour or mobility. Movement over rugged terrain has been invoked as a key influence in modulating limb bone morphology in response to mobility (Ruff 1995b; Marchi 2008). The body shape of individuals may also influence the distribution of rigidity throughout the lower limb diaphyses, and this may explain the very high values of limb bone rigidity towards the proximal and distal ends of each diaphysis observed among the Yahgan and Sadlermiut, who both present a relatively wider-bodied phenotype than the other populations. Furthermore, Sladek et al. (2010) have highlighted the extent to which there can be high levels of intragroup variation, both in the magnitude of cross-sectional properties and in the patterning of variation expressed throughout diaphyses. Whilst attributing variation in the structuring of bone rigidity from proximal to distal limb segments is difficult to assign to specific causes, the present study highlights variation across Holocene foragers and illustrates future potential to examine variability between groups of similar subsistence modes (for example, variation among marine foragers), as further populations are studied.

Comparing results for *TA* to those for J_{solid}, there are a number of differences that are difficult to interpret in terms of mobility levels. Cross-sectional areas are

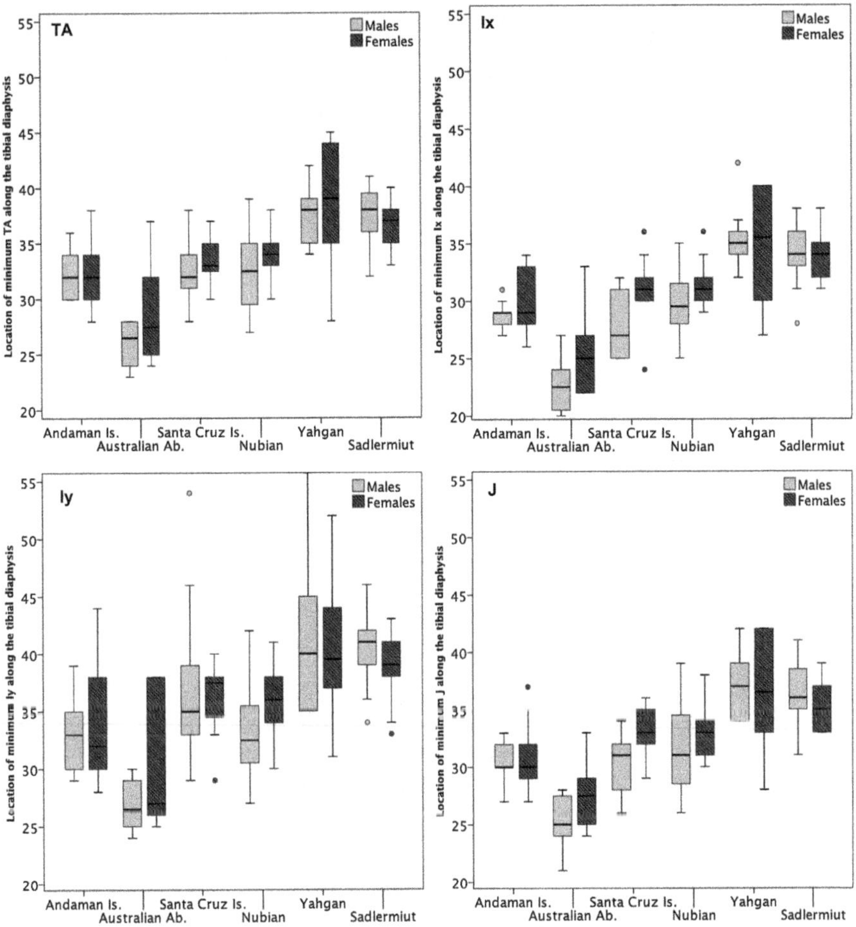

Fig. 5.8 Box-plots for location of minimum cross-sectional geometric properties in the tibia for the six populations. *Boxes* indicate median, lower and upper quartiles, *whiskers* indicate extreme values within 1.5 times inter-quartile range, and *outliers* are plotted as circles. *TA* total subperiosteal area; I_x second moment of area, x-axis; I_y second moment of area, y-axis; J_{solid} polar second moment of area. All properties estimated from periosteal contour only

generally thought of as providing an indication of compressive strength, whilst second moments of area are more biomechanically relevant measures of bending and/or torsional rigidity. However, the observed differences in properties among the populations observed for *TA* as opposed to J_{solid} are difficult to interpret given the close mathematical relationship between *TA* and *J* (Sparacello and Pearson 2010). Part of the disparity in results between *TA* and J_{solid} may reflect contrasting methods applied in standardizing the properties, and whilst these are justified biomechanically (Ruff 1994) for populations of such contrasting physique (in which body mass and limb lengths are poorly correlated), the inclusion of limb length terms only for second moments of areas (I_x, I_y and J_{solid}) may account for some of the resultant differences in properties observed among the populations.

TIBIA	Population	N	MALES Australian	Santa Cruz Is.	Kerma	Yahgan	Sadlermiut
TA^1	Andaman Is.	13	*			**	**
	Australian	8		**	**	**	
	Santa Cruz Is.	13				**	*
	Kerma	12					
	Yahgan	9					
	Sadlermiut	16					
I_x^2	Andaman Is.	13	**			***	***
	Australian	8			**	**	
	Santa Cruz Is.	13				***	***
	Kerma	12				*	
	Yahgan	9					
	Sadlermiut	16					
I_y^3	Andaman Is.	13	*				***
	Australian	8		**	*		*
	Santa Cruz Is.	13					
	Kerma	12					**
	Yahgan	9					
	Sadlermiut	16					
J_{solid}^4	Andaman Is.	13	**			***	***
	Australian	8		*	*		
	Santa Cruz Is.	13				***	***
	Kerma	12					
	Yahgan	9					
	Sadlermiut	16					

[1]TA = total subperiosteal area; [2]I_x = second moment of area, x-axis; [3]I_y = second moment of area, y-axis; [4]J_{solid} = polar second moment of area. All properties estimated from periosteal contour only.

Fig. 5.9 Results of pairwise comparisons of median differences in the location of minimum cross-sectional property throughout the tibia

This study applied a solid section model, as it was not practical to quantify variation in the relative size of the medullary cavity in such a large sample. Therefore, observed variation in the patterning of limb rigidity from proximal to distal sites throughout diaphyses cannot account for differences in the relative size of the medullary cavity throughout the limb bones either between individuals or populations. The extent of variation in the patterning of *%CA* from proximal to distal ends of bone shafts is relatively poorly known for such a wide range of human morphological diversity. However, Sladek et al. (2010) observed considerable variation in related properties, e.g. *CA*, across Bronze Age samples from Europe. For second moments of area, however, the periosteal contour has the greatest role in determining bone rigidity (Sparacello and Pearson 2010), and observed patterns in solid cross-sectional properties are known to closely reflect those of true properties at bone midshafts (Stock and Shaw 2007), and across much of the diaphysis in certain samples (Macintosh et al. 2013).

5.4.2 Location of Minimal Properties Within the Tibia

Results obtained here imply that at least two processes act to set the pattern from proximal to distal locations of minimum rigidity in the tibial shaft. Compared to other populations, Yahgan and Sadlermiut populations possess large distal tibial cross-sectional properties (i.e. at the ankle) relative to body mass, leading to more proximally located minimal properties within their tibiae. This suggests a possible role for epiphyseal integration with diaphyseal morphology, which has been previously discussed with regards to other regions of the limb (Ruff 1999; Weaver 2003). Secondly, Australian Aborigines display a more distally situated location of minimum rigidity relative to other populations. This result is consistent with an influence of high mobility among the populations leading to increased deposition of bone in the midshaft region of the diaphysis, and consequently a shift in the location of minimum rigidity to a more distal position. As discussed in the above section, there is some evidence for increased *TA* in the midshaft region of the tibia among Australian Aborigines (Fig. 5.2), relative to the other populations studied, which may support this interpretation. The unique periosteal geometry of the tibial diaphysis displayed by Australian Aborigines may also be associated with better structural optimization of the bone under greater selective pressures for tissue economy (Stock 2006). Further analysis of this result in a larger number of tibiae, and with respect to regional variation in behaviour across the continent, is required to further isolate the extent to which mobility patterns may be contributing to this observed contrast in the morphology. However, it is possible that more distally located minima observed for Australian Aborigines rather reflect localized remodelling of the mid-distal region of the tibia in response to mobility (and not at locations along the diaphysis closer to the distal epiphysis). Subsequent studies are required to differentiate causes of variability in traits such as the location of minimal rigidity along bone diaphyses. Finally, this study has focused only on indicators of diaphyseal rigidity, and whether the same holds true for cross-sectional shape is the subject of further exploration.

5.5 Conclusions

This chapter set out to assess Holocene variability in cross-sectional properties throughout the femur and tibia with a particular focus on insights into mobility that may be obtained by comparing populations displaying diversity in body size and shape. The results suggest that within particular regions of the lower limb diaphyses, most notably the mid-diaphyseal region of the tibia, signals of high mobility may override confounding effects of differing body size and shape. In other diaphyseal regions, however, particularly towards their ends, alternative factors may need to be sought to explain observed variation. The results suggest that limb rigidity does not reflect simple classifications of populations as either "marine" or "terrestrial", "agricultural" or "hunter-gatherer", but rather reflects population-specific distributions of properties that perhaps ought to be matched to better ethnographic or archaeological indicators of underlying mobility levels. Further study is required in a range of high mobility groups to ascertain the extent to which high mobility is the causal explanation for relatively high mid-tibial rigidity among Australian Aborigines, or whether the high mid-tibial rigidity is a unique Australian Aborigine trait. The finding that the Australian Aborigines display a more distally located position of minimal cross-sectional properties, perhaps reflecting structural optimization of the distal limb in response to high levels of terrestrial mobility, also requires verification on individuals from a wider range of populations.

Acknowledgements We would like to thank two anonymous reviewers, as well as the editors, for comments provided, which have greatly improved the manuscript. We thank the following individuals for access to skeletal collections under their care, or for their assistance: Maggie Bellatti, Mercedes Okumura, Margaret Clegg, Robert Kruszynski, Heather Bonney, Giorgio Manzi, Monica Zavattaro, Jerry Cybulski, Janet Young, Stacey Girling and Megan Gardiner. Thanks also to the following institutions: Duckworth Collection, University of Cambridge; Natural History Museum, London; Canadian Museum of Civilization, Gatineau; Museo di Storia Naturale, Firenze; Museo di Antropologia "G. Sergi", Sapienza, Universita di Roma. Funding for this research was provided by the Natural Environment Research Council, UK, and Churchill College, Cambridge, UK.

References

Bridges EL (1948) Uttermost part of the Earth: a history of Tierra Del Fuego and the Fuegians. Hodder & Stoughton, London

Carlson KJ, Grine FE, Pearson OM (2007) Robusticity and sexual dimorphism in the postcranium of modern hunter-gatherers from Australia. Am J Phys Anthropol 134:9–23

Churchill SE (1994) Human upper body evolution in the Eurasian Later Pleistocene. Ph.D. thesis, University of New Mexico

Churchill SE (1996) Particulate versus integrated evolution of the upper body in late pleistocene humans: a test of two models. Am J Phys Anthropol 100:559–583

Collier S (1989) The influence of economic behaviour and environment upon robusticity of the postcranial skeleton: a comparison of Australian Aborigines and other populations. Archaeol Ocean 24:17–30

Colten RH, Arnold JE (1998) Prehistoric marine mammal hunting on California's Northern Channel Islands. Am Antiquity 63(4):679–701

Coltrain JB (2009) Sealing, whaling and caribou revisited: additional insights from the skeletal isotope chemistry of eastern Arctic foragers. J Archaeol Sci 36:764–775

Davies TG (2013) Cross-sectional variation in the human femur and tibia: the influence of physique and habitual mobility on diaphyseal morphology. Ph.D. thesis, University of Cambridge

Davies TG, Shaw CN, Stock JT (2012) A test of a new method and software for the rapid estimation of cross-sectional geometric properties of long bone diaphyses from 3D laser surface scans. Archaeol Anthropol Sci 4:277–290

Frost HM (1997) Why do marathon runners have less bone than weight lifters? A vital-biomechanical view and explanation. Bone 20:183–189

Glassow MA (1993) Changes in subsistence on marine resources through 7,000 years of prehistory on Santa Cruz Island. In: Glassow MA (ed) Archaeology on the Northern Channel Islands of California, studies of subsistence, economics, and social organization. Coyote, Salinas, pp 75–94

Gosman JH, Stout SD, Larsen CS (2011) Skeletal biology over the life span: a view from the surfaces. Am J Phys Anthropol 54:86–98

Higgins RW, Ruff CB (2011) The effects of distal limb segment shortening on locomotor efficiency in sloped terrain: implications for neandertal locomotor behavior. Am J Phys Anthropol 146:336–345

Jones HH, Priest JD, Hayes WC, Tichenor CC, Nagel DA (1977) Humeral hypertrophy in response to exercise. J Bone Joint Surg Am 59:204–208

Lieberman DE, Polk JD, Demes B (2004) Predicting long bone loading from cross-sectional geometry. Am J Phys Anthropol 123:156–171

Macintosh AA, Davies TG, Ryan TM, Shaw CN, Stock JT (2013) Periosteal versus true cross-sectional geometry: a comparison along humeral, femoral, and tibial diaphyses. Am J Phys Anthropol 150(3):442–452

Marchi D (2008) Relationships between lower limb cross-sectional geometry and mobility: the case of a Neolithic sample from Italy. Am J Phys Anthropol 137:188–200

Marchi D, Sparacello VS, Holt BM, Formicola V (2006) Biomechanical approach to the reconstruction of activity patterns in Neolithic Western Liguria, Italy. Am J Phys Anthropol 131:447–455

Merbs CF (1983) Patterns of activity-induced pathology in a Canadian Inuit population. National Museums of Canada, Ottawa

Ohman JC, Lovejoy CO (2001) The shape of a long bone's shaft: bending stress or growth plate form? Am J Phys Anthropol 114(S32):S115

Pearson OM (2000) Activity, climate, and postcranial robusticity—implications for modern human origins and scenarios of adaptive change. Curr Anthropol 41:569–607

Perry JE (2005) Early period resource use on eastern Santa Cruz Island, California. Institute for Wildlife Studies, Arcata, CA

Roy TA, Ruff CB, Plato CC (1994) Hand dominance and bilateral asymmetry in the structure of the second metacarpal. Am J Phys Anthropol 94:203–211

Ruff CB (1987) Sexual dimorphism in human lower limb bone structure: relationship to subsistence strategy and sexual division of labor. J Hum Evol 16:391–416

Ruff CB (1994) Morphological adaptation to climate in modern and fossil hominids. Am J Phys Anthropol 37:65–107

Ruff CB (1995a) Limb bone structure: influence of sex, subsistence, and terrain [abstract]. Am J Phys Anthropol Suppl. 38(S1):186

Ruff CB (1995b) Biomechanics of the hip and birth in early Homo. Am J Phys Anthropol 98:527–574

Ruff CB (1999) Skeletal structure and behavioural patterns of prehistoric Great Basin populations. In: Hemphill BE, Larsen CS (eds) Understanding prehistoric lifeways in the Great Basin Wetlands: bioarchaeological reconstruction and interpretation. University of Utah, Salt Lake City, pp 290–320

Ruff CB (2000) Body size, body shape, and long bone strength in modern humans. J Hum Evol 38:269–290

Ruff CB (2002) Long bone articular and diaphyseal structure in old world monkeys and apes. I: locomotor effects. Am J Phys Anthropol 119:305–342

Ruff CB (2003) Ontogenetic adaptation to bipedalism: age changes in femoral to humeral length and strength proportions in humans, with a comparison to baboons. J Hum Evol 45:317–349

Ruff CB (2008) Biomechanical analyses of archaeological human skeletal samples. In: Katzenberg M, Saunders A (eds) Biological Anthropology of the Human Skeleton. New York: Wiley, p183–206

Ruff CB, Hayes WC (1983) Cross-sectional geometry of Pecos Pueblo femora and tibiae—a biomechanical investigation: II. Sex, age, side differences. Am J Phys Anthropol 60:383–400

Ruff CB, Holt BM, Trinkaus E (2006a) Who's afraid of the big bad Wolff?: "Wolff's law" and bone functional adaptation. Am J Phys Anthropol 129:484–498

Ruff CB, Holt BM, Sladek V, Berner M, Murphy WA Jr, Zur Nedden D et al (2006b) Body size, body proportions, and mobility in the Tyrolean "Iceman". J Hum Evol 51:91–101

Ruff CB, Larsen CS, Hayes WC (1984) Structural changes in the femur with the transition to agriculture on the Georgia coast. Am J Phys Anthropol 64(2):125–136

Ruff CB, Trinkaus E, Walker A, Larsen CS (1993) Postcranial robusticity in Homo. I: Temporal trends and mechanical interpretation. Am J Phys Anthropol 91:21–53

Ruff CB, Trinkaus E, Holliday TW (1997) Body mass and encephalization in Pleistocene Homo. Nature 387(6629):173–176

Shaw CN, Stock JT (2009a) Habitual throwing and swimming correspond with upper limb diaphyseal strength and shape in modern human athletes. Am J Phys Anthropol 140(1): 160–172

Shaw CN, Stock JT (2009b) Intensity, repetitiveness, and directionality of habitual adolescent mobility patterns influence the tibial diaphysis morphology of athletes. Am J Phys Anthropol 140(1):149–159

Shaw CN, Stock JT (2011) The influence of body proportions on femoral and tibial midshaft shape in hunter-gatherers. Am J Phys Anthropol 144:22–29

Shaw CN, Stock JT (2013) Extreme mobility in the late Pleistocene? Comparing limb biomechanics among fossil *Homo*, varsity athletes and Holocene foragers. J Hum Evol 64(4):242–249

Shaw CN, Stock JT, Davies TG, Ryan TM (2014) Does the distribution and variation in cortical bone along lower limb diaphyses reflect selection for locomotor economy? In: Carlson KJ, Marchi D (eds) Reconstructing mobility: environmental, behavioral, and morphological determinants. Springer, New York

Sladek V, Berner M, Galeta P, Friedl L, Kudrnova S (2010) Technical note: the effect of midshaft location on the error ranges of femoral and tibial cross-sectional parameters. Am J Phys Anthropol 141:325–332

Sparacello V, Marchi D (2008) Mobility and subsistence economy: a diachronic comparison between two groups settled in the same geographical area (Liguria, Italy). Am J Phys Anthropol 136:485–495

Sparacello V, Pearson OM (2010) The importance of accounting for the area of the medullary cavity in cross-sectional geometry: a test based on the femoral midshaft. Am J Phys Anthropol 143(4):612–624

Stock JT (2006) Hunter-gatherer postcranial robusticity relative to patterns of mobility, climatic adaptation, and selection for tissue economy. Am J Phys Anthropol 131:194–204

Stock JT, O'Neill MC, Ruff CB, Zabecki M, Shackelford LL, Rose J (2011) Body size, skeletal biomechanics, mobility and habitual activity from the late Palaeolithic to the mid-dynastic Nile Valley. In: Pinhasi R, Stock JT (eds) Human bioarchaeology of the transition to agriculture. Wiley, Hoboken, NJ, pp 347–367

Stock JT, Pfeiffer S (2001) Linking structural variability in long bone diaphyses to habitual behaviors: foragers from the southern African Later Stone Age and the Andaman Islands. Am J Phys Anthropol 115:337–348

Stock JT, Shaw CN (2007) Which measures of diaphyseal robusticity are robust? A comparison of external methods of quantifying the strength of long bone diaphyses to cross-sectional geometric properties. Am J Phys Anthropol 134(3):412–423

Trinkaus E, Churchill SE, Ruff CB (1994) Postcranial robusticity in Homo II: humeral bilateral asymmetry and bone plasticity. Am J Phys Anthropol 93(1):1–34

Walker PL (1986) Porotic hyperostosis in a marine-dependent California Indian population. Am J Phys Anthropol 69:345–354

Weaver TD (2003) The shape of the Neandertal femur is primarily the consequence of a hyperpolar body form. Proc Natl Acad Sci U S A 100:6926–6929

Weiss E (2003) Effects of rowing on humeral strength. Am J Phys Anthropol 121:293–302

Wescott DJ (2006) Effect of mobility on femur midshaft external shape and robusticity. Am J Phys Anthropol 130:201–213

Chapter 6
The Importance of Considering Fibular Robusticity When Inferring the Mobility Patterns of Past Populations

Vitale S. Sparacello, Damiano Marchi, and Colin N. Shaw

Abstract In this chapter we investigate the lower limb structural rigidity (using cross-sectional geometric properties of the diaphyseal midshaft) within a sample of 124 individuals from the Late Upper Paleolithic, Neolithic and Iron Age from Italy, Medieval Germany, and twenty-first Century Britain (long distance runners, field hockey players, and sedentary controls). Late Upper Paleolithic, Neolithic and Iron Age samples were settled in rugged areas, whereas the other samples inhabited plain areas. The aim of this study is to assess whether fibular diaphyseal properties reflect mobility patterns or terrain properties in past populations. Both fibular rigidity and relative fibular rigidity ratio (fibula/tibia) have been analyzed.

Results reveal that Late Upper Paleolithic, Neolithic and Iron Age samples show high fibular rigidity and have values of relative fibular rigidity that are most similar to modern hockey players. The relative fibular diaphyseal rigidity of hockey players has been previously explained as the consequence of their dynamic and repetitive change of direction. Late Upper Paleolithic and Neolithic individuals are thought to have been highly terrestrially mobile, while Iron Age people were probably fairly sedentary. However, all of the three groups lived in areas of uneven terrain. We conclude that fibular rigidity and relative fibular rigidity are influenced by factors

V.S. Sparacello
Department of Archaeology, Durham University, Durham, UK

Department of Anthropology, University of New Mexico, Albuquerque, NM, USA

D. Marchi (✉)
Department of Biology, University of Pisa, Via Derna 1, Pisa 56126, Italy

Evolutionary Studies Institute, University of the Witwatersrand, Johannesburg, South Africa
e-mail: damiano.marchi@unipi.it

C.N. Shaw
McDonald Institute for Archaeological Research, University of Cambridge, Cambridge, UK

PAVE Research Group, Department of Archaeology and Anthropology, University of Cambridge, Cambridge, UK

K.J. Carlson and D. Marchi (eds.), *Reconstructing Mobility: Environmental, Behavioral, and Morphological Determinants*, DOI 10.1007/978-1-4899-7460-0_6,

that increase foot eversion/inversion such as frequent directional changes and uneven terrain. The results of this study suggest that inclusion of the fibula provides a valuable additional perspective that complements traditional predictions of mobility patterns based on the femur or the tibia alone.

Keywords Fibula • Tibia • Bioarchaeology • Cross-sectional geometry • Terrain conformation

6.1 Introduction

Although it constitutes one of the parameters that are most often investigated in bioarchaeological research, mobility has not yet received a univocal definition. For research aimed at the reconstruction of past activity patterns, mobility is usually broadly defined as the habitual amount of traveling (either through walking or running) that characterized a population. However, it is difficult to numerically quantify mobility in bioarchaeology, for example, through the reconstruction of home ranges or trade networks. Qualitative assessments such as "high" or "low" mobility are then usually used, and are linked to the logistic requirements of different subsistence strategies (see Wescott 2014) based on ethnographic analogies.

Since walking and running are the main causes of anteroposterior (AP) repetitive mechanical loading on lower limbs, the amount of physical activity associated with traversing the landscape can be indirectly quantified through its effects on long bone diaphyses. Research that has investigated the relationship between mobility and long bone diaphyseal structure has generally considered AP-oriented loads as the main cause of lower limb loading (Lovejoy et al. 1976; Ruff 1999, 2000a; Shaw and Stock 2009; see Wescott 2014 and references therein). However, experimental and nonexperimental studies conducted on mammals—including primates—have demonstrated that different "types" of mobility may considerably influence the amount of mediolateral-oriented loads to which the lower limb is subjected (Carlson et al. 2005; Demes et al. 2006; Carlson and Judex 2007; Marchi 2007; Marchi and Shaw 2011; Marchi et al. 2011). Following these findings, here mobility is considered as having two components, both of which influence lower limb morphology. The first component is the amount of traveling that people undertake to have access to resources: this is the traditional definition of mobility. The second component is the type of movement, particularly focused on how movement influences mediolateral (ML) loading (e.g., uphill/downhill and/or on even/uneven terrain). Untangling the concomitant effect of the two components may improve the interpretation of lower limb skeletal properties for behavioral reconstructions.

Cross-sectional geometry (CSG) is a biomechanical technique that studies the plastic behavior of long bone diaphyses to adapt to mechanical loads (for reviews, see Pearson and Lieberman 2004; Ruff et al. 2006b; and references therein). Various bioarchaeological studies have described a correspondence between femoral diaphyseal shape and levels of mobility dictated by subsistence patterns. In particular, highly mobile hunter-gatherers generally show higher values of femoral shape ratios (I_x/I_y and I_{max}/I_{min})[1] than sedentary agriculturalists (e.g., Ruff and Hayes 1983; Ruff 1987, 1999; Larsen 1995; Stock and Pfeiffer 2001; Holt 2003; Ruff et al. 2006a).

However, research suggests that distance traveled is not the sole agent of lower limb remodeling; other factors should be taken into account, and possibly factored out when comparing skeletal samples and inferring habitual behavior patterns (Ruff 1999, 2000a; Sparacello and Marchi 2008). One of the confounding factors to consider when interpreting lower limb robusticity is the potential influence of topography. Ruff (1999) found that Native American groups characterized by different subsistence economies (preagricultural and agricultural, with assumed differences in patterns of mobility) were not differentiated in femoral diaphyseal robusticity. However, groups coming from more rugged areas showed significantly more robust femora. Based on these results, Ruff (2000a) proposed that once terrain is factored out, the influence of subsistence strategies on lower limb bone robusticity greatly declines.

Diaphyseal cross-sectional robusticity may therefore be influenced by both the activity performed and the topography upon which the activity is performed. Further support to this hypothesis came from the comparison of two groups that adopted different subsistence economies (herding and fishing-agriculture, respectively) settled in the same rugged area (Sparacello and Marchi 2008). Despite differences in presumed mobility levels, comparisons did not reveal significant differences in femoral robusticity, but showed significant differences in diaphyseal shape (I_x/I_y). This was interpreted as reflecting the influence of mobility levels and terrain properties on femoral midshaft robusticity and shape. The high femoral robusticity of both samples was interpreted as dictated by terrain ruggedness. The more elliptical diaphyseal shape (greater I_x/I_y ratios) of the Neolithic sample was interpreted as a consequence of higher mobility levels (Sparacello and Marchi 2008). A similar pattern was observed for the tibia in a comparison among several groups that adopted different subsistence economies (hunter-gatherers and herders) who had settled in both plain and mountainous regions (Sparacello et al. 2008). These results suggest that, when comparing groups dwelling in areas with similar topographies, different subsistence economies seem to generate less dramatic differences in lower limb robusticity.

Differently from the femur, there is not consistent correspondence between tibial shape (as revealed by I_{max}/I_{min} ratio) and mobility patterns (cf. Stock and Pfeiffer 2001; Holt 2003; Marchi 2008; Marchi et al. 2011). A comparison of upper and lower limb CSG properties in Andaman Islanders and Later Stone Age southern African foragers (Stock and Pfeiffer 2001) showed that Andamanese people had significantly more robust upper limbs than Later Stone Age people. Andamanese people incorporated a significant degree of marine mobility into their behavioral pattern that included swimming and canoe paddling, which probably led to this result. Later Stone Age people were highly mobile terrestrial hunter-gatherers and were characterized by significantly more

[1] I_x is the AP bending rigidity; I_y is the ML bending rigidity; I_{max} is the maximum moment of area; and I_{min} is the minimum moment of area of a bone at a cross section (Ruff and Hayes 1983). Femoral I_x/I_y is normally referred to as a mobility index.

robust lower limbs, and more AP-strengthened femoral cross sections. However, Later Stone Age people did not display statistically significant differences (more) in platycnemic tibiae relative to the Andamanese people. It therefore appears that femoral cross-sectional shape shows a more strict correspondence with inferred mobility levels than the tibia. A similar pattern was observed in European skeletal samples ranging from the Early Upper Paleolithic to the Bronze Age (Holt 2003; Sládek et al. 2006a, b; Marchi 2008; Marchi et al. 2011; see also Pearson et al. 2014, for an analysis of the weak correlation between femoral and tibial shape indices).

In a study of the associated influence of activity, climate, and mechanical constraints for tissue economy in the lower limb, Stock (2006) found that tibial diaphyseal robusticity was less correlated than femoral diaphyseal robusticity with climate and body shape, and showed less inherent variance in the samples. Stock (2006) concluded that the strongest morphological correlates of terrestrial mobility are femoral cross-sectional shape and tibial diaphyseal robusticity. A recent study on modern athletes and a bioarchaeological sample also found a good correspondence between tibial robusticity (relative to humeral robusticity) and mobility patterns (Shaw and Stock 2013).

Recently, Shaw and Stock (2009) compared three modern human samples: cross-country runners, field hockey players, and sedentary controls. Results revealed that the two highly mobile athlete samples had significantly more robust tibiae compared to controls. Further, diaphyseal shape differences between runners and hockey players were significant, with runners showing higher values (more elliptical diaphyseal cross sections). This result was interpreted as reflecting a greater degree of AP-oriented bending stress in runners—whose mode of locomotion was mainly linear—compared to field hockey players who performed frequent changes of direction (Spencer et al. 2004, and references therein). The results of Shaw and Stock (2009) suggest that tibial diaphyseal robusticity provides information about the level of mobility, while tibial cross-sectional shape is more informative about the directionality of loading.

Despite the abundance of work focused on the relationship between bone structure and mobility, the fibula has been often overlooked in anthropological studies because most of the body weight is supported by the tibia (Marchi and Shaw 2011 and references therein). Moreover, in bioarchaeological skeletal samples the fibula is less frequently found intact compared to the tibia. Nevertheless, the fibula may provide valuable information on mobility patterns. Studies on living hominoids have demonstrated that the relative fibular diaphyseal robusticity (tibia/fibula) corresponds with variation in locomotor patterns (Marchi 2007). Specifically, primarily arboreal hominoids possess a relatively more robust fibula compared to primarily terrestrial hominoids. These differences have been interpreted as a likely consequence of a more mobile fibula (Barnett and Napier 1953) and more adducted hindlimb position in the former (Carlson et al. 2005), which is necessary for travel in arboreal environments.

Marchi and Shaw (2011) analyzed fibular robusticity and tibio-fibular ratios in university varsity athletes and controls, in order to assess whether fibular properties are influenced by the intensity and type of mobility (i.e., straight line movement or with frequent changes of direction). Results showed a trend of increased fibular diaphyseal robusticity from controls to runners to field hockey players, with a significant difference ($P<0.05$) between field hockey players and controls. Moreover, relative fibular robusticity (fibula/tibia ratio) was significantly greater in hockey players compared to runners. The authors concluded that fibular robusticity and relative fibula/tibia robusticity may reflect adaptation to patterns of mobility that incorporate high degrees of foot eversion and inversion. In field hockey players, foot eversion/inversion is likely to have been caused by constant and abrupt changes of direction. When studying mobility patterns in bioarchaeological research, frequent foot eversion and inversion may have been caused by mobility in highly uneven terrains. Comparison of Italian Neolithic and Iron Age skeletal series from individuals dwelling in mountainous terrains versus medieval and modern samples seems to provide support for this interpretation (Marchi et al. 2011; see also Higgins 2014, who found a similar effect of terrain properties on ML bending of bovid metacarpi).

In another study, Rantalainen et al. (2010) investigated the influence of locomotor patterns on tibial and fibular rigidity. The authors found that the repetitive loadings associated with running appear to primarily influence tibial robusticity. According to their model, runners would show low levels of relative fibular robusticity. This prediction was empirically supported by the results of Marchi and Shaw (2011). By contrast, Rantalainen et al. (2010) suggested that high impact activities involving jumping influence the mechanical strength of both the tibia and the fibula. The authors concluded that although the tibia and the fibula are spatially close, they experience substantially different loading environments. Analyzing both the fibula and the tibia may therefore help in obtaining a more thorough understanding of mobility patterns.

In the present study we analyze tibial and fibular diaphyseal robusticity, and tibio-fibular ratios within bioarchaeological and modern skeletal samples. The samples are characterized by different levels of mobility: in the bioarchaeological samples, mobility is inferred from archaeological evidence and femoral shape indices, while modern samples include varsity athletes and sedentary controls (Shaw and Stock 2009; Marchi and Shaw 2011). Three bioarchaeological samples come from mountainous areas, while one bioarchaeological sample and the modern samples come from areas associated with relatively flat terrain (see below). The aim of this research is to assess whether fibular CSG properties can successfully be integrated with the information drawn from femoral and tibial data to provide a more accurate reconstruction of mobility levels and types in bioarchaeological populations. In particular, the presence of samples characterized by varying degrees of mobility coming from both relatively flat and rugged areas gives the opportunity to untangle the possible concomitant influence of mobility and terrain properties on the tibio-fibular complex. On the basis of the above research on tibial and fibular robusticity, we hypothesize that tibial robusticity will be influenced by both mobility levels and terrain ruggedness, while fibular robusticity and relative fibular proportions will mainly reflect foot eversion/inversion, associated with the terrain ruggedness. In particular, we hypothesize that (1) when comparing groups dwelling in areas with similar terrain properties, tibial diaphyseal robusticity will be higher in more mobile groups; (2) when comparing groups with similar levels of mobility, tibial diaphyseal robusticity will be higher in groups settled in mountainous (more rugged) terrain; and (3) fibular diaphyseal robusticity and the fibula/tibia robusticity ratio will be higher in skeletal series drawn from more mountainous/uneven areas, independent of mobility levels.

6.2 Materials and Methods

6.2.1 The Sample

The skeletal series analyzed here include four bioarchaeological and three modern samples. Only male individuals were included in this study, given that, cross-culturally, most of the mobility-oriented activities were performed by males, at least beginning with the Neolithic (Ehrenberg 1989). Bioarchaeological skeletal series include 7 Late Upper Paleolithic, 15 Neolithic, 33 Iron Age, and 14 medieval individuals (Table 6.1). Mobility levels for these samples are presumed based on archaeological information and ethnographic analogy with modern or recent groups (e.g., Hudson and Hudson 1980; Kelly 1983, 1995; Larsen 1995; Carlson et al. 2007).

The Late Upper Paleolithic sample (12,000–10,000 BP) (Alessio et al. 1967; Martini et al. 2004; Paoli et al. 1980) consists of individuals from the sites of Arene Candide (Liguria, Northwestern Italy) and Romito (Calabria, Southern Italy).

Table 6.1 Bioarchaeological skeletal samples composition

Period	*N*	Necropolis	Terrain	Subsistence	Mobility level
Late Upper Paleolithic	7	Arene Candide 2, 4, 5, 10 Romito 3, 7, 8	Mountainous	Hunting	High
Neolithic	15	Arene Candide 2 Tinè, E VI, 7, 8, IX, XIII Arma dell'Aquila II Bergeggi 2, 3 Boragni 2 Pollera 10, 13, 30, 32, 6246	Mountainous	Herding	High
Iron Age	35	Alfedena 1, 4, 5, 6, 9, 18, 19, 21, 40, 41, 53, 66, 67, 68, 73, 82, 83, 84, 88, 89, 90, 91, 97, 98, 102, 105, 109, 114, 115, 116, 119, 121, 126, 130, 132	Mountainous	Agriculture-Herding	Moderate-Low
Medieval	14	Neuburg 24, B21, B26, 61, 65_57, 75, 80, 101_73, 109, 111, 167, 175, 176, 189	Plain	Agriculture	Low

Subsistence for these peoples was based on hunting mid-sized ungulates such as red deer, roe deer, and ibex (Mussi 2001; Martini et al. 2009), an activity pattern that required a high level of mobility (Kelly 1983, 1995).

Neolithic individuals date to 6,000–5,500 BP (Maggi 1997) and were unearthed from a series of neighboring caves including Arene Candide (Liguria, Northwestern Italy). The main subsistence activity for Neolithic people was sheep herding, although agriculture played a minor role (Marchi et al. 2006, 2011, and references therein). In general, pastoral systems rely on both seasonal movements among various pasture zones (ranging from 20 to more than 300 km), as well as daily dispersal from encampments (Niamir-Fuller 1999). Herder mobility is predicated on the availability of pasture and water. For example, in arid areas cattle herders may walk 8–9 km per day (Coppolillo 2000; Turner and Hiernaux 2002) and, during the dry season, up to 17 km per day (Adriansen and Nielsen 2005). Therefore, the most mobile herders perform both high logistic and high residential mobility. This level of mobility overlaps with that of modern hunter-gatherers (Kelly 1983, 1995; Marlowe 2005). The Ligurian Neolithic people were part of a small-scale transhumance system in a region that virtually lacked pastures (Marchi et al. 2006, 2011), a subsistence strategy that likely required logistic mobility. High mobility of the Ligurian Neolithic people was supported by previous analysis performed on their femoral CSG (Marchi et al. 2006).

Iron Age individuals date back to 2,600–2,400 BP and were unearthed from the necropolis of Alfedena in Abruzzo (central Italy). The economy at Alfedena was based on agriculture, while a small subset of the population was involved in herding. This subsistence strategy would have required lower population level mobility levels (Sparacello et al. 2011). Accordingly, the Iron Age sample is associated with a relatively circular femoral shape.

Medieval individuals (1,300 BP) were mainly agriculturalists (Marchi 2007; Benjamin Höke, pers. comm.) and were unearthed from the necropolis of Neuburg in Bavaria (Southern Germany). Historical studies indicate that early medieval peasants had little or no opportunity for residential mobility and lived the majority of their lives close to the field (Le Goff 1988, 1990). Furthermore, previous comparisons of hunter-gatherer and agricultural skeletal remains suggest decreased levels of mobility in agricultural populations (Larsen 1995; Ruff et al. 2006a).

The modern samples include 15 field-hockey players, 15 cross-country runners, and 21 sedentary control individuals (Shaw and Stock 2009). Two additional individuals practicing rugby were included in the field hockey players' sample, given the similarity of the movements involved in the two sports (Marchi and Shaw 2011). The two athlete samples are characterized by high levels of mobility. However, in general, runners travel in a relatively straight-ahead direction, while hockey players perform frequent and abrupt changes of direction (Shaw and Stock 2009; Marchi and Shaw 2011).

Bioarchaeological samples come from areas that are easy to categorize topographically, being either flat or fairly rugged. Modern samples performed their sports in mainly flat terrain. However, we (Sparacello et al. 2008) developed a protocol to assess terrain ruggedness in an objective way using the freeware program

Google Earth™. A circle with a diameter of 5 km is drawn, with the archaeological site at the center. An altimetry profile is calculated using a function of Google Earth™ for the four paths drawn along the directions N-S, E-W, NW-SE, SW-NE of the circle (Sparacello et al. 2008). The altimetry profile provides the sum of the elevation gain and loss along the path. We consider the average of this value among the four paths as an effective measure of terrain ruggedness. In fact, the value provides a standardized assessment of the amount of vertical traveling (either uphill or downhill) imposed by traversing a landscape. Moreover, the value is not dependent on the altitude of the starting point. After testing the method on several landscapes, it was decided to consider "flat" terrain as having an average value between 0 and 500 m for the sum of elevation gain and loss. "Moderately hilly" was defined as a sum between 500 and 1,000 m; "hilly-mountainous" as a sum between 1,000 and 1,500 m. Finally, we consider "mountainous" territory to have a sum above 1,500 m. For example, the Black Hills in Wyoming (United States) and the iconic landscape of the Tuscany hills (Italy) both average ~1,000 m, while the Himalayan village of Chukhung, Nepal, at the fringes of Mount Everest, averages 2,512 m. Using this method, the site of Neuburg falls in the "flat" category, averaging 202.5 m, while Alfedena (average 1,580 m), Arene Candide (average 1,868 m), and Romito (average 1,875 m) fall in the "mountainous" category.

6.2.2 *Methods*

Cross-sectional properties were calculated at 50 % bone length, using three different methods: (1) polysiloxane molds and measurements of biplanar radiographs of the diaphysis for the Late Upper Paleolithic sample and the majority of the Ligurian Neolithic sample (O'Neill and Ruff 2004); (2) polysiloxane molds of the cortical contour and regression equations for some Ligurian Neolithic individuals and the Iron Age sample (Sparacello and Pearson 2010); (3) pQCT scans for the modern athlete and control samples (Shaw and Stock 2009). Previous research has demonstrated the compatibility of results obtained using different techniques (Stock 2002; Stock and Shaw 2007; Sparacello and Pearson 2010; Davies et al. 2012). For the first two methods, dry bones were positioned following Ruff (2002) and Marchi (2004, 2007); for the third method, limbs of the living individuals where held in place using purpose-designed clamping devices as described in Shaw and Stock (2009).

The cross-sectional variable Z_p (section modulus) is used here to evaluate overall bone rigidity in both the tibia and the fibula. Z_p is calculated by raising the polar second moment of area (*J*) to the power of 0.73 (Ruff 1995, 2000b). Mechanical loading of long bones is a function of physical activity, bone length, and body mass (Ruff 2000b). Thus, to identify behaviorally significant differences in robusticity, it is necessary to control for the effects of body size. Z_p was scaled for body size by dividing it by bone mechanical length and body mass (Ruff 2000a, b). Body mass was calculated from femoral head superoinferior (SI) diameter by averaging the values obtained using equations in Grine et al. (1995), McHenry (1992), and Ruff et al. (1991).

Diaphyseal shape in CSG refers to the ratios of second moment of areas (SMAs), which are proportional to bending rigidity. For the tibia, I_{max}/I_{min} (ratio of the maximum and minimum SMA) was used, while for the femur, I_x/I_y (ratio of SMAs calculated about ML and AP planes) was used. Relative fibular robusticity was calculated as $100\times(J$ fibula/J tibia). Shape indices and relative fibular robusticity are derived from unstandardized data.

Statistical analysis consisted of a one-way ANOVA for each variable considered in this study, and both Fisher LSD and Tukey HSD post-hoc tests. Using Fisher LSD with seven groups increases the risk of Type I errors, because it does not correct for multiple comparisons. Tukey HSD corrects for multiple comparisons, but given the small sample size of several samples included here, this test may be too restrictive for the purposes of this study. We present results for both tests and base our discussion on the LSD test. However, we note instances for which LSD and HSD tests provide different results. In those cases, results should be further verified using a larger sample size. All statistical analyses were carried out with STATISTICA 10 (Statsoft Inc. 2011).

6.3 Results

Table 6.2 shows the mean, standard deviations, Fisher LSD, and Tukey HSD post-hoc results for comparisons of femoral, tibial, and fibular CSG variables. Figure 6.1 displays the femoral shape index (I_x/I_y) of those bioarchaeological samples for which the femur was available (Late Upper Paleolithic, Neolithic, and Iron Age). The Late Upper Paleolithic and Neolithic samples show midshaft femora that are more elliptical and AP-oriented, while the Iron Age sample displays significantly more circular sections (Fig. 6.1 and Table 6.2).

Figure 6.2 displays variation in tibial Z_p across all samples. Tibial Z_p is higher in the bioarchaeological samples settled in mountainous areas (Late Upper Paleolithic, Neolithic, and Iron Age samples) when compared with the sedentary sample settled in a flat terrain (medieval individuals). However, only the comparison between the Neolithic and medieval sample is significant after correcting for multiple comparisons. Within the samples settled in a rugged terrain, the less mobile Iron Age individuals have the lowest average value of tibial Z_p, and the difference is significant when compared with Neolithic individuals. Late Upper Paleolithic, Neolithic, and Iron Age individuals are not significantly different from runners (which have the highest values among modern samples) and have significantly higher tibial Z_p than the sedentary control sample (Table 6.2).

Figure 6.3 displays variation in tibial shape (I_{max}/I_{min}). Tibial shape reveals a diachronic decreasing trend from the Late Upper Paleolithic sample to the medieval sample, and all pairwise comparisons are significant according to LSD post-hoc analyses, but not according to the Tukey HSD post-hoc analyses (Table 6.2). When compared with the modern athlete samples, the Late Upper Paleolithic sample shows a significantly higher shape index than runners, while the Neolithic and

Table 6.2 Temporal differences in diaphyseal geometric properties of the femur, tibia, and fibula

I_x/I_y *Femur*	*N*	*Mean*	*SD*	*NEOL*	*IRONAGE*				
LUP	7	1.32	0.21	NS[a]	***				
NEOL	13	1.34	0.20		***				
IRONAGE	27	1.03	0.14						
Z_p *Tibia*	*N*	*Mean*	*SD*	*NEOL*	*IRONAGE*	*MEDGER*	*HOCKEY*	*RUNNERS*	*CONTROL*
LUP	7	101.39	52.08	NS	NS	(NS)*	NS	NS	(*)**
NEOL	15	110.22	14.48		(*)***	***	***	NS	***
IRONAGE	33	98.39	10.59			(NS)**	NS	NS	***
MEDGER	14	88.5	12.71				NS	(**)***	NS
HOCKEY	17	92.27	9.33					(NS)**	NS
RUNNERS	15	103.4	10.1						***
CONTROL	21	85.28	9.7						
I_{max}/I_{min} *Tibia*	*N*	*Mean*	*SD*	*NEOL*	*IRONAGE*	*MEDGER*	*HOCKEY*	*RUNNERS*	*CONTROL*
LUP	7	3.55	0.57	***	***	***	***	***	***
NEOL	15	2.55	0.44		(NS)*	***	(NS)*	NS	(NS)*
IRONAGE	33	2.27	0.37			(NS)**	NS	(NS)**	NS
MEDGER	14	1.93	0.29				(NS)*	***	(NS)*
HOCKEY	17	2.22	0.26					(NS)**	NS
RUNNERS	15	2.61	0.50						(NS)**
CONTROL	21	2.26	0.28						

Z_p *Fibula*	*N*	*Mean*	*SD*	*NEOL*	*IRONAGE*	*MEDGER*	*HOCKEY*	*RUNNERS*	*CONTROL*
LUP	7	14.58	3.66	NS	NS	(**)***	(NS)*	(NS)**	(**)***
NEOL	15	15.17	3.10		NS	***	(**)***	(**)***	***
IRONAGE	21	13.39	3.12			(**)***	(NS)*	(NS)*	(**)***
MEDGER	14	10.01	2.01				NS	NS	NS
HOCKEY	17	11.6	2.40					NS	NS
RUNNERS	15	11.13	2.61						NS
CONTROL	21	10.1	2.23						
(J Fibula/ J Tibia)×100	*N*	*Mean*	*SD*	*NEOL*	*IRONAGE*	*MEDGER*	*HOCKEY*	*RUNNERS*	*CONTROL*
LUP	7	7.45	2.57	NS	NS	(NS)*	NS	(*)**	(NS)*
NEOL	15	7.04	1.77		NS	(NS)*	NS	(*)**	(NS)*
IRONAGE	21	7.04	2.00			(NS)**	NS	(**)***	(NS)*
MEDGER	14	5.45	1.32				NS	NS	NS
HOCKEY	17	6.18	1.72					(NS)*	NS
RUNNERS	15	4.91	1.23						NS
CONTROL	21	5.70	1.72						

[a]Fisher LSD post-hoc test and Tukey HSD post-hoc test (in parentheses) are both reported in cases where results were different. *NS* nonsignificant; $*p<0.05$; $**0.01<p<0.001$; $***p<0.001$

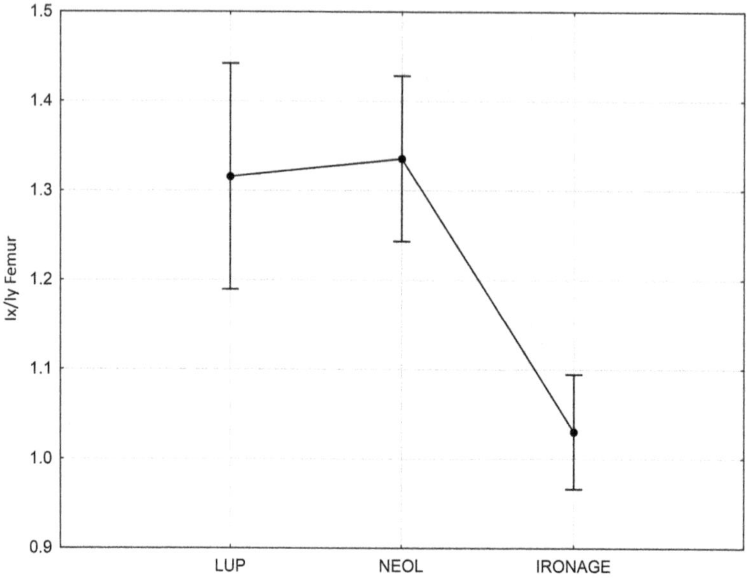

Fig. 6.1 Femoral shape index I_x/I_y. I_x = anteroposterior bending rigidity; and I_y = mediolateral bending rigidity. *LUP* Late Upper Paleolithic, *NEOL* Neolithic, *IRONAGE* Iron Age

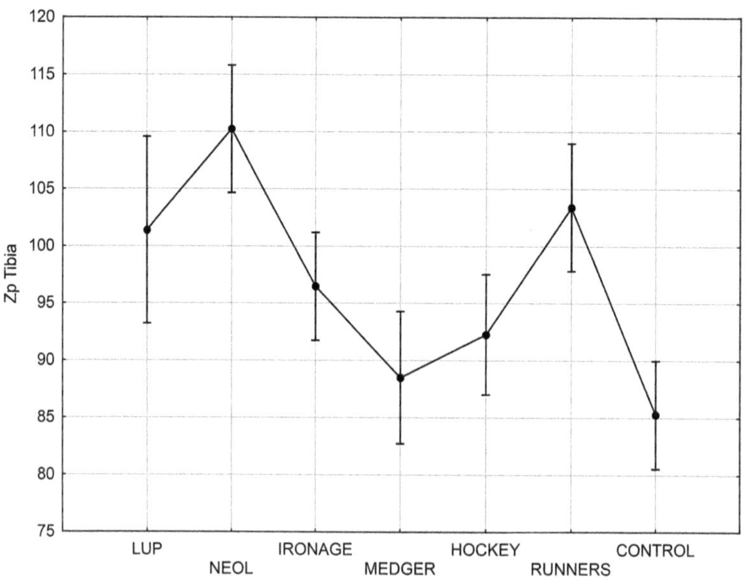

Fig. 6.2 Tibial section modulus Z_p: size-standardized diaphyseal torsional rigidity. *LUP* Late Upper Paleolithic, *NEOL* Neolithic, *IRONAGE* Iron Age, *MEDGER* Medieval, *HOCKEY* field hockey players, *RUNNERS,* cross-country runners, *CONTROL* sedentary control

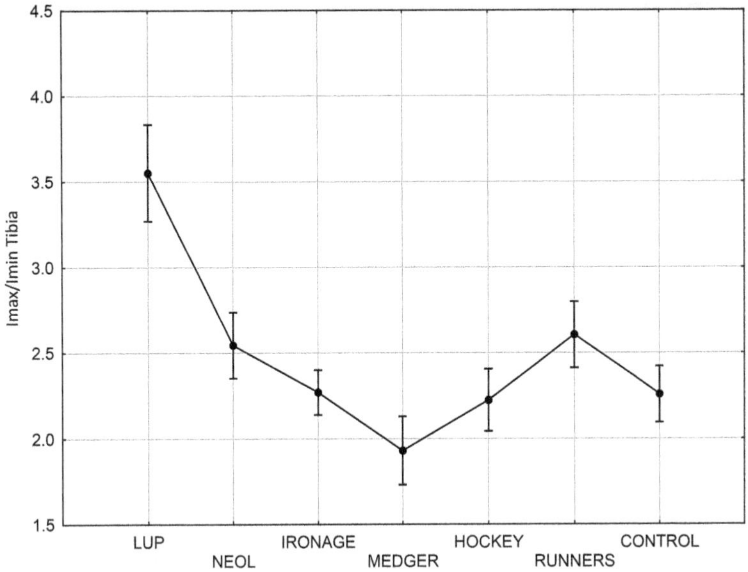

Fig. 6.3 Tibial shape index I_{max}/I_{min}: I_{max}=maximum bending rigidity; I_{min}=minimum bending rigidity. *LUP* Late Upper Paleolithic, *NEOL* Neolithic, *IRONAGE* Iron Age, *MEDGER* Medieval, *HOCKEY* field hockey players, *RUNNERS* cross-country runners, *CONTROL* sedentary control

runner samples show comparable values. Iron Age individuals have a shape index that is comparable with hockey players and the control sample.

Figure 6.4 displays variations in fibular Z_p. As seen for tibial Z_p, within bioarchaeological samples, fibular Z_p of the samples settled in mountainous areas (Late Upper Paleolithic, Neolithic, and Iron Age) is significantly higher than values in the sedentary control sample and those settled in flat terrain (medieval sample). These results are still significant after correcting for multiple comparisons with the Tukey HSD test. The bioarchaeological samples settled in mountainous areas also display higher values of fibular Z_p than most of the modern samples. After correcting for multiple comparisons by using the Tukey HSD test, comparisons of the Late Upper Paleolithic and Iron Age samples with hockey players and runners are not significant at the 0.05 level. Among the samples settled in rugged terrains, the less mobile Iron Age individuals display the lowest average fibular Z_p, but differences from other samples settled in rugged terrains are not statistically significant.

Figure 6.5 displays variations in relative fibular rigidity [100×(fibula *J*/ tibia *J*)]. All of the bioarchaeological samples settled in mountainous areas (Late Upper Paleolithic, Neolithic, and Iron Age samples) have a relatively more robust fibula when compared to the medieval, runner, and control samples. However, after correcting for multiple comparisons, the same groups show a significantly higher value of relative fibular rigidity only in comparison to runners. No significant difference is present when compared with hockey players. Finally, no differences in relative fibular robusticity are present within samples settled in a rugged terrain.

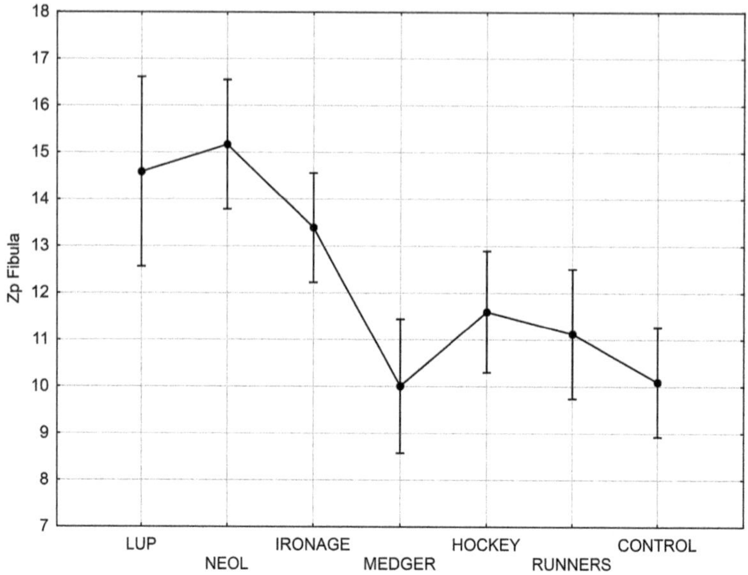

Fig. 6.4 Fibular section modulus Z_p: size-standardized diaphyseal torsional rigidity. *LUP* Late Upper Paleolithic, *NEOL* Neolithic, *IRONAGE* Iron Age, *MEDGER* Medieval, *HOCKEY* field hockey players, *RUNNERS* cross-country runners, *CONTROL* sedentary control

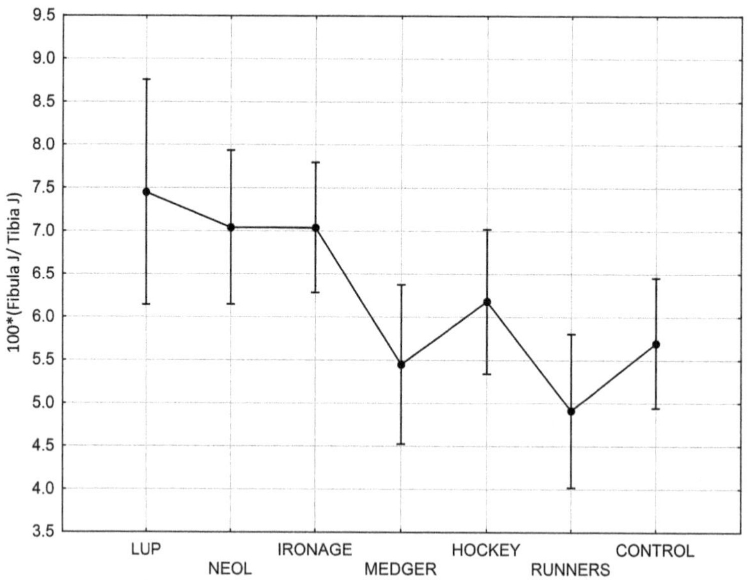

Fig. 6.5 Fibular relative robusticity: ratio between fibular J (polar moment of area) and tibial J. *LUP* Late Upper Paleolithic, *NEOL* Neolithic, *IRONAGE* Iron Age, *MEDGER* Medieval, *HOCKEY* field hockey players, *RUNNERS* cross-country runners, *CONTROL* sedentary control

6.4 Discussion and Conclusion

The purpose of this research was to investigate the concomitant effects of mobility level and mobility type on lower limb mechanical properties, and in particular, fibular robusticity and tibio-fibular robusticity ratios. We compared bioarchaeological and modern samples, each with different levels of mobility (known or inferred on the basis of subsistence), and with or without factors influencing ML loadings (sport-induced changes in direction or terrain ruggedness). Overall, the results suggest that including the fibula in bioarchaeological behavioral reconstruction may provide insights on the "type" of mobility performed.

Femoral shape indices could be calculated for Late Upper Paleolithic, Neolithic, and Iron Age individuals and confirm the expectations based on previous research: Late Upper Paleolithic and Neolithic individuals show similarly elliptical and AP-oriented femoral midshaft cross sections that are likely the result of high mobility levels, while Iron Age people display a significantly more circular midshaft shape. This finding is in agreement with Ruff's work (1999, 2000a), which concluded that femoral shape indices are good indicators of mobility levels after terrain is factored out. Tibial cross-sectional properties provide a less clear correspondence with mobility levels. Given the same terrain, tibial Z_p is generally higher in more mobile groups, as evidenced by the comparison between Neolithic and Iron Age individuals, and between modern athletes and controls. However, if mobility was the only factor responsible for tibial diaphyseal robusticity, we would expect the Late Upper Paleolithic individuals to be significantly more robust than Iron Age individuals and that Iron Age individuals should not be significantly more robust than medieval individuals. Instead, we did not find any significant difference between Late Upper Paleolithic and Iron Age individuals, while the latter showed significantly more robust tibiae than medieval individuals (although the comparison is nonsignificant after correcting for multiple comparisons). For the comparison between Late Upper Paleolithic and Iron Age samples, the small sample size of the Late Upper Paleolithic sample could have played a role. We propose that terrain plays a major role in determining tibial diaphyseal robusticity. As Ruff (1999, 2000a) suggested for the femur, when comparing groups settled in similar terrains, the influence of different mobility levels seems to decline. This would also explain why the Iron Age individuals, who we assume were not very mobile but were settled in a mountainous area, show tibial Z_p values significantly higher than medieval individuals and sedentary modern controls, and comparable with the ones shown by hockey players.

Previous research hypothesized that tibial shape may be influenced by both mobility level (increasing AP bending rigidity, and thus I_{max}) and frequent inversion/eversion of the foot caused by frequent changes of direction or terrain unevenness (increasing the ML bending rigidity, and thus I_{min}) (Marchi et al. 2011; see also Higgins 2014, for a comparable result in bovid metacarpals). Taking into account the influence of both mobility and terrain conformation on tibial shape, we would predict that, when comparing groups settled in areas with similar topographies, more mobile groups will show higher shape indices (less circular diaphyseal cross sections). Our results partially support these expectations, but some pairwise differences are difficult to explain in this framework and call for more investigation on the reliability of tibial shape as an indicator of mobility levels. Within groups settled in a mountainous terrain, the more mobile Late Upper Paleolithic and Neolithic individuals show a higher shape index than Iron Age individuals. However, Late Upper Paleolithic individuals have extremely platycnemic tibiae (Fig. 6.3); if tibial shape was strictly correlated with mobility, this would signal that these individuals were much more mobile than Neolithic individuals. Although this explanation may be possible, the signal should have been similar when comparing femoral shape and tibial robusticity. Even more problematic is the result of comparisons involving medieval and control individuals. It is difficult to imagine a more sedentary lifestyle than the one performed by modern college students who work out less than one hour a week (Shaw and Stock 2009). The medieval agricultural lifestyle required at least some degree of mobility due to farming activities. Yet, medieval individuals show significantly less elliptical tibial cross-sectional shape than control individuals. Both samples dwelled on flat terrain, which excludes the possibility that the higher shape in medieval individuals is due to traversing rugged terrains. It is more likely that, as Stock (2006) suggested, tibial shape is influenced by factors in addition to mobility and terrain, causing the extreme values found here in Late Upper Paleolithic and medieval German individuals (Fig. 6.3).

Mobility as generally implied in bioarchaeological studies, i.e., the amount of traveling due to subsistence activities, is probably only one of the factors that characterize lower limb robusticity and shape. The type of substratum, different intensity and repetitiveness of activity, and the linearity or nonlinearity of the movement should be taken into account when analyzing mobility (Carlson and Judex 2007; Shaw and Stock 2009; Carlson 2014). For the Late Upper Paleolithic and medieval samples, activities such as long distance running or plowing, or other factors hitherto not investigated, may have had an influence on shaping lower limb properties. However, it is difficult to incorporate information on the type of movements performed by past populations for subsistence tasks. It appears that the inclusion of the fibula in the study of lower limb bone structure can provide useful insights when developing behavioral interpretations in bioarchaeological contexts.

While tibial Z_p is significantly higher in the Neolithic sample when compared with the Iron Age sample (a difference that we interpreted as due to different levels of mobility in similar terrains), the groups settled in mountainous areas show more robust fibulae compared with non-mountainous samples, regardless of the assumed level of mobility (although some of the pairwise comparisons would not be significant after correcting for multiple comparisons). Furthermore, while runners have the highest tibial rigidity among modern samples (Fig. 6.2), no significant difference in fibular robusticity is present among modern groups, and the highest value is displayed by hockey players (Fig. 6.4). Fibular Z_p appears, therefore, not significantly influenced by the level of mobility, but mainly correlated with terrain properties and with sport-dictated frequent changes of direction.

The pattern described above for fibular diaphyseal rigidity is more apparent when considering the ratio between fibular and tibial diaphyseal rigidity. All the groups settled in a mountainous terrain show significantly higher fibula/tibia ratios than all other groups (with the exception of hockey players). This ratio appears not to be influenced by mobility levels given equivalent terrain conditions. In fact, Late Upper Paleolithic, Neolithic, and Iron Age individuals display similar values, and also medieval, modern runners, and control individuals are not significantly different from each other. Hockey players show the highest fibula/tibia ratio among groups settled in plain areas, and the result is significant when compared to the ratio of runners. It therefore appears that what drives the increase in relative (to the tibia) fibular robusticity may be either terrain ruggedness or sport-related abrupt changes of direction, i.e., activities that have in common high levels of foot eversion/inversion. It should be noted, however, that after correcting for multiple comparisons the bioarchaeological samples settled in mountainous areas show significantly higher fibula/tibia ratios only in comparison to runners, whose ratio is low due to high tibial robusticity (Fig. 6.2). This calls for further verification of the results found here using a larger sample size.

The above results suggest a clear and coherent correspondence between fibular cross-sectional properties, relative fibular proportions, and factors increasing the frequency of foot eversion/inversion, such as frequent and abrupt changes of direction (Marchi and Shaw 2011) and traveling on uneven surfaces (Marchi et al. 2011). Rugged terrain may also increase fibular loading using a different mechanism than increasing the frequency of foot eversion/inversion: traveling downhill on particularly rugged terrain may increase the frequency of high-impact ground reaction forces that enhance fibular robusticity compared to traveling on level rugged terrain (Rantalainen et al. 2010). The apparent specificity of the response of the tibio-fibular complex should be further verified through experimental studies and larger sample sizes. However, the study of the tibio-fibular complex in bioarchaeology may integrate additional inferences about past population mobility. For example, in areas with mixed relief, with plains and mountains, a robust fibula with a high fibula/tibia ratio may indicate a preferential subsistence-related exploitation of mountainous areas. The same properties can be used to assess degree of exploitation of inland resources by coastal hunter-gatherers, provided that the inland region is mountainous.

Femoral shape and, to a lesser extent, tibial robusticity are integral to inferences of mobility patterns in past populations. Results presented here suggest that fibular analyses also have the potential to improve these inferences by providing anatomical information that may reflect variation in loading directionality and ankle mobility.

References

Adriansen HK, Nielsen TT (2005) The geography of pastoral mobility: a spatio-temporal analysis of GPS data from Sahelian Senegal. Geogr J 64:177–188

Alessio M, Bella F, Cortesi C (1967) University of Rome carbon-14 dates V. Radiocarbon 9:357–358

Barnett CH, Napier JR (1953) The rotatory mobility of the fibula in eutherian mammals. J Anat 87:11–21

Carlson KJ (2014) Linearity in the real world—an experimental assessment of non-linearity in terrestrial locomotion. In: Carlson KJ, Marchi D (eds) Reconstructing mobility: environmental, behavioral, and morphological determinants. Springer, New York

Carlson KJ, Judex S (2007) Increased non-linear locomotion alters diaphyseal bone shape. J Exp Biol 17:3117–3125

Carlson KJ, Demes B, Franz TM (2005) Mediolateral forces associated with quadrupedal gaits of lemurids. J Zool (Lond) 266:261–273

Carlson KJ, Grine FE, Pearson OM (2007) Robusticity and sexual dimorphism in the postcranium of modern hunter gatherers from Australia. Am J Phys Anthropol 9:23

Coppolillo PB (2000) The landscape ecology of pastoral herding: spatial analysis of land use and livestock production in east Africa. Hum Ecol 28:527–560

Davies TG, Shaw CN, Stock JT (2012) A test of a new method and software for the rapid estimation of cross-sectional geometric properties of long bone diaphyses from 3D laser surface scans. Archaeol Anthropol Sci 4:277–290

Demes B, Carlson KJ, Franz TM (2006) Cutting corners: the dynamics of turning behaviors in two primate species. J Exp Biol 209:927–937

Ehrenberg M (1989) Women in prehistory. University of Oklahoma Press, Norman

Goff L (1988) Medieval civilization, 400-1500. Blackwell, Oxford

Goff L (1990) The medieval world. Parkgate, London

Grine FE, Jungers WL, Tobias PV, Pearson OM (1995) Fossil Homo femur from Berg Aukas, northern Namibia. Am J Phys Anthropol 26:67–78

Higgins RW (2014) The effects of terrain on long bone robusticity and cross-sectional shape in lower limb bones of bovids, Neandertals, and Upper Paleolithic modern humans. In: Carlson KJ, Marchi D (eds) Reconstructing mobility: environmental, behavioral, and morphological determinants. Springer, New York

Holt BM (2003) Mobility in Upper Paleolithic and Mesolithic Europe: evidence from the lower limb. Am J Phys Anthropol 122:200–215

Hudson RD, Hudson ND (1980) Nomadic pastoralism. Ann Rev Anthropol 9:15–16

Kelly RL (1983) Hunter-gatherer mobility strategies. J Anthropol Res 39:277–306

Kelly RL (1995) The foraging spectrum: diversity in hunter-gatherer lifeways. Smithsonian Institution Press, Washington, DC

Larsen CS (1995) Biological changes in human populations with agriculture. Annu Rev Anthropol 24:185–213

Lovejoy CO, Burstein H, Heiple K (1976) The biomechanical analysis of bone strength: a method and its application to platycnemia. Am J Phys Anthropol 44:489–506

Maggi R (1997) The radiocarbon chronology. Mem Ist Ital Paleontol Um 5:31–52

Marchi D (2004) Cross-sectional geometry of the limb bones of the Hominoidea: its relationships with locomotion and posture. Dissertation, University of Pisa, Pisa

Marchi D (2007) Relative strength of the tibia and fibula and locomotor behavior in hominoids. J Hum Evol 53:647–655

Marchi D (2008) Relationships between limb cross-sectional geometry and mobility: the case of a Neolithic sample from Italy. Am J Phys Anthropol 137:188–200

Marchi D, Shaw CN (2011) Variation in fibular robusticity reflects variation in mobility patterns. J Hum Evol 61:609–616

Marchi D, Sparacello VS, Holt BM, Formicola V (2006) Biomechanical approach to the reconstruction of activity patterns in Neolithic Western Liguria, Italy. Am J Phys Anthropol 131:447–455

Marchi D, Sparacello VS, Shaw CN (2011) Mobility and lower limb robusticity of a pastoralist Neolithic population from North-Western Italy. In: Pinhasi R, Stock J (eds) Human bioarchaeology of the transition to agriculture. Wiley-Liss, New York, pp 317–346

Marlowe FW (2005) Hunter-gatherers and human evolution. Evol Anthropol 14:54–67

Martini F, Bisconti M, Casciarri S, Fabbri PF, Leonini V, Lo Vetro D, et al (2004) La nuova sepoltura epigravettiana 'Romito 7' a Papasidero. Atti della XXXVII Riunione Scientifica dell'Istituto Italiano di Preistoria e Protostoria: 'Preistoria e Protostoria della Calabria'; Sept 29–Oct 4 2002; Tortora, Italy. Istituto Italiano di Preistoria e Protostoria, Florence, pp 101–111

Martini F, De Curtis O, Di Giuseppe Z, Ghinassi M, Lo Vetro D, Nannini L, et al (2009) Humans, climate and environment in Calabria during the Second Pleniglacial: new data from Grotta del Romito. Atti del 6° Convegno Nazionale di Archeozoologia; May 21-24 2009; Lucca, Italy. Soprintendenza al Museo Nazionale Preistorico Etnografico 'Luigi Pigorini', Rome, p 17

McHenry HM (1992) Body size and proportions in early hominids. Am J Phys Anthropol 87:407–431

Mussi M (2001) Earliest Italy. An overview of the Italian Paleolithic and Mesolithic. Kluwer, New York

Niamir-Fuller M (1999) Managing mobility in African rangelands. Intermediate Technology, London

O'Neill MC, Ruff CB (2004) Estimating human long bone cross-sectional geometric properties: a comparison of noninvasive methods. J Hum Evol 47:221–235

Paoli G, Parenti R, Sergi S (1980) Gli scheletri mesolitici della caverna delle Arene Candide (Liguria). Mem Ist Ital Paleontol Um 3:33–154

Pearson OM, Lieberman DE (2004) The aging of Wolff's "law": ontogeny and responses to mechanical loading in cortical bone. Yearb Phys Anthropol 47:63–99

Pearson OM, Petersen TR, Sparacello VS, Daneshvari SR, Grine FE (2014) Activity, "body shape", and cross-sectional geometry of the femur and tibia. In: Carlson KJ, Marchi D (eds) Reconstructing mobility: environmental, behavioral, and morphological determinants. Springer, New York

Rantalainen T, Nikander R, Heinonen A, Suominen H, Sievänen H (2010) Direction-specific diaphyseal geometry and mineral mass distribution of tibia and fibula: a pQCT study of female athletes representing different exercise loading types. Calcif Tissue Int 86:447–454

Ruff CB (1987) Sexual dimorphism in human lower limb bone structure: relationship to subsistence strategy and sexual division of labor. J Hum Evol 16:391–416

Ruff CB (1995) Biomechanics of the hip and birth in early Homo. Am J Phys Anthropol 98:527–574

Ruff CB (1999) Skeletal structure and behavioral patterns of prehistoric Great Basin populations. In: Hemphill BE, Larsen CS (eds) Understanding prehistoric lifeways in the Great Basin Wetlands: bioarchaeological reconstruction and interpretation. University of Utah Press, Salt Lake City, pp 290–320

Ruff CB (2000a) Biomechanical analyses of archaeological human skeletons. In: Katzenberg MA, Saunders SR (eds) Biological anthropology of the human skeleton. Wiley, New York, pp 71–102

Ruff CB (2000b) Body size, body shape, and long bone strength in modern humans. J Hum Evol 38:269–290

Ruff CB (2002) Long bone articular and diaphyseal structure in old world monkeys and apes. I: locomotor effects. Am J Phys Anthropol 119:305–342

Ruff CB, Hayes WC (1983) Cross-sectional geometry of Pecos Pueblo femora and tibiae—a biomechanical investigation. II. Sex, age, and side differences. Am J Phys Anthropol 60:383–400

Ruff CB, Scott WW, Liu AY (1991) Articular and diaphyseal remodeling of the proximal femur with changes in body mass in adults. Am J Phys Anthropol 86:397–413

Ruff CB, Holt BM, Sládek V, Berner M, Murphy W, zur Nedden D et al (2006a) Body size, body proportions, and mobility in the Tyrolean 'Iceman'. J Hum Evol 51:91–101

Ruff CB, Holt B, Trinkaus E (2006b) Who's afraid of the big bad Wolff? 'Wolff 's law' and bone functional adaptation. Am J Phys Anthropol 129:484–498

Shaw CN, Stock JT (2009) Intensity, repetitiveness, and directionality of habitual adolescent mobility patterns influence the tibial diaphysis morphology of athletes. Am J Phys Anthropol 140:149–159

Shaw CN, Stock JT (2013) Extreme mobility in Late Pleistocene? Comparing limb biomechanics among fossil Homo, varsity athletes and Holocene foragers. J Hum Evol 64:242–249

Sládek V, Berner M, Sailer R (2006a) Mobility in Central European Late Eneolithic and Early Bronze Age: femoral cross-sectional geometry. Am J Phys Anthropol 130:320–332

Sládek V, Berner M, Sailer R (2006b) Mobility in Central European Late Eneolithic and Early Bronze Age: tibial cross-sectional geometry. J Archaeol Sci 33:470–482

Sparacello VS, Marchi D (2008) Mobility and subsistence economy: a diachronic comparison between two groups settled in the same geographical area (Liguria, Italy). Am J Phys Anthropol 136:485–495

Sparacello VS, Pearson OM (2010) The importance of accounting for the area of the medullary cavity in cross-sectional geometry: a test based on the femoral midshaft. Am J Phys Anthropol 143:612–624

Sparacello VS, Pearson OM, Petersen TR (2008) Untangling the effects of terrain and mobility on the cross-sectional geometry of femur and tibia. Am J Phys Anthropol 46:199

Sparacello VS, Pearson OM, Coppa A, Marchi D (2011) Changes in robusticity in an Iron Age agropastoral group: the Samnites from the Alfedena necropolis (Abruzzo, Central Italy). Am J Phys Anthropol 144:119–130

Spencer M, Lawrence S, Rechichi C, Bishop D, Dawson B, Goodman C (2004) Time motion analysis of elite field hockey, with special reference to repeated-sprint activity. J Sports Sci 2:843–850

Statsoft Inc. (1984–2011) Statistica. Statsoft, Tulsa

Stock J (2002) A test of two methods of radiographically deriving long bone cross-sectional properties compared to direct sectioning of the diaphysis. Int J Osteoarchaeol 12:335–342

Stock J (2006) Hunter-gatherer postcranial robusticity relative to patterns of mobility, climatic adaptation, and selection for tissue economy. Am J Phys Anthropol 131:194–204

Stock J, Pfeiffer S (2001) Linking structural variability in long bone diaphyses to habitual behaviors: foragers from the southern African Later Stone Age and the Andaman Islands. Am J Phys Anthropol 115:337–348

Stock JT, Shaw CN (2007) Which measures of skeletal robusticity are robust? A comparison of external methods of quantifying diaphyseal strength to cross-sectional geometric properties. Am J Phys Anthropol 134:412–423

Turner MD, Hiernaux P (2002) The use of herders' accounts to map livestock activities across agropastoral landscapes in Semi-Arid Africa. Landsc Ecol 17:367–385

Wescott DJ (2014) The relationship between femur shape and terrestrial mobility patterns. In: Carlson KJ, Marchi D (eds) Reconstructing mobility: environmental, behavioral, and morphological determinants. Springer, New York

Chapter 7
The Relationship Between Femur Shape and Terrestrial Mobility Patterns

Daniel J. Wescott

Abstract Femoral diaphyseal cross-sectional shape is commonly used to interpret levels of terrestrial logistic mobility (TLM; daily distance covered on land by individuals or groups) in human archaeological populations. However, variation in femoral diaphyseal shape can be influenced by factors other than TLM, such as other lower limb habitual activities, differences in body physique (especially body breadth), age of onset of activities, terrain type, and other environmental and cultural factors. Therefore, similarly shaped femora can occur in populations with different levels of TLM depending on whether the shape differences are due to changes in anteroposterior or mediolateral bending strength. In this chapter, I discuss factors that influence femoral diaphyseal shape and robusticity by comparing diaphyseal shape between individuals with normal mobility and limited or impaired ambulatory ability, examining temporal trends in Native American and modern US populations, and examining ontogenetic factors and non-ambulatory activities on femoral diaphyseal cross-sectional shape. I also discuss and summarize why using multiple biomechanical properties and several bones can provide a clearer picture of the pattern of activity obtained from long bone morphology. I argue that basic femoral diaphyseal shape ratios can be used to estimate levels of TLM when all variables are carefully considered, and that the use of multiple bones and indicators provides a more robust understanding of the mechanical loads that cause similarities and differences in long bone morphology than shape alone.

Keywords Femur • Logistic mobility • Bone biomechanics • Diaphyseal cross-sectional shape • Bioarchaeology • Skeletal biology • Secular change • Reduced mobility

---

D.J. Wescott (✉)
Department of Anthropology, FACTS, Texas State University,
601 University Drive, San Marcos, TX 78666, USA
e-mail: dwescott@txstate.edu

K.J. Carlson and D. Marchi (eds.), *Reconstructing Mobility: Environmental, Behavioral, and Morphological Determinants*, DOI 10.1007/978-1-4899-7460-0_7,

7.1 Introduction

Mobility patterns provide a wealth of information about prehistoric and historic populations including subsistence strategy, demography, sexual division of labor, trade, and territoriality (Kelly 1992). Mobility is a broad term and there are many types of mobility, but logistic mobility is the type of mobility that is most likely to leave its mark on the human skeleton and therefore of interest here. For simplification, logistic mobility can be divided into terrestrial and aquatic mobility (Stock and Pfeiffer 2001; Weiss 2003). Aquatic logistical mobility involves the use of watercraft and its effects are primarily increased strength of the clavicles and humeri and a larger upper/lower limb robusticity index (Stock and Pfeiffer 2001; Weiss 2003). Terrestrial logistic mobility (TLM) involves walking or running over land and impacts the strength and shape of the femora, tibiae, fibulae, and metatarsals (Stock and Pfeiffer 2001). In this chapter, mobility is defined as the daily walking and/or running activities of an individual to move from one location to another, unless otherwise noted.

7.1.1 Femur Midshaft Shape as an Indicator of TLM

Cross-sectional geometric properties of long bones provide a generalized measure of in vivo loading when comparing skeletal remains from genetically similar species (Ruff et al. 2006a). Therefore, the examination of lower limb cortical bone distribution using cross-sectional geometry provides one of the most direct methods of interpreting TLM in human populations (Ruff and Larsen 2014). The ratio of femoral diaphyseal midshaft anteroposterior (AP) and mediolateral (ML) external dimensions and the maximum (I_{max}) and minimum (I_{min}) or AP (I_x) and ML (I_y) second moments of area from cross-sectional geometry are commonly used for comparing levels of TLM (Bridges 1989, 1995; Brock and Ruff 1988; Holt 2003; Marchi et al. 2006; Nikita et al. 2011; Ruff 1987; Stock 2006; Stock and Pfeiffer 2001). In fact, Larsen (1997) labeled the ratio of femoral second moments of area (I_{max}/I_{min} and I_x/I_y) as the "mobility index" because of its usefulness in interpreting mobility patterns in populations and individuals. Trinkaus and colleagues (1991) argued that diaphyseal shape provides a good estimator of the *type* of activity, while robusticity (strength relative to length) helps explain the *intensity* of activities. Furthermore, numerous experimental (Carter and Beaupré 2001; Chamay and Tschantz 1972; Jones et al. 1977; Lanyon et al. 1982; Lieberman et al. 2001; Martin 2003; Shaw and Stock 2009) and correlational (Marchi 2008; Ruff 1987; Ruff and Hayes 1983; Ruff et al. 2006b; Sládek et al. 2006; Sparacello and Marchi 2008; Stock 2006; Stock and Pfeiffer 2001, 2004) studies have demonstrated the relationship between long bone morphology and mechanical loading due to habitual activities, especially those like TLM that are weight-bearing (Duncan et al. 2002; Shaw and Stock 2009). These studies provide validity to inferences about habitual patterns from archaeological skeletal remains.

While long bone diaphyseal shape and strength provide valuable information for reconstructing habitual activities in past populations, variation in lower limb bones can be influenced by both intrinsic and extrinsic factors other than TLM, including non-ambulatory habitual activities involving the lower limb (Wescott 2001, 2008, 2013; Wescott and Cunningham 2006), body size and shape (i.e., body mass, body breadth, and limb length) (Gruss 2007; Weaver 2003), climate (Pearson 2000; Ruff 1993, 1994; Ruff et al. 2006b; Stock 2006; Weaver 2003), terrain (Marchi 2008; Ruff 1999; Ruff et al. 1993; Wescott 2001, 2006a), ontogeny (Cowgill 2010; Cowgill et al. 2010; Ruff 2003a, b; Ruff et al. 1994; Wescott 2006b), hormones (Devlin 2011; Devlin and Lieberman 2007; Devlin et al. 2010), and epigenetic and genetic factors (Arnsdorf et al. 2010; Buckwalter et al. 1995; Lovejoy et al. 2003; Marks and Popoff 1988). In addition, second moments of area only theoretically describe a beam if its material components are homogenous and isotropic, the stress–strain curve is linearly elastic, and the beam is straight (Vogel 2013). Long bones do not meet perfectly the requirements of a beam. Finally, long bones are routinely subject to bending stress in planes other than their direction of greatest bending rigidity (Demes et al. 2001; Lanyon and Rubin 1985; Lieberman et al. 2004). As a result, there is not always a straightforward relationship between the cross-sectional properties of bone and the orientation of loads placed on them during life (Demes et al. 1998; Lovejoy et al. 2003; Pearson and Lieberman 2004), and similarities in femoral shape can occur in populations with different levels of TLM (Wescott 2006a). Anthropologists must be cautious not to simply apply femoral shape or the "mobility index" to interpret TLM from long bones. Instead, before making interpretations of mobility, researchers must closely examine the potential factors that might cause a high femoral shape ratio as well as other biological indicators of activity and the supporting archaeological information. It is also necessary to critically evaluate long bone cross-sectional data within an archaeological and biological context when making interpretations about mobility in past human populations.

Femoral diaphyseal shape has been shown to be a useful indicator of TLM (Stock 2006), but there are several examples of where using femoral diaphyseal shape alone is misleading. The mobility index is only useful for examining changes in mobility when the reasons for the shape changes are fully understood since a high mobility index can result from either an increase in AP bending strength or a reduction in ML bending strength. In this chapter, I will discuss the influence of impaired mobility, secular change, non-ambulatory habitual activities, and growth and development patterns on femoral diaphyseal robusticity and shape. Understanding these factors and how they affect femoral cross-sectional morphology, especially shape, is necessary to interpreting behavior, including TLM, from femoral diaphyseal cross-sectional geometry. In addition, while the cross-sectional shape of the femur can provide a wealth of information about activity and behavior, the examination of cross-sectional properties and shape of additional bones, especially in combination with other biological information (e.g., osteoarthritis, entheses, functional angles, articular surface morphology, histomorphometry, and trabeculae orientation) and archaeological data, aids interpretations of activity. Therefore, I will also discuss how using multiple biomechanical properties of several bones can provide a clearer picture of the pattern of activity obtained from long bone morphology, which in the long run will strengthen the validity of interpretations of mobility levels from human skeletal remains.

7.2 Mobility Impairment

One method of examining the usefulness of the mobility index is to compare individuals with long-term limited mobility associated with conditions such as cerebral palsy, paraplegia, and other impairments to those with various other levels of unimpaired mobility. Numerous researchers have discussed how high activity levels in athletes affect long bone cross-sectional properties (Nilsson and Westin 1971; Jones et al. 1977; Ruff et al. 1994; Shaw and Stock 2009; Trinkaus et al. 1994), but the effects of long-term mobility impairment have received far less attention in the anthropological literature. For this comparison, I used highly terrestrial mobile Paleolithic European hunter-gatherers, terrestrially mobile Arikara horticulturalists from South Dakota, marine mobile Inuit, sedentary modern American Blacks and Whites, and five modern mobility-impaired individuals. The mobility-impaired sample is heterogeneous and includes a female in her early thirties with cerebral palsy who encountered mobility impairment as a teenager and became wheelchair-bound by her late twenties, an older adult male who became paraplegic as the result of an accident in adulthood, two adolescent males with an unknown, but probably long-term, period of mobility impairment, and a 21-year-old male with known long-term impaired mobility. Cross-sectional properties of femora were calculated using *MomentMacro* (Ruff 2013) in *ImageJ* (National Institute of Health 2008) from computed tomography scans that were standardized using appropriate bone lengths and body weight estimates (Ruff 2008). Moments of area were standardized by body weight multiplied by bone length. Body weight was estimated from femoral head diameter (Auerbach and Ruff 2004).

While the sample sizes are small and heterogeneous, the most informative result of the comparison between the mobility-impaired sample and a variety of human populations with different levels of TLM is that the individuals with impaired mobility exhibit an I_{max}/I_{min} ratio mean of 1.34 (range 1.17–1.67), which is equal to or greater than the mean for Inuits, Arikara, and American Blacks and Whites (Fig. 7.1). Ratios greater than 1.0 indicate that the femur diaphysis is AP elongated. Using only the I_{max}/I_{min} ratio, these five individuals have a higher mobility index than relatively mobile Native Americans, and all of them fall within the range of Paleolithic hunter-gatherers (1.05–1.88). However, when standardized I_{max} and I_{min} values are examined separately, and not combined into an index, the mobility-impaired individuals have much lower bending rigidity than the mobile Arikara (Fig. 7.2). An examination of the relationship between I_{max}/I_{min} ratio and standardized I_{max} shows a moderate-to-weak positive correlation ($R^2=0.38$) among Arikara males. This indicates that among this group the greater I_{max}/I_{min} is the result of

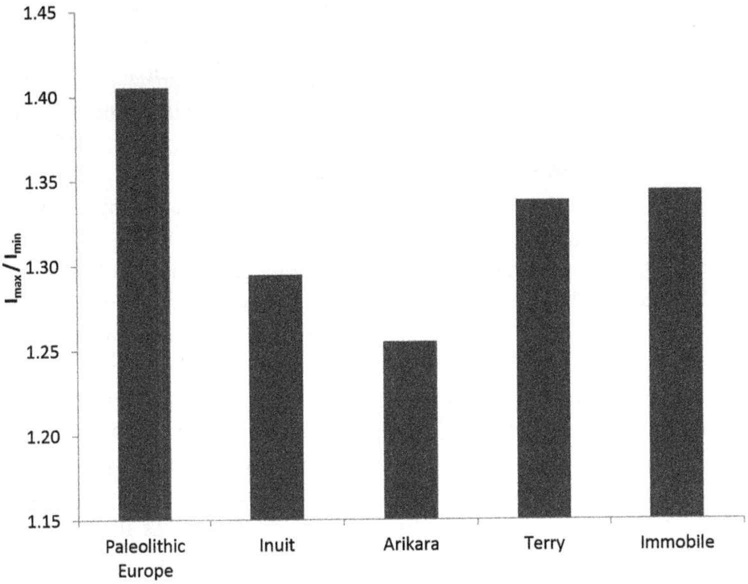

Fig. 7.1 Mean femoral midshaft I_{max}/I_{min} ratio for male European hunters ($n=39$), Inuit ($n=20$), Arikara ($n=127$), modern Americans (Terry; $n=58$), and mobility-impaired individuals (4 m, 1 f)

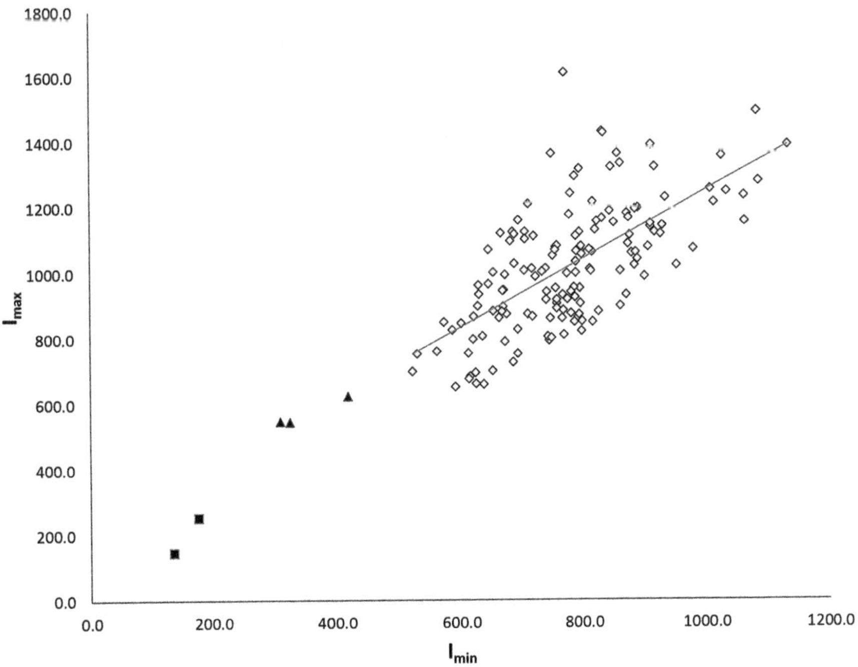

Fig. 7.2 Relationship between I_{max} and I_{min} for Arikara males (*open diamonds*), two long-term mobility-impaired individuals (*black squares*) and three shorter-term mobility-impaired individuals (*black triangles*) illustrating the general decrease in bending rigidity among mobility-impaired individuals. The regression line is for the Arikara

expansion in the AP bending strength. The Arikara show a very weak negative correlation ($R^2=0.03$) between I_{max}/I_{min} ratio and I_{min}. Among the immobile individuals there is also a very slight positive correlation between I_{max}/I_{min} and I_{max}, but the correlation is much weaker ($R^2=0.09$).

Fascinating, but not wholly unexpected, are the differences manifested between individuals who became immobile as adults and those who had mobility impairment since childhood. In Fig. 7.2, the individuals with long-term immobility (squares) are in the lower left and the individuals who became immobile later in life (triangles) are closest to the Arikara (diamonds). In addition to the differences in bone rigidity, these individuals exhibit differences in other morphological features of the femur. There is no observable effect on femoral head diameter or length in individuals who became immobile later in life. The primary change these three individuals exhibit is a reduction in the I_{min} value (Fig. 7.2; Carlson et al. 2008). But, individuals who had impaired mobility from childhood exhibited clear wasting and more circular femoral diaphyses. For the long-term immobile individuals, there was also a large reduction in I_{max}, femur length, and noticeably smaller femoral head diameter, frequent coxa valga (angle of inclination greater than 135), and antetorsion (high angle of torsion) (Dionyssiotis et al. 2007; Megyesi et al. 2009; Robin et al. 2003).

The take-home message here is that a simple examination of the I_{max}/I_{min} ratio does not separate mobile and immobile individuals, but the sample size is very small and heterogeneous for the mobility-impaired group and must be viewed with caution. The combination of a reduction in bending and torsional strength, as well as morphological features such as coxa valga and antetorsion do, however, clearly separate the mobile and immobile individuals. The main factor affecting the I_{max}/I_{min} ratio of the immobile individuals is a decrease in ML bending rigidity while the I_{max}/I_{min} ratio of the mobile individuals is associated with an increase in AP bending rigidity. Since the I_{max}/I_{min} ratio is not standardized for body size, the ratio used alone does not allow researchers to differentiate between the different causes of change in the ratio.

7.3 Secular Change in Femur Shape Among Modern Americans

By examining Fig. 7.1, it becomes clear that the mobility index does not perfectly reflect mobility levels. The Paleolithic hunter-gatherers have the greatest TLM and the greatest mean I_{max}/I_{min} ratio, as would be predicted. The modern Americans (Terry Collection) have the lowest level of TLM, but they exhibit a greater I_{max}/I_{min} value than either the Inuit or the Arikara (Fig. 7.1). This is contrary to expectations and brings up another example where using diaphyseal shape ratios of the femur alone can be misleading when inferring mobility.

In the past two centuries, Americans have undergone significant secular changes in activity levels, body mass, and stature, all of which affect lower limb morphology by altering the magnitude and direction of forces placed on the limbs. During this same period, the femoral midshaft diaphyseal shape of modern Americans has changed from relatively circular to more AP-elongated (increase in external AP/ML and I_{max}/I_{min} ratios; Rockhold 1998; Wescott and Zephro 2012). A more AP-elongated femoral midshaft is conventionally associated with increased TLM. However, in recent Americans TLM has greatly decreased due to technological advances. This suggests that other factors are causing femur shape to mimic the effects of greater mobility.

Documented secular changes that are relevant to femoral morphology include an increase in long bone length (stature) (Meadows and Jantz 1995; Meadows Jantz and Jantz 1999; Wescott and Zephro 2012; Harrington 2013), increase in body mass (Ogden et al. 2006), slight decrease in bi-iliac breadth (Driscoll 2010), and an increase in the AP diameter of the pelvic inlet (Driscoll 2010). There has been little or no significant increase in femoral head diameter (Cridlin 2007; Wescott and Zephro 2012), which most likely reflects lean body mass (Auerbach and Ruff 2004). Most of the secular changes in skeletal morphology among Americans can be attributed to decreases in activity, improvements in nutrition and healthcare (reduced early childhood disease load) and climate (Jantz 2001; Meadows Jantz and Jantz 1999).

Several studies have examined secular trends in femoral diaphyseal external measurements among Americans. Trotter and colleagues (1968) were the first to observe a significant secular trend in femoral midshaft dimensions. They examined changes in femoral length and ML (transverse) diameter at midshaft among individuals born between 1840 and 1949 in the Terry Collection (Hunt and Albanese 2005). They observed a positive trend in length and a negative trend in ML diameter. Later, Rockhold (1998) examined secular change in external measurements of the femur among American Blacks and Whites of both sexes using a sample drawn from the Terry Collection and the Forensic Data Bank (Jantz and Moore-Jansen 1988). She also observed a decrease in the ML diameter at midshaft, nonsignificant changes in the AP diameter, and that subperiosteal area is increasing at a slower rate than femur length. The results of a study conducted by Meadows Jantz et al. (2012) on modern Americans over a 140-year period showed that limb bones have become longer, especially the distal bones, the femoral midshaft shape index (AP/ML × 100) increased, and the cnemic index (ratio of maximum to minimum diameter at the nutrient foramen of the tibia) exhibits a sharp increase in the late nineteenth and early twentieth century. Wescott and Zephro (2012) examined secular trends using 492 adult Black and White males and females drawn from the Forensic Data Bank, Terry Collection, and historic cemeteries with birthdates ranging from 1814 to 1983. They examined trends in femur length, femoral head diameter, diaphyseal diameters, cross-sectional area, robusticity, and torsional rigidity at subtrochanteric and midshaft. The results of their study showed that femoral head diameter and subtrochanteric rigidity did not change significantly over the 169-year period in any group. Femoral midshaft shape, however, showed a significant positive trend in all groups with the femoral midshaft becoming more AP-elongated (Fig. 7.3). From 1814 to about 1890 the femoral midshaft became relatively more circular due mainly to an increase in the ML diameter (Fig. 7.3). The femoral midshaft shape index (AP/ML)

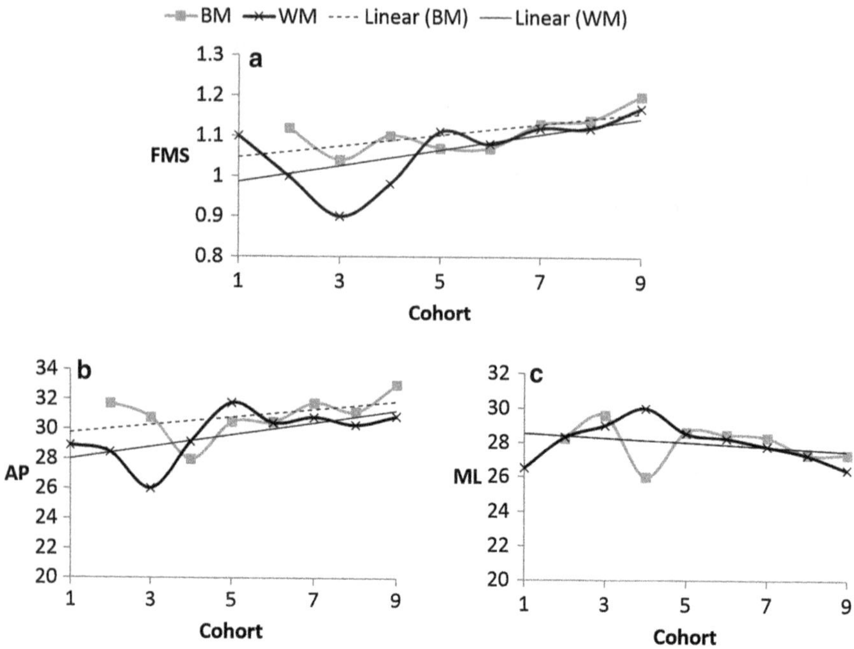

Fig. 7.3 Secular change in femoral midshaft shape (**a**), AP diameter (**b**), and ML diameter (**c**) for modern American black (BM) and white males (WM). The sample mean for 20-year cohorts was used as points for illustrative purposes (Cohorts—1: ≤1829, 2: 1830–1849, 3: 1850–1869; 4: 1870–1889, 5: 1890–1909; 6: 1910–1929; 7: 1930–1929; 8:1950–1969, 9: 1970–1989)

then increased steadily through time due to a slight increase in the AP diameter and a decrease in the ML diameter. The AP diameter, but not the ML diameter, showed a significant age effect, suggesting that some of the changes seen in the AP diameter are associated with age-related expansion of the femur and not secular change. Therefore, the primary driver of the increased femoral midshaft shape index was caused by a decrease in the ML diameter, at least since the early 1900s. The decrease in the ML diameter probably reflects a general decrease in activity levels while the AP diameter is maintained due to increased body mass and femur length (see below). While the mobility index increased among Americans, midshaft robusticity declined slightly, although not significantly, in all groups. The decline in robusticity reflects decreased mobility among modern Americans as expected.

7.3.1 Obesity

One factor that also affects the morphology of the femur in recent modern Americans is a secular change in body weight. The prevalence of obesity has rapidly increased among Americans in the past three decades. Prior to 1980 the prevalence of obesity was relatively stable, but the percentage of Americans adults who are obese has more than doubled, and the number of obese children has tripled, in the past 30 years (Flegal et al. 2002, 2010; Ogden et al. 2006). Obesity in females seems to have reached a plateau but it is still increasing among males (Ogden et al. 2006).

The effect of obesity on femoral size and shape has only recently been investigated. Increases in body weight are known to affect cross-sectional properties of the femur due to increased biomechanical load (Demes et al. 1991; Moore 2008; Petit et al. 2005; Shaw and Stock 2011). Obese individuals have a greater ML ground reaction force, an increased stance length, and greater ankle eversion (DeVita and Hortobagyi 2003). Petit and colleagues (2005) showed that American overweight and obese children and adolescents had 11 % greater femoral shaft section modulus compared to those who were of normal weight due to the increased lean body mass associated with obesity. Agostini and Ross (2011) examined the effect of obesity on femoral diaphyseal shape of American White males. They found a significant positive relationship between the body mass index (BMI) and femoral midshaft ML diameter after controlling for age. However, Moore and Schaefer (2011) found no significant relationship between cross-sectional shape (I_{max}/I_{min} or I_y/I_x) and obesity among modern Americans in the Bass Donated Collection at the University of Tennessee (Shirley et al. 2011). Most recently, Harrington (2013) examined the effects of obesity on the distal femur and proximal tibia. She found significant differences between obese (BMI ≥ 30) and normal weight (BMI ≤ 25) individuals in femoral AP and ML external dimensions and midshaft shape using individuals from the Bass Donated Collection and the Texas State University Donated Skeletal Collection. Obese individuals exhibit a more circular femoral midshaft primarily due to expansion of the ML diameter (Harrington 2013).

Studies of obese individuals indicate there is a positive relationship between body weight and ML bending rigidity of the femoral midshaft. Obese individuals may also be less mobile than normal weight individuals, but the more circular femoral midshaft shape (smaller mobility index) is not due to a decrease in AP diameter, but rather due to increases in ML bending forces associated with gait posture (Agostini and Ross 2011; Seung-uk et al. 2010). As seen in the mobility-impaired individuals and secular changes among recent Americans, mechanical forces not related to TLM levels appear to primarily affect the ML diameter in obese individuals.

7.3.2 Femur Length and Pelvic Breadth

Increases in femur length and decreases in pelvic breadth may also have a minor effect on femoral diaphyseal shape in modern Americans (Pearson et al. 2014; Wall-Scheffler 2014). Gruss (2007) found that AP bending moments are greater among individuals with longer lower limbs, which could lead to increased AP bending strength. Shaw and Stock (2011) examined this question by controlling for body weight associated with increased stature. They found that among Pleistocene and

Holocene skeletons, limb length and bone length did not affect the shape of either the femur or the tibia, but bi-iliac breadth was associated with greater ML bending rigidity. Wescott and Zephro (2012) examined the correlation of femoral shaft properties with femoral head diameter (used as a proxy for body weight) and femur length (used as a proxy for stature). They found that midshaft AP diameter significantly correlates with femur length in all groups and with femoral head diameter in males. In females, there was also a significant correlation between femoral head diameter and femoral midshaft shape. Based on a biomechanical model, we would expect a decrease in the femoral midshaft AP diameter among relatively sedentary modern Americans. The lack of a decrease in the AP diameter is possibly prevented by the observed increase in femur length (Gruss 2007). The observed decreases in ML diameter probably reflect decreases in activity intensity, but may also be linked to a narrowing of the pelvis.

The decrease in bi-iliac breadth among modern Americans would support a slight decrease in the ML diameter and robusticity at the subtrochanteric level, and possibly at midshaft as seen in the secular trend data (Harrington 2013; Meadows Jantz et al. 2012; Wescott and Zephro 2012). This suggests that the bi-iliac or bi-acetabular breadth must be considered when using femoral shape to interpret mobility patterns (Ruff 2008; Pearson et al. 2014). Even so, Shaw and Stock (2011) convincingly argue that while pelvic breadth influences femoral shape it does not rule out the use of lower limb diaphyseal shape when inferring mobility patterns from human remains. For example, Ruff et al. (2006b) demonstrated that femoral shape of the Tyrolean Iceman, which is relatively circular due to ML strengthening, is associated with his broad body but that the shape of the tibia reflects his high mobility over rough terrain.

The main point of this section on secular trends in Americans is that femoral midshaft AP/ML ratio can increase in a population that is becoming increasingly sedentary. However, while researchers must use caution when using femoral shape to estimate mobility, by examining the cause of the shape change, it becomes clear that the changes among American Whites and Blacks are not associated with an increase in AP bending strength, as would be predicted, since they are relatively sedentary. Furthermore, the slight decrease in midshaft rigidity does track the decrease in mobility as expected using a biomechanical model.

7.4 Non-mobility Habitual Activities

Studies involving non-mobility activities are usually restricted to the upper limbs, but a multitude of non-mobility activities also contribute to adult lower limb morphology. Wescott and Cunningham (2006) examined within- and between-sex temporal (sixteenth to nineteenth centuries) variation in femoral cross-sectional morphology and asymmetry among the Arikara. The Arikara are a Native American tribe centered along the Missouri River in South and North Dakota. They had a mixed subsistence strategy that included maize horticulture, bison hunting, and plant gathering. Females conducted all activities associated with horticulture, and it is known that historic Arikara females increased their crop output compared to earlier females to have surplus for trade (Wescott and Cunningham 2006). This would have resulted in intensification of activities associated with preparing fields, planting, and harvesting. Wescott and Cunningham (2006) observed that increased femoral strength in the left thigh caused asymmetry to increase among females in the historic period. They argued that the increased femoral subtrochanteric and midshaft rigidity seen in historic females was due to the greater mechanical loads placed on the left lower limb when conducting right-handed activities such as hoeing and raking.

Wescott's (2013) examination of femoral and humeral asymmetry in Kennewick Man, a 9,500-year-old skeleton from the Columbia River Plateau in northwestern United States, has also shown the effect of non-mobility-related activities on the long bones. Kennewick Man exhibits significant asymmetry of both the upper and lower limbs. The right humerus has 47 % greater torsional rigidity than the left. For the lower limbs, the left femur is stronger and more AP-elongated (greater I_{max}/I_{min} ratio) than the right. The asymmetry in femoral strength and shape indicates the effects of non-ambulatory activities. Wescott (2013) argued that the greater relative strength of the left femur was caused by the elevated ground reaction force experienced by the left lower limb used to decelerate the body while throwing a spear or harpoon with the right upper limb.

Another possible cause of femoral diaphyseal shape change is associated with functional angles resulting from other non-ambulatory activities. Child (2013) examined the relationship between diaphyseal cross-sectional properties and functional angles (neck-shaft and torsion) in 64 males from the Terry Collection and 33 males from Indian Knoll (Archaic hunter-gatherers). Her results show that in both populations neck-shaft and torsional angles significantly affect the cross-sectional properties of the diaphysis. There was a negative trend between neck-shaft angle and cortical area, second moments of area, and polar moments of area at both subtrochanteric and midshaft. The I_x/I_y ratio also increased at midshaft with a higher neck-shaft angle due to a decrease in ML (I_y) bending rigidity. High angles of torsion significantly affect midshaft variation in polar moments of area and AP bending rigidity. In other studies, Basgall (2008) and Wescott and Cunningham (2013) examined femoral neck torsion among Great Plains groups. Basgall (2008) found significant asymmetry in femoral torsion that she attributed to sitting posture. Wescott and Cunningham (2013) examined temporal changes in femoral torsion among the Arikara. They observed no temporal change among males, significant sexual dimorphism in the angle of torsion in all temporal groups, and significant changes in femoral torsion asymmetry between historic Arikara females and earlier groups. They attributed the observed change in asymmetry and sexual dimorphism to changes in sitting posture among females. Historic Arikara females habitually side-sit with one thigh medially rotated, while males sit cross-legged (Wescott and Cunningham 2013).

The main argument of this section is that copious non-ambulatory habitual activities apply loads to the lower limbs that clearly have an effect on cross-sectional strength and shape of the femur. This does not mean that femoral midshaft shape does not reflect mobility levels, but anthropologists must consider the possible effect of other non-ambulatory activities when trying to reconstruct TLM patterns in past populations using femoral cross-sectional shape. As with the Arikara females and Kennewick Man, changes in functional angles and asymmetry may provide clues about the non-ambulatory activities reflected in femur diaphyseal shape.

7.5 Growth and Development

Several studies have demonstrated that many of the differences among populations in adult femoral shape and strength are established very early during growth (Cowgill 2010; Cowgill and Hager 2007; Wescott 2006b). Wescott (2006b) examined diaphyseal shape at subtrochanteric level among an ontogenetic sample of American Whites, American Blacks, and Native Americans. He observed rapid diaphyseal shape changes between 1 and 5 years of age as gait patterns develop, but no significant changes after this age (Wescott 2006b). This shows that the adult shape in each population was established by approximately 5 years of age. Cowgill (2010) also analyzed changes in diaphyseal strength during growth in immature humans from seven populations. The results of her study show that differences between populations in diaphyseal strength and shape occur extremely early in growth, possibly even during the first year of life. The results of these ontogenetic studies indicate that factors other than adult TLM are involved in establishing the femoral variation observed in different populations. Genetic propensities and childhood nutritional status, among other factors, are likely to have effects on children, thereby influencing their adult long bone diaphyseal strength and shape.

Furthermore, ontogenetic studies suggest that bone's reaction to loading caused by habitual activities is not consistent throughout life (Cowgill 2010; Cowgill et al. 2010; Pearson and Lieberman 2004; Ruff 2003a, b; Wescott 2006b). Mechanical loading has a greater influence on bone during growth than during adulthood (Lieberman et al. 2003; Pearson and Lieberman 2004). However, this does not mean that adult bone is unresponsive to mechanical loading (Pearson and Lieberman 2004; Ruff et al. 2006a). But, the age at which individuals begin to participate in particular activities greatly influences their adult long bone morphology and robusticity (Bass et al. 2002; Haapasalo et al. 1998; Ruff et al. 1994). Numerous studies have demonstrated that long bones will exhibit greater external dimensions if mechanical loading occurs prior to skeletal maturity (see references in Pearson and Lieberman 2004). After skeletal maturity, bone strength primarily increases by reducing the medullary cavity rather than increasing the periosteal area (Pearson and Lieberman 2004).

What the ontogenetic studies of femoral diaphyseal strength and shape have shown is that the adult pattern is established very early in life and habitual activities that commence post-adolescence have less of an effect on the strength and shape of the bone than those occurring during growth. Therefore, researchers must consider genetic, nutrition, and growth patterns of a population when comparing femoral strength and shape among heterogeneous populations. Even when examining temporal changes in relatively homogeneous populations, activities occurring during childhood and the age at which individuals begin adult activities may be playing a significant role in adult femur morphology. As Pearson and Lieberman (2004) point out, adult activities rather than childhood activities, may be reflected in the long bones of adults because in many societies, children often become involved in strenuous adult activities during their adolescent growth spurt when bone is rapidly modeling. However, this also suggests that temporal changes in femoral strength and shape could reflect a change in intensity and/or a change in the age at which adult activities commence in a population.

7.6 Use of Multiple Bones and Properties

While long bone cross-sectional shape and rigidity provide generalized information about the loading regime and therefore a means by which to investigate the adaptive response of bone, there is enough evidence suggesting that inferring activity patterns from long bone cross-sectional properties is not a simple endeavor. Biomechanical analyses should also take into consideration other indicators of activity (e.g., osteoarthritis, entheses, histomorphometry) within an archaeological context. There are many good examples of this process in the literature (e.g., Carlson et al. 2007; Larsen and Ruff 2011; Marchi 2008; Ruff et al. 2006b; Stock 2006; Stock and Pfeiffer 2001). For example, Larsen and Ruff (2011) show that the combined use of osteoarthritis and long bone cross-sectional properties allows for a clearer picture of regional variation in the transition from hunting-gathering to agriculture in North America. In an earlier study, Robling (1998) used a novel approach to study mobility in Peru that combines long bone cross-sectional geometry and histomorphometrics to control for systematic remodeling. Here, I will provide an example from my own work on Kennewick Man that illustrates the use of multiple bones and indicators in interpreting activity patterns.

Archaeological studies suggested that Kennewick Man's lifeway probably included high TLM in pursuit of large terrestrial animals (Chatters 2000; Chatters et al. 2012). However, isotopic data suggest that a majority of the protein in his diet came from marine mammals and fish (Taylor et al. 1998). Wescott (2013) examined the major long bones of Kennewick Man and compared him to Paleolithic Europeans and Native Americans from various temporal periods to address the issue of the types of habitual activities in which Kennewick Man may have been involved.

Kennewick Man was relatively tall and broad-bodied with a greater than average body weight for prehistoric populations on the Northwest Coast (Wescott 2014). He exhibits very strong upper and lower limb bones that suggest participation in strenuous habitual activities. Based on his stable isotope values (Taylor et al. 1998), it is possible that Kennewick Man had a lifestyle similar to historic populations in Alaska and the Aleutian Islands. Maritime mobile hunters have relatively high humeral-to-femoral strength and very little upper limb asymmetry (Stock and Pfeiffer 2001; Weiss 2003). Kennewick Man, however, has low size-standardized section moduli humeral-to-femoral midshaft ratios and high humeral asymmetry. These patterns are similar to inland hunters and unlike highly mobile ocean-rowing populations, indicating that Kennewick Man had a mobility pattern very different from the Inuit.

If Kennewick Man was a large game terrestrial hunter, he would be expected to have strong and AP-elongated femoral diaphyseal cross sections. Compared to Archaic foragers and Great Plains bison-hunters, Kennewick Man had very strong and AP-expanded, elliptically-shaped femoral diaphyses. Based solely on femoral cross-sectional shape, the hypothesis that Kennewick Man was a terrestrial hunter would seem very plausible. When other information about Kennewick Man is used to reconstruct his habitual activities, however, the picture of his lifeway changes dramatically. Kennewick Man had extremely strong (high rigidity) tibiae, but a tibial index (I_{max}/I_{min}) of only 1.62. Shaw and Stock (2009) observed an average tibial I_{max}/I_{min} ratio of 2.6 in British runners. When compared to highly mobile Paleolithic and Mesolithic European hunters (Holt 2003; Ruff et al. 2006b), Kennewick Man's femoral shape index exceeds that of Mesolithic males and most Paleolithic males, but his tibial index is well below averages of all Paleolithic, Mesolithic, and Neolithic groups (Holt 2003; Ruff et al. 2006a, b; Wescott 2010). This would suggest that while Paleolithic hunters were placing unidirectional forces on their tibia, presumably due to running or walking (Shaw and Stock 2009), Kennewick Man was placing multidirectional or torsional stress on his tibia (Nikander et al. 2010; Shaw and Stock 2009). Likewise, Kennewick Man's tibial-to-femoral polar section modulus ratio (tibia Z_p/femur Z_p) is high compared to most European hunters. Therefore, while the strength of the lower limbs and the shape of the femora are consistent with high terrestrial mobility, the shape of the tibia suggests that he was not involved in relatively unidirectional running or walking like Paleolithic Europeans. Overall, Kennewick Man's lower limb morphology is more consistent with terrestrial activities involving high- or odd-impact loading (Nikander et al. 2010; Shaw and Stock 2009; Wallace et al. 2012) caused by activities such as rabbit hunting that involve explosive acceleration with rapid turns. However, rabbit hunting is not consistent with his known diet (Taylor et al. 1998) or other archaeological evidence (Butler 1993; Butler and Campbell 2004). Using all the available information, it is more likely that Kennewick Man was involved in a subsistence regime that included habitual harpooning of fish and mammals while walking along the ocean coast or through flowing rivers (Wescott 2014). Using only the femoral mobility index would have resulted in incorrect interpretations of Kennewick Man's lifeway.

The primary message from this section is that the use of multiple bones and cross-sectional properties provides a clearer picture of the habitual activities causing the observed bone morphology. By examining multiple bones and comparing the results to other indicators of activity, we can gain a better understanding of the activities of individuals and populations in the past.

7.7 Discussion and Conclusions

Mobility is critical to our understanding of the archaeological and human evolutionary records. Understanding mobility forms and levels are necessary for interpreting cultural resource strategies, and the cost and benefits of movement are the primary attribute that affect the type, frequency, and length of movement (Kelly 1992). These costs and benefits are tied to food density, terrain, transport technology, and many other energetic and non-energetic factors. Diminishing returns of logistic mobility are reached at shorter distances for plant foods than large game, and therefore subsistence-related movement of groups is primarily affected by food gathering resources (Kelly 1992). But, as Kelly (1992:60) correctly points out "The dimensions of movement need to be disentangled and studied independently so that we can understand how factors altering one component affect other areas of behavior and culture." We also must go beyond the simple dichotomy of "mobile" versus "sedentary" and away from interpretations based on long bone shape alone. To truly understand logistic mobility in past societies, we need to understand the range of activities that influence osteogenic responses in long bones and ultimately cause differences in their shape, size, and strength. We also need to interpret the data within a biological, ecological, archaeological, and cultural context. All of the evidence needs to be in agreement before we make broad conclusions about individuals or populations.

Bone is a complex tissue that can increase strength "by adding bone mass, by changing bone geometry to redistribute the force (stress) that it must resist, or by alterations of its microstructure via processes such as Haversian remodeling" (Pearson and Lieberman 2004:65). As a result, the examination of osteological material is one of the most direct methods of inferring mobility patterns and levels in past societies. While femoral shape can provide valuable information when interpreting mobility patterns, several studies have found discrepancies between levels of mobility and femoral cross-sectional shape. When examining changes in femoral shape as an indicator of increased or decreased mobility through time, it is important that anthropologists determine if the shape change is due to a change primarily in AP (I_x) or ML (I_y) planes, or more importantly the I_{max} or I_{min}, and how overall robusticity changes. Demes and colleagues (1991) found that among indriids (Strepsirrhine) ML diameter is affected by body mass more than AP diameter. This also appears to be true for obese Americans (Agostini and Ross 2011; Harrington 2013; Wescott and Zephro 2012). Body breadth relative to height also influences femoral diaphyseal shape. Increases in biacetabular breadth relative to lower limb length will cause ML buttressing of the femur resulting in a decreased mobility index unrelated to mobility levels (Ruff et al. 2006b). Preliminary studies of mobility-impaired individuals and secular changes in modern Americans also indicate that the ML bending plane may be more sensitive to some types of mechanical loading than is the AP bending plane. Therefore, it is imperative that anthropologists examine the underlying causes of changes in the mobility index when interpreting mobility patterns in past populations.

Young's modulus (measure of the stiffness of an elastic material) and second moments of area provide good estimates of a beam's bending resistance. Young's modulus probably does not differ appreciably from one bone to the next, but, second moments of area only accurately describe a beam if its material component is homogenous and isotropic, the stress–strain curve is linearly elastic, and the beam is straight (Vogel 2013). Unfortunately, long bones do not perfectly meet these criteria. Since the cortical wall at the femoral midshaft is generally thickened posteriorly and sometimes anteriorly, not buttressed medially or laterally, and the femoral diaphysis is AP curved, it makes mechanical sense that the ML diameter is more sensitive to reduced mechanical loading than the AP diameter (Trinkaus and Ruff 2012). However, activities such as long-distance running are going to place greater AP than ML bending loads on the femoral diaphyses (Holt 2003; Ruff et al. 2006a). Therefore, highly mobile groups should have a higher mobility index due to greater AP bending strength than more sedentary groups (Stock 2006; Shaw and Stock 2009).

Femoral diaphyseal shape should not be used alone to reconstruct activity patterns. Other indicators of bone strength and rigidity along with the analysis of several long bones are necessary to correctly interpret TLM levels and patterns (Shaw et al. 2014; Davies and Stock 2014; Sparacello et al. 2014). Data on mobility-impaired individuals and modern Americans demonstrate that the use of multiple cross-sectional properties will provide much greater information regarding the types of loads placed on the lower limbs during life. Examination of other bones in the lower limbs is also necessary (Sparacello et al. 2014). Tibial diaphyseal shape and strength may actually provide better information about mobility than the femur (Ruff and Hayes 1983; Ruff et al. 2006b; Shaw and Stock 2009; Stock 2006). Tibial diaphyseal shape and robusticity correlate significantly with mobility patterns (Shaw and Stock 2009). Since the longitudinal axis of the tibia is nearly parallel with the center of gravity while standing, the tibial diaphyseal shape ratio should be less affected by body shape than femoral diaphyseal ratios (Sládek et al. 2006; Stock 2006; Ruff et al. 2006b). But, like the femur, tibial diaphyseal shape ratios should not be used exclusively to reconstruct TLM. Tibial morphology is very likely to be influenced by activities that generate multidirectional loads due to locomotor patterns and/or rugged terrains (Marchi et al. 2011; Shaw and Stock 2009). Marchi (Marchi 2008; Marchi et al. 2011; Sparacello et al. 2014) has also demonstrated that the fibula can provide valuable information when interpreting mobility, especially in regions with rugged terrain. Even examination of the upper limb bones can help provide clues about overall activity patterns that may be useful when inferring mobility patterns. Finally, other biological indicators of activity, such as patterns of osteoarthritis, will provide corroborating evidence for interpretations of TLM (Larsen and Ruff 2011).

In conclusion, the examination of osteological material is one of the most direct methods available for examining patterns and levels of activity in past societies, and femoral diaphyseal shape ratios based on external dimensions or cross-sectional moments of area are a valuable tool for interpreting TLM when used cautiously by anthropologists. Demes and colleagues (1991) argue that "inferences derived solely from cross-sectional shape about the likely mode of locomotion in fossils should be stated carefully" (Demes et al. 1991:544). This same caution should also be applied when trying to reconstruct the level of mobility patterns in archaeological populations.

References

Agostini GM, Ross AH (2011) The effect of weight on the femur: a cross-sectional analysis. J Forensic Sci 56:339–343

Arnsdorf EJ, Tummala P, Castillo AB, Zhang F, Jacobs CR (2010) The epigenetic mechanism of mechanically induced osteogenic differentiation. J Biomech 42:2881–2886

Auerbach BM, Ruff CB (2004) Human body mass estimation: a comparison of "morphometric" and "mechanical" methods. Am J Phys Anthropol 125:331–342

Basgall AL (2008) Northwestern Plains Indian women: a bioarchaeological analysis of changing roles and status [thesis]. University of Wyoming, Laramie

Bass SL, Saxon L, Daly RM, Turner CH, Robling AG, Seeman E, Stuckey S (2002) The effect of mechanical loading on the size and shape of bone in pre-, peri- and postpubertal girls: a study of tennis players. J Bone Miner Res 17:2274–2280

Bridges PS (1989) Changes in activities with the shift to agriculture in the southeastern United States. Curr Anthropol 30:385–394

Bridges PS (1995) Skeletal biology and behavior in ancient humans. Evol Anthropol 4:112–120

Brock SL, Ruff CB (1988) Diachronic patterns of change in structural properties of the femur in the prehistoric American Southwest. Am J Phys Anthropol 75:113–127

Buckwalter JA, Glimcher MJ, Cooper RR, Recker R (1995) Bone biology. J Bone Joint Surg Am 77:1276–1289

Butler VL (1993) Natural versus cultural Salmonid remains: origins of the Dalles Roadcut bones, Columbia River, Oregon, U.S.A. J Archaeol Sci 20:1–24

Butler VL, Campbell SK (2004) Resource intensification and resource depression in the Pacific Northwest of North America: a zooarchaeological review. J World Prehist 18:327–405

Carlson KJ, Grine FE, Pearson OM (2007) Robusticity and sexual dimorphism in the postcranium of modern hunter-gatherers from Australia. Am J Phys Anthropol 134:9–23

Carlson KJ, Summer DR, Morbeck ME, Nishida T, Yamanaka A, Boesch C (2008) Role of nonbehavioral factors in adjusting long bone diaphyseal structure in free-ranging Pan troglodytes. In J Primatol 29:1401–1420

Carter DR, Beaupré GS (2001) Skeletal function and form: mechanobiology of skeletal development, aging, and regeneration. Cambridge University Press, New York

Chamay A, Tschantz P (1972) Mechanical influences in bone remodeling: experimental research on Wolff's Law. J Biomech 5:173–180

Chatters JC (2000) The recovery and first analysis of an early Holocene human skeleton from Kennewick, Washington. Am Antiq 65:291–316

Chatters JC, Hackenberger S, Prentiss AM, Thomas JL (2012) The Paleoindian to Archaic transition in the Pacific Northwest: in situ development or ethnic replacement? In: Bousman CB, Vierra BJ (eds) From the Pleistocene to the Holocene: human organization and cultural transformation in prehistoric North America. Texas A&M Press, College Station, pp 37–66

Child SL (2013) Relationship between angles and cross-sectional geometry in the femur across subsistence economies of North America: implications for behavior and activity patterns. [Dissertation research proposal]. University of Missouri, Columbia

Cowgill LW (2010) The ontogeny of Holocene and late Pleistocene human postcranial strength. Am J Phys Anthropol 141:16–37

Cowgill LW, Hager LD (2007) Variation in the development of postcranial robusticity: an example from Çatalhöyük, Turkey. Int J Osteoarchaeol 17:235–252

Cowgill L, Warrener A, Pontzer H, Ocobock C (2010) Waddling and toddling: the biomechanical effects of an immature gait. Am J Phys Anthropol 143:52–61

Cridlin S (2007) Examination of secular change in the vertical head diameter of the human femur in American males and females [dissertation]. University of Tennessee, Knoxville

Davies TG, Stock JT (2014) Human variation in the periosteal geometry of the lower limb: signatures of behaviour among human Holocene populations. In: Carlson KJ, Marchi D (eds) Reconstructing mobility: environmental, behavioral, and morphological determinants. Springer, New York

Demes B, Jungers WL, Selpien K (1991) Body size, locomotion, and long bone cross-sectional geometry in indriid primates. Am J Phys Anthropol 86:537–547

Demes B, Stern JT, Hausman MR, Larson SG, McLeod KJ, Rubin CT (1998) Patterns of strain in the macaque ulna during functional activity. Am J Phys Anthropol 106:87–100

Demes B, Qin Y, Stern JT, Larson SG, Rubin CT (2001) Patterns of strain in the Macaque tibia during functional activity. Am J Phys Anthropol 116:257–265

DeVita P, Hortobagyi T (2003) Obesity is not associated with increased knee joint torque and power during level walking. J Biomech 36:1355–1362

Devlin MJ (2011) Estrogen, exercise, and the skeleton. Evol Anthropol 20:54–61

Devlin MJ, Lieberman DE (2007) Variation in estradiol level affects cortical bone growth in response to mechanical loading in sheep. J Exp Biol 210:602–613

Devlin MJ, Stetter CM, Lin HM, Beck RS, Petit MA, Lieberman DE, Lloyd T (2010) Peripubertal estrogen levels and physical activity affect femur geometry in young adult women. Osteoporos Int 21:609–617

Dionyssiotis Y, Trovas G, Galanos A, Raptou P, Papaioannou N, Papagelopoulos P, Petropoulou K, Lyritis GP (2007) Bone loss and mechanical properties of tibia in spinal cord injured men. J Musculoskelet Neuronal Interact 7:62–68

Driscoll KRD (2010) Secular change of the modern human bony pelvis: examining morphology in the United States using metric and geometric morphometry [dissertation]. University of Tennessee, Knoxville

Duncan CS, Blimkie C Jr, Kemp A, Higgs W, Cowell CT, Woodhead H, Briody JN, Howman-Giles R (2002) Mid-femur geometry and biomechanical properties in 15- to 18-yr-old female athletes. Med Sci Sports Exerc 34:673–681

Flegal KM, Carroll MD, Ogden CL, Johnson CL (2002) Prevalence and trends in obesity among US adults 1999–2000. J Am Med Assoc 288:1723–1727

Flegal KM, Carroll MD, Ogden CL, Curtin LR (2010) Prevalence and trends in obesity among US adults, 1999–2008. J Am Med Assoc 303:235–241

Gruss LT (2007) Limb length and locomotor biomechanics in the genus Homo: an experimental study. Am J Phys Anthropol 134:106–116

Haapasalo H, Kannus P, Sievänen H, Pasanen M, Uusi-Rasi K, Heinonen A, Oja P, Vuori I (1998) Effects of long-term unilateral activity on bone mineral density of female junior tennis players. J Bone Miner Res 13:310–319

Harrington KI (2013) Secular change in the knee joint and the effects of obesity [thesis]. Texas State University, San Marcos

Holt BM (2003) Mobility in Upper Paleolithic and Mesolithic Europe: evidence from the lower limb. Am J Phys Anthropol 122:200–215

Hunt DR, Albanese J (2005) History and demographic composition of the Robert J. Terry Anatomical Collection. Am J Phys Anthropol 127:406–417

Jantz RL (2001) Cranial change in Americans: 1850-1975. J Forensic Sci 46:784–787

Jantz RL, Moore-Jansen PH (1988) A data base for forensic anthropology: structure, content, and analysis. Report of investigations no. 47, Department of Anthropology, University of Tennessee, Knoxville

Jones HN, Priest JD, Hayes WC, Tichenor CC, Nagel DA (1977) Humeral hypertrophy in response to exercise. J Bone Joint Surg Am 59A:204–208

Kelly RL (1992) Mobility/sedentism: concepts, archaeological measures, and effects. Annu Rev Anthropol 21:43–66

Lanyon LE, Rubin CT (1985) Static vs dynamic loads as an influence on bone remodeling. J Biomech 15:142–154

Lanyon LE, Goodship AE, Pye CJ, MacFie JH (1982) Mechanical adaptive bone remodeling. J Biomech 15:142–154

Larsen CS (1997) Bioarchaeology: interpreting behavior from the human skeleton. Cambridge University Press, Cambridge

Larsen CS, Ruff C (2011) 'An external agency of considerable importance': the stresses of agriculture in the foraging-to-farming transition in eastern North America. In: Pinhasi R, Stock JT (eds) Human bioarchaeology of the transition to agriculture. Wiley-Blackwell, New York, pp 293–315

Lieberman DE, Devlin MJ, Pearson OM (2001) Articular area responses to mechanical loading: effects of exercise, age, and skeletal location. Am J Phys Anthropol 116:266–277

Lieberman DE, Pearson OM, Polk JD, Demes B, Crompton AW (2003) Optimization of bone growth and remodeling in response to loading in tapered mammilian limbs. J Exper Biol 206:3125–3138

Lieberman DE, Polk JD, Demes B (2004) Predicting long bone loading from cross-sectional geometry. Am J Phys Anthropol 123:156–171

Lovejoy CO, McCollum MA, Reno PL, Rosenman BA (2003) Developmental biology and human evolution. Annu Rev Anthropol 32:85–109

Marchi D (2008) Relationships between lower limb cross-sectional geometry and mobility: the case of a Neolithic sample from Italy. Am J Phys Anthropol 137:188–200

Marchi D, Sparacello VS, Holt BM, Formicola V (2006) Biomechanical approach to the reconstruction of activity patterns in Neolithic Western Liguria, Italy. Am J Phys Anthropol 131:447–455

Marchi D, Sparacello V, Shaw C (2011) Mobility and lower limb robusticity of a pastoralist Neolithic population from North-Western Italy. In: Pinhasi R, Stock JT (eds) Human bioarchaeology of the transition to agriculture. Wiley-Blackwell, New York, pp 317–346

Marks SC, Popoff SN (1988) Bone cell biology: the regulation of development, structure, and function in the skeleton. Am J Anat 183:1–44

Martin RB (2003) Functional adaptation and fragility of the skeleton. In: Agarwal SC, Stout SD (eds) Bone loss and osteoporosis: an anthropological perspective. Kluwer, New York, pp 121–138

Meadows Jantz L, Jantz RL (1999) Secular change in long bone length and proportion in the United States, 1800-1970. Am J Phys Anthropol 110:57–67

Meadows Jantz L, Jantz RL, Devlin JB (2012) Changes in postcranial morphology in modern American whites. Am J Phys Anthropol 147(S54):212

Meadows L, Jantz RL (1995) Allometric secular change in the long bones from the 1800s to the present. J Forensic Sci 40:762–767

Megyesi MS, Tubbs RM, Sauer NJ (2009) An analysis of human skeletal remains with cerebral palsy: associated skeletal age delay and dental pathologies. J Forensic Sci 54:270–274

Moore MK (2008) Body mass estimation from the human skeleton [dissertation]. University of Tennessee, Knoxville

Moore MK, Schaefer E (2011) A comprehensive regression tree to estimate body weight from the skeleton. J Forensic Sci 56:1115–1122

National Institute of Health (2008) ImageJ. Public domain image processing software. https://rsb.info.nih.gov.

Nikander R, Kannus P, Rantalainen T, Uusi-Rasi K, Heinonen A, Sievanen H (2010) Cross-sectional geometry of weight-bearing tibia in female athletes subjected to different exercise loadings. Osteoporos Int 21:1687–1694

Nikita E, Siew YY, Stock J, Mattingly D, Lahr MM (2011) Activity patterns in the Sahara Desert: an interpretation based on cross-sectional geometric properties. Am J Phys Anthropol 146:423–434

Nilsson BE, Westin NE (1971) Bone density in athletes. Clin Orthop 77:179–182
Ogden C, Carroll MD, Curtin LR, McDowell MA, Tabak CJ, Flegal KM (2006) Prevalence of overweight and obesity in the United States, 1999-2004. J Am Med Assoc 295:1549–1555
Pearson OM (2000) Activity, climate, and postcranial robusticity. Curr Anthropol 41:569–607
Pearson OM, Lieberman DE (2004) The aging of Wolff's "law": ontogeny and responses to mechanical loading in cortical bone. Yearbk Phys Anthropol 47:63–99
Pearson OM, Petersen TR, Sparacello VS, Daneshvari SR, Grine FE (2014) Activity, "body shape", and cross-sectional geometry of the femur and tibia. In: Carlson KJ, Marchi D (eds) Reconstructing mobility: environmental, behavioral, and morphological determinants. Springer, New York
Petit MA, Beck TJ, Shults J, Zemel BS, Foster BJ, Leonard MB (2005) Proximal femur bone geometry is appropriately adapted to lean mass in overweight children and adolescents. Bone 36:568–576
Robin J, Kerr Graham H, Selber P, Dobson F, Smith K, Baker R (2003) Proximal femur geometry in cerebral palsy: a population-based cross-sectional study. J Bone Joint Surg Br 90B:1372–1379
Robling AG (1998) Histomorphometric assessment of mechanical loading history from human skeletal remains: the relationship between micromorphology and macromorphology at the femoral midshaft [dissertation]. University of Missouri, Columbia
Rockhold LA (1998) Secular change in external femoral measures from 1840 to 1970 [thesis]. University of Tennessee, Knoxville
Ruff CB (1987) Sexual dimorphism in human lower limb bone structure: relationship to subsistence strategy and sexual division of labor. J Hum Evol 16:391–416
Ruff CB (1993) Climate adaptation and hominid evolution: the thermoregulatory imperative. Evol Anthropol 2:53–60
Ruff CB (1994) Morphological adaptation to climate in modern and fossil hominids. Yearbk Phys Anthropol 37:65–107
Ruff CB (1999) Skeletal structure and behavioral patterns of prehistoric Great Basin populations. In: Hemphill BE, Larsen CS (eds) Prehistoric lifeways in the Great Basin wetlands: bioarchaeological reconstruction and interpretation. University of Utah Press, Salt Lake City, pp 290–320
Ruff CB (2003a) Ontogenetic adaptation to bipedalism: age changes in femoral to humeral length and strength proportions in humans, with a comparison to baboons. J Hum Evol 45:317–349
Ruff CB (2003b) Growth in bone strength, body size, and muscle size in a juvenile longitudinal sample. Bone 33:317–329
Ruff CB (2008) Biomechanical analyses of archaeological human remains. In: Katzenberg MA, Saunders SR (eds) Biological anthropology of the human skeleton. Wiley, New York, pp 183–206
Ruff CB (2013) Momentmacro. http://www.hopkinsmedicine.org/FAE/mmacro.htm
Ruff CB, Hayes WC (1983) Cross-sectional geometry of Pecos Pueblo femora and tibiae—a biomechanical investigation. II. Sex, age, and side differences. Am J Phys Anthropol 60:383–400
Ruff CB, Larsen CS (2014) Long bone structural analyses and the reconstruction of past mobility: a historical review. In: Carlson KJ, Marchi D (eds) Reconstructing mobility: environmental, behavioral, and morphological determinants. Springer, New York
Ruff CB, Trinkaus E, Walker A, Larsen CS (1993) Postcranial robusticity in Homo, I: temporal trends and mechanical adaptation. Am J Phys Anthropol 91:21–53
Ruff CB, Walker A, Trinkaus E (1994) Postcranial robusticity in Homo, III: ontogeny. Am J Phys Anthropol 93:35–54
Ruff CB, Holt B, Trinkaus E (2006a) Who's afraid of the big bad Wolff?: "Wolff's Law" and bone functional adaptation. Am J Phys Anthropol 129:484–498
Ruff CB, Holt BM, Sládek V, Berner M, Murphy WA Jr, zur Nedden D, Seidler H, Recheis W (2006b) Body size, body proportions, and mobility in the Tyrolean "Iceman". J Hum Evol 51:91–101

Seung-uk K, Stenholm S, Ferrucci L (2010) Characteristic gait patterns in older adults with obesity—results from the Baltimore Longitudinal Study of Aging. J Biomech 43:1104–1110

Shaw CN, Stock JT (2009) Intensity, repetitiveness, and directionality of habitual adolescent mobility patterns influence the tibial diaphysis morphology in athletes. Am J Phys Anthropol 140:149–159

Shaw CN, Stock JT (2011) The influence of body proportions on femoral and tibial midshaft shape in hunter-gatherers. Am J Phys Anthropol 144:22–29

Shaw CN, Stock JT, Davies TG, Ryan TM (2014) Does the distribution and variation in cortical bone along lower limb diaphyses reflect selection for locomotor economy? In: Carlson KJ, Marchi D (eds) Reconstructing mobility: environmental, behavioral, and morphological determinants. Springer, New York

Shirley NR, Wilson RJ, Jantz LM (2011) Cadaver use at the University of Tennessee's Anthropological Research Facility. Clin Anat 24:372–380

Sládek V, Berner M, Sailer R (2006) Mobility in Central European Late Eneolithic and Early Bronze Age: tibial cross-sectional geometry. J Archaeol Sci 33:470–482

Sparacello VS, Marchi D (2008) Mobility and subsistence economy: a diachronic comparison between two groups settled in the same geographical area Liguria, Italy. Am J Phys Anthropol 136:485–495

Sparacello VS, Marchi D, Shaw CN (2014) The importance of considering fibular robusticity when inferring the mobility patterns of past populations. In: Carlson KJ, Marchi D (eds) Reconstructing mobility: environmental, behavioral, and morphological determinants. Springer, New York

Stock JT (2006) Hunter-gatherer postcranial robusticity relative to patterns of mobility, climatic adaptation, and selection for tissue economy. Am J Phys Anthropol 131:194–204

Stock J, Pfeiffer S (2001) Linking structural variability in long bone diaphyses to habitual behaviors: foragers from southern African Later Stone Age and the Andaman Islands. Am J Phys Anthropol 115:337–348

Stock J, Pfeiffer S (2004) Long bone robusticity and subsistence behavior among Later Stone Age foragers of the forest and fynbos biomes of South Africa. J Archaeol Sci 31:999–1013

Taylor RE, Kimer DL, Southon JR, Chatters JC (1998) Radiocarbon dates of Kennewick Man. Science 280:1171–1172

Trinkaus E, Ruff CB (2012) Femoral and tibial diaphyseal cross-sectional geometry in Pleistocene Homo. PaleoAnthropology 2012:13–62

Trinkaus E, Churchill SE, Villemeur I, Riley KG, Heller JA, Ruff CB (1991) Robusticity versus shape: the functional interpretation of Neandertal appendicular skeleton. J Anthropol Soc Nippon 99:257–278

Trinkaus E, Churchill SE, Ruff CB (1994) Postcranial robusticity in Homo, II: humeral bilateral asymmetry and bone plasticity. Am J Phys Anthropol 93:1–34

Trotter M, Peterson RR, Wette R (1968) The secular trend in the diameter of the femur of American whites and negroes. Am J Phys Anthropol 28:65–74

Vogel S (2013) Comparative biomechanics: life's physical world. Princeton University Press, Princeton

Wallace IJ, Tommasini SM, Judex S, Garland T, Demes B (2012) Genetic variations and physical activity as determinants of limb bone morphology: an experimental approach using a mouse model. Am J Phys Anthropol 148:24–35

Wall-Scheffler CM (2014) The balance between burden carrying, variable terrain and thermoregulatory pressures in assessing morphological variation. In: Carlson KJ, Marchi D (eds) Reconstructing mobility: environmental, behavioral, and morphological determinants. Springer, New York

Weaver TD (2003) The shape of the Neanderthal femur is primarily the consequence of a hyperpolar body form. Proc Natl Acad Sci U S A 100:6926–6929

Weiss E (2003) The effects of rowing on humeral strength. Am J Phys Anthropol 121:293–302

Wescott DJ (2001) Structural variation in the humerus and femur in the American Great Plains and adjacent regions: differences in subsistence strategy and physical terrain [dissertation]. University of Tennessee, Knoxville

Wescott DJ (2006a) Effect of mobility on femur midshaft external shape and robusticity. Am J Phys Anthropol 130:201–213

Wescott DJ (2006b) Ontogeny of femur subtrochanteric shape in Native Americans and American Blacks and Whites. J Forensic Sci 51:1–6

Wescott DJ (2008) Biomechanical analysis of humeral and femoral structural variation in the Great Plains. Plains Anthropol 53:333–355

Wescott DL (2010) Reconstructing habitual activities in Paleoamericans. Am J Phys Anthropol 42(S49):317–318

Wescott DL (2014) Reconstructing habitual activities by biomechanical analysis of long bones. In: Owsley DW, Jantz RL (eds) Kennewick Man: scientific interpretations of an ancient American skeleton. Texas A&M Press, College Station

Wescott DJ, Cunningham DL (2006) Temporal changes in Arikara humeral and femoral cross-sectional geometry associated with horticultural intensification. J Archaeol Sci 33:1022–1036

Wescott DJ, Cunningham DL (2013) Temporal changes in Arikara femoral torsion. Am J Phys Anthropol 150(S56):289

Wescott DJ, Zephro L (2012) Secular change in the femur. Am J Phys Anthropol 147(S54):308

Chapter 8
Activity, "Body Shape," and Cross-Sectional Geometry of the Femur and Tibia

Osbjorn M. Pearson, Timothy R. Petersen, Vitale S. Sparacello, Shamsi R. Daneshvari, and Frederick E. Grine

Abstract Both bi-iliac breadth and stature are considered key aspects of "body shape," vary ecogeographically, and have been proposed to influence femoral midshaft shape, complicating interpretations of "activity." This chapter explores patterns of variation in cross-sectional geometry [especially shape, as measured by I_{max}/I_{min} or midshaft anteroposterior (AP) and mediolateral (ML) diameters] in the femur and tibia using three data sets that comprise a large amount of external measurements and some data from cross-sectional geometry. These data show that the midshaft shapes of the femur and tibia are only weakly correlated: $r=-0.12$ for AP/ML

O.M. Pearson (✉)
Department of Archaeology, Durham University,
Durham, UK

Department of Anthropology, MSC 01-1040, University of New Mexico,
Albuquerque, NM 87121, USA
e-mail: ompear@unm.edu

T.R. Petersen
Department of Anthropology, MSC 01-1040, University of New Mexico,
Albuquerque, NM 87121, USA

Department of Anesthesiology & Critical Care Medicine, University of New Mexico
Hospital, Albuquerque, NM 87131, USA

V.S. Sparacello
Department of Anthropology, MSC 01-1040, University of New Mexico,
Albuquerque, NM 87121, USA

S.R. Daneshvari
Department of Anthropology, MSC 01-1040, University of New Mexico,
Albuquerque, NM 87121, USA

Health Sciences Library and Informatics Center, MSC 09-5100, University of New Mexico,
Albuquerque, NM 87131, USA

F.E. Grine
Departments of Anthropology and Anatomical Sciences, Stony Brook University,
Stony Brook, NY 11794, USA

K.J. Carlson and D. Marchi (eds.), *Reconstructing Mobility: Environmental, Behavioral, and Morphological Determinants*, DOI 10.1007/978-1-4899-7460-0_8,

diameters; $r=0.33$ for I_{max}/I_{min} ratios. Femoral midshaft shape is weakly, but significantly, associated with bi-iliac breadth and the ratio of bi-iliac breadth to femoral length in some, but not all, data sets. The results indicate that variation in "body shape" does not drive the low correlations observed between femoral and tibial midshaft shapes. We should look to other factors to explain the mismatch.

Keywords Cross-sectional geometry • Mobility • Femur • Tibia • Bi-iliac breadth • Body shape

8.1 Introduction

This chapter investigates the effect of "body shape," as gauged by bi-iliac breadth, as well as the combination of bi-iliac breadth and stature (or long bone length, a proxy for stature) on cross-sectional shape of the midshaft femur and tibia. Researchers generally consider that walking and running produce high anteroposterior (AP) bending loads on the femur and tibia and that, given a high amount of mobility, both bones will grow (model) to deposit more bone in their anterior and posterior cortices to resist the increased loading. Since the mid-1990s (Ruff 1995), most of the researchers working on cross-sectional geometry have accepted the idea that body shape influences the shape of the midshaft femur, and perhaps—although much less markedly—that of the tibia. According to the conventional view, populations with relatively wider hips are likely to have proportionately larger mediolateral (ML) second moments of area in the midshaft femur than proportionately narrower-hipped populations. As a result, one would expect a highly mobile, high-latitude and thus wide-hipped population to have a rounder average midshaft femoral section than an equally mobile narrower-hipped, low-latitude population. Thus, consideration of body shape would be crucial in attempts to infer mobility patterns from the cross-sectional geometry of the midshaft femur. As we show here, this received wisdom turns out to be wrong. We also trace the historical development of the idea, which perhaps explains why it seemed logical and proved so influential, and offer some alternative hypotheses for why contrasts between populations and individuals in the cross-sectional shapes of lower limb bones exist.

At this juncture, it is useful to offer a definition of "mobility," and to critique what has often been the tautological or overly vague definitions of mobility that have plagued the anthropological literature. We define mobility as the sum total of locomotor activities performed using the lower limb. We should note that we have not actually quantified the mobility level of any group. However, neither have almost any of the studies that have sought to evaluate the mobility of prehistoric populations. Instead, the default mode of operation has been to adopt a tautological definition of mobility: groups with highly developed femoral pilasters and platycnemic tibial shafts are assumed to have been highly mobile and the magnitude of the mobility is inferred from the degree of flattening of the femoral and tibial shafts. This circular reasoning is clearly a problem for anthropology. Experimental approaches that measure the degree of morphological response in these bones to strains of known magnitude, strain rate, and loading cycles provide the best route out of this impasse.

To return to the main subject matter of this chapter, in the mid-1970s through early 1980s, anthropologists became interested in deducing patterns of biomechanical strength and mobility from the cross-sectional geometry of long bones of the lower limb (Endo and Kimura 1970; Amtmann 1971; Kimura 1971, 1974; Lovejoy et al. 1976; Lovejoy and Trinkaus 1980; Kimura and Takahashi 1982; Ruff and Hayes 1983; Kimura and Amtmann 1984; Ruff et al. 1984). These early works noted marked differences between populations, and sometimes between sexes within populations, that demanded explanation. These contrasts largely mirrored patterns in the subtrochanteric and pilastric indices of the femur and the cnemic index of the tibia that earlier anthropologists had described and pondered (e.g., the long lists of population means for the meric, pilastric, and cnemic indices in Martin and Saller 1956).

Two of the most striking differences between populations lay in the ratios of maximum to minimum second moments of area of the femoral and tibial midshafts. In both cases, hunter-gatherer populations often had an exaggeratedly large maximum (located in an approximately AP direction) second moment of area compared to the minimum (approximately ML) second moment of area in comparison to presumably more sedentary populations such as horticulturalists and especially modern, industrialized populations. This difference has been widely considered to be a direct result of the body's modeling response to the greater AP bending strains imposed by more mobile lifestyles of hunters and gatherers (e.g., Ruff et al. 1984; Ruff 1994a, b, 1999; Larsen et al. 2002; Stock and Pfeiffer 2001, 2004; but see Bridges 1989 and Carlson et al. 2007 for exceptions), and especially so in hunter-gatherer males given the strongly developed sexual division of labor in almost all hunter-gatherer societies (Ruff 1987).

In our view, two key developments prepared the way for the belief that body shape helps make sense of variation in midshaft femoral shape. The first of these was the recognition of a paradox. Work on the cross-sectional geometry of Neandertals and the early near-modern humans from Skhul and Qafzeh in Israel showed that both had a fairly platycnemic tibial midshaft, but the Skhul–Qafzeh males had some of the highest ratios of I_x/I_y and I_{max}/I_{min} in their femoral midshaft of any population that had been studied, while Neandertal males (and females) tended to have a very round, albeit massive, femoral midshaft that resembled sedentary industrialized populations (Trinkaus and Ruff 1999a, b, 2012). Both populations were presumably highly mobile hunter-gatherers whose lifestyle should have created high AP bending loads on their lower limbs, so how could one explain the difference between the two populations?

The second development grew out of Ruff's long, productive, and influential interest in the role of pelvic width, as gauged most readily by bi-iliac breadth, in bipedal locomotion and in human ecogeographic variation (e.g., Ruff 1991, 1994a, b, 1995. In a carefully argued paper, Ruff (1995) proposed that bi-iliac breadth imposed its own ML bending load on the lower limb, which helped make sense of variation in the cross-sectional geometry of femoral subtrochanteric shape in Plio-Pleistocene hominins, as well as in various populations of modern humans. Ruff argued that the same moments should act on the femoral midshaft, although with less magnitude. This insight eventually offered a breakthrough for resolving the paradox of different shapes of the femoral midshaft in the Skhul–Qafzeh and Neandertal samples: the

Neandertals had a rounder femoral midshaft shape because they had absolutely and proportionately wider hips, and thus their body shape imposed a higher ML bending load on the femoral midshaft necessitating localized deposition of extra bone to resist those forces.

While the implications of Ruff's insight gestated, differences in femoral midshaft cross-sectional shapes continued to be attributed to variation in mobility. For example, in surveying changes in the cross-sectional geometry of the lower limb from the Early Upper Paleolithic through the Mesolithic in Europe, Holt (2003) noted that both males and females showed a decline over time in the ratio of midshaft femoral I_x/I_y, a pattern she attributed to decreasing mobility over time. Several years later, Shackleford (2007) noted that the trend toward rounder femoral midshaft sections was also present in late Upper Paleolithic populations in the Mediterranean (combined samples from the Maghreb and Israel), the Nile Valley, and Asia (Minatogawa and Tam Hang), a pattern that she also attributed to decreases in mobility. Holt (2003) did not find the same trend in the shape of the tibia over time; instead tibial shape did not vary significantly over time in either males or females, although the sample sizes were too small to detect subtle differences, especially in the females. Why femoral and tibial midshafts would produce differing pictures of changes in habitual loading, which was generally equated with mobility, remained unsolved for many researchers.

The paradigm that body shape provided an important influence on femoral midshaft shape was articulated fully in an analysis of the cross-sectional geometry of the Tyrolean Iceman (Ruff et al. 2006), who also had a relatively round femoral midshaft section but an AP-elongated tibial midshaft. The shape of the Iceman's femoral section was attributed to his high bi-iliac breadth, and the analysis presented support for the conclusion from a sample of 21 anatomically modern European males, ranging in antiquity from the Gravettian to the Bronze Age. In their analysis, Ruff et al. (2006) noted a positive correlation between two ratios: the ratio of femoral to tibial ratios of ML/AP section moduli vs. the ratio of bi-iliac breadth to estimated stature. In point of fact, the relationship was weak ($r=0.420$, corresponding to $R^2=0.176$) and not statistically significant ($p=0.058$). Nevertheless, the authors chose to emphasize the conclusion that "The greater the relative pelvic breadth, the greater the increase in ML bending strength of the femur relative to the tibia" (Ruff et al. 2006, p. 96), and this notion became a highly influential perspective for the last half-decade (Stock 2006; Shang et al. 2007; Ruff 2008, 2010; Sparacello and Marchi 2008; Trinkaus 2009, 2011; Shaw and Stock 2011; Marchi et al. 2011; Larsen and Ruff 2011; Trinkaus and Ruff 2012; Wescott 2014).

8.2 Criticisms

The evolution of Ruff and colleagues' ideas into the conclusion that body shape was responsible for the mismatch between femoral and tibial midshaft shape did not occur in isolation from other research, or from a series of criticisms, only some of which informed the development of the theory. Fairly early on, Pearson and Grine

(1996) presented analyses that showed that the ratios of I_{max}/I_{min} in the midshaft femur and tibia were more weakly correlated ($r=0.34$, using the same data as in Data Set 3 in this chapter) than expected if both indices responded to AP (or quasi-AP if the plane of the neutral axis of the maximum second moment of area does not lie in the ML plane) bending stresses from locomotion. Ruff (personal communication) countered that it was essential to consider the second moments of area around the AP and ML axes in both bones rather than the maximum and minimum second moments of area (which would often not be aligned in the same planes). This proposal is itself questionable for reasons explained in Pearson et al. (2006), and which will be explained in more detail at the end of this chapter. Suffice it to say for now that if the ratio of I_{max}/I_{min} results from the sum of a bone's responses to its loading history, and the greatest stresses in the lower limb result from locomotion, why would it make any sense to consider an index (I_x/I_y) that is *less* reflective of those adaptive responses?

Another key caveat arose recently when Stock (2006) published a series of analyses aiming to untangle the interactions between climate (or its correlate, ecogeographic variation in human body form) and activity on cross-sectional geometry of the bones of the lower limb. Stock (2006) found that influence of climate appeared to be dominant in the midshaft femur, but the influence of activity dominated in the midshaft tibia. This result underscores work by Lieberman demonstrating that the midshafts of different long bones in a limb have dissimilar modeling and remodeling responses to stresses and also shows profound differences in the magnitude of response across ontogeny (Lieberman et al. 2003, 2004; Pearson and Lieberman 2004).

Pearson et al. (2006) published a preliminary version of the analyses presented in this chapter in response to criticism that body shape had to be considered when assessing mobility or activity levels from the shape of the femoral midshaft. Pearson et al. (2006) concluded there was only a weak correlation between midshaft femoral and tibial shape (as assessed from external measurements) and that in partial correlation design, controlling bi-iliac breadth and femoral length did not improve this weak correlation.

Most recently, Shaw and Stock (2011) investigated the relationships among bone lengths in the lower limb, bi-iliac breadth, and femoral and tibial midshaft cross-sectional geometry. They found that bone lengths standardized for body mass had no significant correlations with diaphyseal shape [except for a weak ($r=0.13$, $p \leq 0.01$, $n=71$) association between tibia length and midshaft shape in males], but that bi-iliac breadth standardized by estimated body mass showed weak but statistically significant associations with femoral midshaft shape in both sexes [for males, $r=-0.224$, $p \leq 0.05$, $n=26$; for females, $r=-0.243$ (misprinted as 0.0243), $p \leq 0.05$, $n=22$]. The correlations between tibial midshaft shape and bi-iliac breadth divided by body mass were even weaker and did not reach statistical significance. On the face of it, then, if one overlooks the low magnitude of the correlations, Shaw and Stock's results appear to provide substantiation for the idea that body breadth affects femoral midshaft shape. However, the study had three flaws that mandate that the issue should be revisited.

First, Shaw and Stock's (2011) sample of individuals with data for both cross-sectional geometry and measures of bi-iliac breadth conflates one group of narrow-hipped, short, and light-weight tropical and subtropical foragers (Khoesan and Andaman Islanders) with a second set of taller, wider-hipped, heavier high-latitude foragers (people from Tierra del Fuego plus hunter-gatherers from around the Great Lakes in North America). The Khoesan individuals have more heavily developed pilasters than any of the other samples, guaranteeing that analyses of this pooled sample will find an association between narrow hips and high values of I_x/I_y in the femoral midshaft. Second, the standardization of bi-iliac breadth by body mass makes no sense from a biomechanical perspective. The bone in a cross section acts to resist stresses from bending moments (defined as force times acceleration times a distance). In a static situation, the force in the bending moment is created by body mass and acceleration from gravity; the geometry of the hip and limb determine the distance in the moment. Thus, the product or interaction between body mass and bi-iliac breadth influences both I_x and I_y: it makes no sense to standardize one mechanical influence by the other. Third, following the logic presented by Ruff (1995), proportionately (i.e., wide when standardized by mass), laterally flaring iliac blades should *reduce* ML bending rather than *increase* it, and thus the correlations for the femoral midshaft found by Shaw and Stock (2011) actually run contrary to what Ruff (1995) had argued.

In sum, the questions remain open of whether variation in body shape—as gauged by bi-iliac breadth or the combination of bi-iliac breadth and long bone length—strongly affects femoral or tibial midshaft shape, and whether ecogeographic variation resolves the potential paradox that femoral and tibial midshaft shapes sometimes lead to conflicting inferences about the habitual activity of past populations. Tests of the predictions below address these questions.

8.3 Materials and Methods

To test these predictions and assess relationships among ratios of I_{max}/I_{min} and I_x/I_y in the femur and tibia, we use three data sets (Table 8.1). The first comprises a worldwide sample of external dimensions of the limb bones (Pearson 1997, 2000; Pearson and Millones 2005). Data Set 1 includes populations that differ dramatically in physique, as well as the shapes of midshafts of lower limb bones, but regrettably it does not include data on bi-iliac breadth of the vast majority of individuals. The second consists of external measurements of documented skeletons from four North American collections (Daneshvari 2011). In addition to being a large sample, Data Set 2 also has the bi-iliac breadth for each individual, which is essential for testing some of the hypotheses about the relationship between "body shape" and midshaft sections of lower limb bones. The third includes cross-sectional properties derived from CT scans of Zulu, African American, and Khoesan samples. Data on bi-iliac breadth was also recorded for all of the African American skeletons with sufficiently well-preserved pelvic bones. Additional information on these samples and

Table 8.1 Sample sizes of the three data sets used in this study

Data set	Subsample	Females	Males	Indeterminate
1—Worldwide populations—external measurements[a]				
	Pooled sample	208	333	19
	Khoesan	25	37	6
	Zulu	31	31	0
	African American	31	41	0
	Jebel Sahaba	13	18	8
	Chinese	0	28	0
	Buriat	2	1	0
	Maori	1	1	0
	Inuit	25	62	0
	Tierra del Fuego	9	21	0
	Mesolithic Europe	13	16	4
	Sami	25	34	0
	European Americans	25	25	0
	Australian aborigines	8	18	1
2—Recent, documented skeletons[b]—external measurements				
	Pooled sample	186	475	0
	Maxwell Museum	25	50	0
	U. Tennessee	41	144	0
	Terry collection	41	156	0
	Hamann–Todd collection	79	125	0
3—Cross sectional geometry (and external measurements)[c]				
	Pooled sample	44	76	1
	Khoesan	14	14	1
	Zulu	24	24	0
	African American	6	38	0
4—Cross-sectional geometry[d]				
	Iron Age Samnites	70	160	0

[a]Collected by Pearson (Pearson 1997, 2000; Pearson and Millones 2005)
[b]Collected by Daneshvari (2011)
[c]Collected by Grine and Pearson (Grine et al. 1995; Carlson et al. 2007)
[d]Collected by Sparacello (2013)

protocols used to CT scan the bones have been published elsewhere (Grine et al. 1995; Carlson et al. 2007). Unfortunately, only data on I_{max}/I_{min} (rather than data for both I_{max}/I_{min} and I_x/I_y) were recorded for these bones, and the way in which they were positioned for scanning (with, for femora, the anterior surface of the proximal femur facing inferiorly and the midshaft in contact with the gantry and, for tibiae, with the ventral edge of the shaft facing superiorly and the distal end elevated to keep the shaft parallel to the gantry) are not clearly equivalent to other protocols (e.g., Ruff and Hayes 1983) for orienting these bones in an AP plane. So long as a bone's diaphysis is parallel to the gantry, the differences in orientation make no difference for values of I_{max}/I_{min} but they directly affect what direction appears to be "anterior" and thus would influence values of I_x/I_y. As a result, only values of I_{max}/I_{min} (and not I_x/I_y) are presented and evaluated for Data Set 3. Data Set 4 consists of a pooled-sex sample of 230 Samnites (70 females and 160 males) from the Iron Age of central Italy (Sparacello 2013). Each individual in Data Set 4 had a femur and tibia that were complete enough for analysis. The bones were orientated following the protocol of Ruff (2002). The external contour of each midshaft section was molded, AP and ML directions were recorded, and then the periosteal contour was digitized. Values of second moments of area and ratios of I_{max}/I_{min} and I_x/I_y were predicted from the periosteal contours using formulae presented in Sparacello and Pearson (2010).

Different variables were available for each data set (Table 8.2). All analyses were performed using JMP 6.0.3 (SAS Institute Inc., 2006) using Pearson product–moment correlations. Additional relevant details are given under specific tests below.

8.4 Predictions

1a. Given that variations in individual "mobility" are expected to produce AP bending moments in both the midshaft femur and midshaft tibia, there should be a reasonably strong and positive correlation between the values of I_{max}/I_{min} (and in external AP to ML diameters) of the femoral and tibial midshaft. Obviously, values for ratios of I_{max}/I_{min} as well as their orientation may depart from values of I_x/I_y.

1b. If one considers that only the maximum bending moments (which approximate but may not coincide with the AP axis) should be considered, and should be scaled for body size, then this prediction can be restated as an expectation for a fairly high correlation between the size-adjusted section moduli in the (quasi-) AP direction of both bones (i.e., femoral I_{max} divided by the product of one-half of the external femoral AP diameter, predicted body mass, and femoral length should correlate positively and fairly strongly with its tibial counterpart).

2. If bi-iliac breadth influences femoral or tibial midshaft shape, a negative correlation should exist between bi-iliac breadth and ratios of I_{max}/I_{min} (and the ratio of external AP to ML diameters) in these sections. The negative correlation should be present between samples of humans, as well as within samples, since its cause should stem from individual variation in geometry.
3. If "body shape," as gauged by the ratio of bi-iliac breadth to bone or limb length, exerts an influence on femoral or tibial midshaft shape, then there should be a negative correlation between the ratio of bi-iliac width to stature and the ratio of I_{max}/I_{min} (and the ratio of external AP to ML diameters) of each bone.
4. A similar relationship to that observed by Ruff et al. (2006)—a negative correlation between ratios of femoral to tibial I_{max}/I_{min} vs. the ratio of bi-iliac breadth to stature (or, in this case, femoral or tibial length)—should be present in this study in the pooled sample, as well as within individual samples subdivided by sex.

Data Set 4 contains values for both the ratio of I_{max}/I_{min} and I_x/I_y, and thus permits them to be compared in order to evaluate the proposal that values of I_x/I_y capture functional relationships in a way that values for I_{max}/I_{min} do not.

Table 8.2 Variables and summary statistics for each data set

		Females	Males
Data set	Variable	Mean ± SD (n)	Mean ± SD (n)
1	Femur maximum length	414.8 ± 27.1 (184)	438.7 ± 31.2 (298)
	Femur midshaft AP	26.5 ± 2.3 (195)	29.3 ± 2.8 (300)
	Femur midshaft ML	24.3 ± 2.1 (195)	26.4 ± 2.5 (301)
	Tibia articular length[a]	323.0 ± 28.2 (177)	342.6 ± 31.4 (281)
	Tibia midshaft AP	26.8 ± 2.4 (190)	30.4 ± 2.8 (288)
	Tibia midshaft ML	19.4 ± 2.0 (190)	21.5 ± 2.2 (289)
	Bi-iliac breadth	268.3 ± 15.8 (6)	271.1 ± 20.0 (85)
2	Femoral length[b]	434.9 ± 24.2 (185)	472.3 ± 26.7 (470)
	Femur midshaft AP	26.9 ± 2.1 (181)	30.2 ± 2.4 (461)
	Femur midshaft ML	26.3 ± 2.2 (181)	29.5 ± 2.3 (461)
	Femur midshaft AP/ML	102.7 ± 10.5 (181)	102.9 ± 10.5 (461)
	Tibia length[c]	357.0 ± 22.5 (185)	389.1 ± 25.9 (461)
	Tibia midshaft AP	27.7 ± 2.1 (180)	32.0 ± 2.2 (458)
	Tibia midshaft ML	20.0 ± 1.5 (180)	22.5 ± 1.9 (458)
	Tibia midshaft ML/AP	72.3 ± 5.4 (180)	70.7 ± 5.8 (458)
	Bi-iliac breadth	265.1 ± 19.8 (186)	274.3 ± 18.3 (473)
	Bi-iliac breadth/Femoral length	0.611 ± 0.051 (185)	0.582 ± 0.046 (468)
3-Pooled	Femur bicondylar length	417.7 ± 21.3 (44)	453.7 ± 31.5 (76)
	Femur midshaft I_{max}/I_{min}	1.426 ± 0.240 (44)	1.442 ± 0.279 (76)
	Tibia articular length[a]	335.7 ± 18.6 (44)	364.9 ± 27.2 (76)
	Tibia I_{max}/I_{min}	2.190 ± 0.420 (44)	2.283 ± 0.488 (76)
	Predicted body mass[d] (kg)	53.6 ± 5.1 (44)	67.0 ± 8.0 (75)
	Femur body-size-adjusted "AP" section modulus[e]	0.061 ± 0.012 (43)	0.068 ± 0.016 (76)
	Tibia body-size-adjusted "AP" section modulus[e]	0.070 ± 0.014 (43)	0.071 ± 0.016 (76)
3-Khoesan	Femur bicondylar length	399.0 ± 11.5 (13)	419.9 ± 24.9 (15)
	Femur midshaft I_{max}/I_{min}	1.413 ± 0.290 (13)	1.682 ± 0.349 (15)
	Tibia articular length[a]	323.9 ± 15.2 (13)	336.5 ± 20.1 (15)
	Tibia I_{max}/I_{min}	2.398 ± 0.475 (13)	2.774 ± 0.483 (15)
	Bi-iliac breadth	–	–
	Predicted body mass[d] (kg)	48.8 ± 4.7 (13)	56.0 ± 5.7 (14)
	Femur body-size-adjusted "AP" section modulus[e]	0.056 ± 0.010 (13)	0.061 ± 0.0132 (14)
	Tibia body-size-adjusted "AP" section modulus[e]	0.065 ± 0.012 (13)	0.080 ± 0.019 (14)
3-Zulu	Femur bicondylar length	423.2 ± 20.4 (24)	451.8 ± 25.6 (24)
	Femur midshaft I_{max}/I_{min}	1.413 ± 0.228 (24)	1.311 ± 0.159 (24)
	Tibia articular length[a]	339.3 ± 17.5 (24)	368.7 ± 26.4 (24)
	Tibia I_{max}/I_{min}	2.062 ± 0.320 (24)	2.227 ± 0.453 (24)
	Bi-iliac breadth	–	–
	Predicted body mass[d] (kg)	55.6 ± 3.6 (24)	67.0 ± 5.9 (24)
	Femur body-size-adjusted "AP" section modulus[e]	0.062 ± 0.011 (24)	0.065 ± 0.009 (24)
	Tibia body-size-adjusted "AP" section modulus[e]	0.074 ± 0.014 (24)	0.078 ± 0.016 (24)

(continued)

Table 8.2 (continued)

		Females	Males
Data set	Variable	Mean ± SD (n)	Mean ± SD (n)
3-African-American	Femur bicondylar length	435.8 ± 14.5 (6)	467.4 ± 27.1 (38)
	Femur midshaft I_{max}/I_{min}	1.433 ± 0.250 (6)	1.489 ± 0.212 (38)
	Tibia articular length[a]	350.4 ± 16.0 (6)	372.4 ± 24.2 (38)
	Tibia I_{max}/I_{min}	2.094 ± 0.340 (6)	2.148 ± 0.411 (38)
	Bi-iliac breadth	–	254.9 ± 17.7 (25)
	Bi-iliac breadth/Femur bicondylar length	–	0.549 ± 0.038 (25)
	Femur/Tibia midshaft I_{max}/I_{min}	–	0.684 ± 0.172 (25)
	Predicted body mass[d] (kg)	56.6 ± 3.8 (6)	70.5 ± 6.9 (38)
	Femur body-size-adjusted "AP" section modulus[e]	0.072 ± 0.012 (6)	0.072 ± 0.019 (38)
	Tibia body-size-adjusted "AP" section modulus[e]	0.065 ± 0.014 (6)	0.065 ± 0.013 (38)
4	Femur I_{max}/I_{min}	1.23 ± 0.138 (230)[f]	
	Femur I_x/I_y	1.060 ± 0.178 (230)[f]	
	Tibia I_{max}/I_{min}	2.150 ± 0.412 (230)[f]	
	Tibia I_x/I_y	1.430 ± 0.315 (230)[f]	

[a]Middle of the distal articular surface for the talar trochlea to the center of the superior surface of the medial condyle (Pearson 1997)
[b]Maximum length (Daneshvari 2011)
[c]Tibia Length excluding intercondylar eminence, but including medial malleolus (M-1b) (Daneshvari 2011)
[d]Body mass predicted by the formula based on the femoral head in Grine et al. (1995)
[e]Calculated as the I_{max} value/[body mass (calculated from the formula in Grine et al. 1995) × bone length × 0.5 × external AP diameter of the midshaft section of the bone]. I_{max} for each section may depart from the true AP plane so this variable represents a pseudo-AP section modulus rather than a true one
[f]Sexes pooled

8.5 Results

Prediction 1a, that there should be a reasonably strong and positive correlation between the values of I_{max}/I_{min} (and in external AP to ML diameters) of the femoral and tibial midshaft, was tested using all three data sets. In Data Set 1, there is a low but statistically significant correlation between the pilastric index of the femur (100 × AP/ML) and the midshaft shape index of the tibia (100 × ML/AP): $r = -0.109$, $p = 0.018$, $n = 477$. If one subdivides the sample by sex, the correlation approximates zero for females and becomes nonsignificant ($r = 0.060$, $p = 0.415$, $n = 187$), but the correlation remains low and statistically significant for males [$r = -0.190$, $p = 0.001$, $n = 285$ (five individuals in the pooled sample were of indeterminate sex and are not included here)]. The pattern present in males indicates that a degree of AP elongation in the femoral midshaft section predicts a measure of AP elongation in the tibial midshaft, as one might expect, but the relationship is clearly weak. In addition, it appears that the pattern present in males drives the overall correlation in this data set.

Table 8.3 Correlations between size-adjusted I_{max} section moduli of femoral and tibial midshafts, with samples subdivided by group and sex

Group	Sex	*r*	*p*	*n*
Pooled	Pooled	0.347	0.0001[a]	120
Females (pooled)	Female	0.517	0.0003[a]	44
Males (pooled)	Male	0.288	0.0122[a]	75
Khoesan	Female	0.418	0.1555	13
Khoesan	Male	0.786	0.0009[a]	14
African American	Female	0.676	0.1408	6
African American	Male	0.475	0.0026[a]	38
Zulu	Female	0.686	0.0002[a]	24
Zulu	Male	0.275	0.1933	24

These results address Prediction 1b

See the text for discussion

[a]Indicates a correlation significant at $p<0.05$ or better

In Data Set 2, the tibial midshaft index (100×ML/AP) is also weakly but statistically significantly correlated with the pilastric index (100×AP/ML) ($r=0.139$, $p=0.001$, $n=623$). As in the worldwide sample, subdividing the sample by sex removes the significant correlation for females ($r=0.044$, $p=0.560$, $n=176$), but not for males ($r=0.178$, $p=0.0002$, $n=447$).

In the pooled data set of cross-sectional geometry (Data Set 3), a positive (albeit weak) and statistically significant correlation exists between the indices of I_{max}/I_{min} in the femoral and tibial midshafts ($r=0.340$, $p=0.0001$, $n=121$). When the samples are subdivided by sex, the significant correlation again disappears for the females ($r=0.157$, $p=0.309$, $n=44$), but remains for males [$r=0.402$, $p=0.0003$, $n=76$ (one individual in the pooled sample was of indeterminate sex and is not considered here)]. In sum, both the external dimensions and cross-sectional geometry show only weak correlations between midshaft shapes (AP/ML external diameters or ratios of I_{max}/I_{min}), and these relationships appear to be driven by the patterns present in males alone.

Prediction 1b, that a fairly high, positive correlation should exist between the size-adjusted section moduli in the direction of I_{max} (roughly AP in each, but with some incongruence between sections and individuals) of each bone (see Table 8.2), was tested using Data Set 3 (cross-sectional geometry). In the pooled data, the correlation between body-size-adjusted "AP" section moduli for the femur and tibia is weak ($r=0.347$), but is statistically significant. Curiously, subdivision by sex strengthens the correlation for females ($r=0.517$), but weakens it for males ($r=0.288$). If one further subdivides by both group and sex, a strange pattern emerges (Table 8.3) in which male Khoesan have a stronger correlation than Khoesan females, but the opposite is true of the Zulus. The upshot of these results may be that patterns in these data may reflect group-specific patterns of sexual division of labor rather than any universal pattern determined by sex or body shape.

Prediction 2, that a negative correlation should exist between bi-iliac breadth and ratios of I_{max}/I_{min} (and the ratio of external AP to ML diameters), was tested using Data Sets 2 and 3. Data Set 2, which consists of external measurements of recent,

Table 8.4 Correlations between bi-iliac breadth and the pilastric and midshaft tibial indices in Data Set 2 (external dimensions of recent skeletons from documented collections) and the African American males in Data Set 3 (cross-sectional geometry and bi-iliac breadth)

Analysis	Sex	*r*	*p*	*n*
Data Set 2	Pooled	–0.047	0.237	640
Pilastric vs. bi-iliac breadth	Female	–0.082	0.274	181
	Male	–0.035	0.454	459
Data Set 2	Pooled	–0.015	0.701	636
Tibial midshaft vs. bi-iliac breadth	Female	0.019	0.797	180
	Male	0.010	0.840	456
Data Set 3 (African Americans)				
Femoral midshaft I_{max}/I_{min} vs. bi-iliac breadth	Male	–0.434	0.030[a]	25
Tibial midshaft I_{max}/I_{min} vs. bi-iliac breadth	Male	–0.248	0.232	25

These results address Prediction 2

[a]Indicates a correlation significant at $p<0.05$ or better

documented skeletons does not support this prediction; neither the pilastric index nor the midshaft tibial index are significantly correlated with bi-iliac breadth, and this does not change with subdivision by sex (Table 8.4). In contrast, a subset of Data Set 3 comprising cross-sectional geometry for 25 African American males supports the prediction. For the femur, the correlation between midshaft I_{max}/I_{min} and bi-iliac breadth is moderately strong ($r=-0.434$) and statistically significant, while the correlation for the tibia is weaker ($r=-0.266$) and nonsignificant, but is still in the predicted direction. Overall, this prediction receives mixed support, especially from cross-sectional geometry (Data Set 3). The prediction can be considered valid, at least in the case of cross-sectional geometry of the femur, but the overall influence of hip breadth on cross-sectional geometry of the femoral midshaft is weak to moderate.

Prediction 3, that there should be a negative correlation between the ratio of bi-iliac width to femoral length and the ratio of I_{max}/I_{min} (as well as the ratio of external AP to ML diameters) of each bone, was tested with Data Set 2 and the African Americans from Data Set 3. The results from Data Set 2 support this prediction for femoral shape in the pooled sample (and in both sexes when subdivided, although the magnitudes of the correlations are weak at $r \sim -0.14$), and in males the correlation reaches statistical significance only for tibial midshaft shape (Table 8.5). Although these correlations point in the expected direction, it is worth reiterating that their weakness does not indicate a robust relationship. Likewise, cross-sectional geometry and external measurements of 25 African American males supports this prediction (Table 8.4), but the relationships are not strong. For femoral midshaft I_{max}/I_{min}, the correlation with bi-iliac breadth divided by femoral length is negative, but weak and nonsignificant ($r=-0.347$), while for the tibia, the correlation with this measure of body shape is weaker ($r=-0.248$) and nonsignificant.

Prediction 4, that a negative correlation should exist between ratios of femoral to tibial I_{max}/I_{min} vs. the ratio of bi-iliac breadth to femoral length, a measure of "body shape," was tested using the African Americans in Data Set 3. These data did not

Table 8.5 Correlations between bi-iliac breadth divided by femoral length (a measure of "body shape") and the pilastric and midshaft tibial indices in Data Set 2 (documented collections, external measurements only) and in 25 African American males from Data Set 3 (cross-sectional geometry and external measurements)

Analysis	Sex	*r*	*p*	*n*
Data Set 2				
Pilastric index vs. body shape	Pooled	−0.142	0.0003[a]	640
	Female	−0.150	0.0439[a]	181
	Male	−0.144	0.0020[a]	459
Data Set 2	Pooled	−0.053	0.1821	631
Tibial midshaft index vs. body shape	Female	−0.081	0.2802	179
	Male	−0.098	0.0371[a]	452
Data Set 3				
Femoral midshaft I_{max}/I_{min} vs. body shape	Male	−0.347	0.089	25
Tibial midshaft I_{max}/I_{min} vs. body shape	Male	−0.248	0.232	25

These results address Prediction 3

[a]Indicates a correlation significant at $p<0.05$ or better

Table 8.6 Correlations among second moments of area and ratios of I_{max}/I_{min} and I_x/I_y in Data Set 4, a pooled-sex sample ($n=230$) of Iron Age Samnites from central Italy

		Femur		Tibia	
		I_{max}/I_{min}	I_x/I_y	I_{max}/I_{min}	I_x/I_y
Femur	I_{max}/I_{min}	1.000	0.345[a]	0.107	−0.026
	I_x/I_y	0.345[a]	1.000	−0.151[a]	−0.067
Tibia	I_{max}/I_{min}	0.107	−0.151[a]	1.000	0.122
	I_x/I_y	−0.026	−0.067	0.122	1.000

[a]Indicates a correlation significant at $p<0.05$ or better

support this prediction. For the 25 African American males in the data set of cross-sectional geometry and external measurements, the correlation between the two ratios was close to zero and nonsignificant ($r=-0.038$, $p=0.856$).

Data Set 4 provides a final opportunity to revisit the relationships among ratios of I_{max}/I_{min} and I_x/I_y in the femur and tibia. These data comprise cross-sectional properties of 230 Iron Age Samnites from central Italy (Sparacello 2013), and allow direct comparisons among these ratios (Table 8.6). Data Set 4 shows a weak ($r=0.107$) and nonsignificant correlation between values of I_{max}/I_{min} in the femur and tibia. If the values for I_x/I_y better capture the bones' responses to AP bending moments that routinely result from walking and running, as has been proposed by Ruff and others, then one would expect the ratios of I_x/I_y for the midshaft femur and tibia to show a stronger correlation. They do not. Among the Samnites, the correlation between I_x/I_y values is even weaker at $r=-0.067$ and nonsignificant. This example, for which both kinds of shape ratios are available, provides no support for the claim that ratios of I_x/I_y are preferable to ratios of I_{max}/I_{min}, but instead lends some support to the idea that ratios of I_x/I_y may obscure rather than clarify functional adaptations visible in midshaft sections of lower limb bones.

8.6 Discussion and Conclusions

Results of the analyses in this chapter support the idea that body shape, whether gauged by bi-iliac breadth alone or by the ratio of bi-iliac breadth to femoral length, influences the shape of the femoral cross section, but the magnitude of that influence is weak (accounting for only between 0 and 25 % of the variance in femoral shape) and sometimes inconsistent across the sexes or data sets. Ruff's (1995; Ruff et al. 2006) insight about the importance of body shape remains valid, but it does not function to provide a strong explanatory model for the variation that researchers observe in femoral midshaft shapes, or for the conflicting inferences about mobility that can sometimes be drawn from femoral and tibial cross sections. We need to explore some alternative hypotheses for these patterns, and several seem promising.

One alternative is that the contrasting patterns sometimes observed in the shapes of femoral and tibial midshafts are a product of the ontogenetic ages at which the individual engaged in activities. More specifically, it is possible that a high amount of early running or walking (in late childhood or during the adolescent growth spurt) induces femoral midshaft changes, while later activity is much less influential. At the same time, the tibia may provide a better gauge of patterns of mobility during adulthood. Pearson and Lieberman (2004) noted profound differences in modeling responses can occur across ontogeny. At least some support for this specific hypothesis comes from some preliminary work (Sparacello et al. 2010) on the ontogeny of cross-sectional shapes of the femur and tibia in Gravettian children, who obtained ~100 % of adult values for the pilastric index by 11.5 years, but only 80–90 % of the adult value of midshaft tibial shape by 14–16 years. Cowgill (2014) provides additional details regarding the ontogeny of cross-sectional shapes in seven groups of immature recent humans. She reports that some shape differences appear by approximately 6 years of age, while many appear only around 12 years of age or later and that those differences seem to conform only loosely to inferred disparities in group mobility.

A second alternative is that perhaps the midshaft femur and tibia record different types of activity; perhaps a substantial amount of relatively fast running or sprinting are required to produce a heavily pilastered femur, but a great deal of slower running or even walking will suffice to produce AP reinforcement of the tibia. As a case in point, Rantalainen et al. (2010) found differential effects of various kinds of athletic activities on bone segments in which the repetitive, low-impact loadings associated with running appear to have mainly influenced tibial robusticity, while high impact activities involving jumping influenced both the tibia and the fibula. There have been a few experimental tests of these effects in the tibia (e.g., Burr et al. 1996; Shaw and Stock 2009a; Rantalainen et al. 2010) and other bones (Shaw and Stock 2009b; Shaw et al. 2012), but much more experimental work is needed. Wescott's chapter (2014) explores the effects of some finer-grained measures of mobility and loss of mobility on femoral shape, and finds that there appear to be several paths to changes in femoral cross-sectional shape. On the subject of what forces may have shaped limb morphology, Wall-Scheffler's chapter (2014) offers the intriguing observation that a wide pelvis makes locomotion more efficient when carrying burdens over appreciable distances.

A third possibility is that walking and/or running produce different patterns (and directions) of bending stress in the midshaft femur and tibia. If so, it would make much more sense to model the bending regimes as accurately as possible rather than simply assume that the midshafts of both the tibia and femur experience bending in the same AP plane. There is already ample reason to doubt that this model of AP bending is accurate. Based on the distribution of section moduli around a midshaft femoral section, Pauwels (1980) showed that this section of the bone was best-adapted to resist bending along an axis from anterolateral to posteromedial. Some bone biologists accept that this is the direction of bending at the femoral midshaft (e.g., Martin et al. 1998), but most anthropologists have not adopted this view.

In the tibia, mechanical models suggest that the direction of maximum bending forces changes across heel-strike, stance, and toe-off (Kimura 1974; Wehner et al. 2009). Moreover, anthropologists have generally considered tibial I_{max} to be the best gauge of AP bending strains for the whole lower leg segment, which actually consists of two bones. The fibula certainly has a role in buffering bending strains (as do the ligaments and muscles of the lower leg), and the ratio between tibial and fibular robusticity as gauged by cross-sectional geometry appears informative about types of activity, i.e., running on a straight line or swerving (Marchi and Shaw 2011). As a complex comprising two bones, the leg segment appears to have more options in how it responds to various activities. This may decrease the strength of the correlations between tibial second moments of area and mobility.

In addition, a substantial amount of experimental work has demonstrated that bones do not always bend in the presumed orientations. Work by Demes and colleagues has shown that in vivo bending regimes in primate limb bones are often not what one would expect from an abstracted consideration of their locomotion, or from the cross-sectional shape of their bones (Demes et al. 1998, 2001; Lieberman et al. 2004; Demes 2007; Demes and Carlson 2009).

In a trenchant critique of the logic and clinical data that underlies inferences about prehistoric activities from osteoarthritis, entheses (or the so-called musculoskeletal stress markers), and cross-sectional geometry, Jurmain (1999) decried the lack of clear and consistent evidence that any of these features actually develop in response to the stimuli that anthropologists suppose they do. In a recent follow-up, Jurmain et al. (2012) noted that there is now more experimental validation for the idea that variations in cross-sectional geometry reflect patterns of activity, but even more experimental work is needed to understand their development. The results of new experimental studies promise to further inform and improve analyses of cross-sectional geometry and improve the factual underpinning of the inferences about the mobility and life ways of prehistoric people that we draw from them.

Acknowledgments Data collection by OMP was supported by an NSF predoctoral fellowship, the Wenner-Gren Foundation, Boise Fund, the University of New Mexico; FEG and OMP's collection of the cross-sectional data was supported by NSF DBS-9120117 to FEG. We are grateful to all of these institutions and to the many curators who facilitated access to their collections. We are grateful for a series of insightful criticisms by two anonymous reviewers and D. Marchi. Their critiques helped to improve this chapter.

References

Amtmann VE (1971) Mechanical stress, functional adaptation and the variation in structure of the human femur diaphysis. Ergeb Anat Entwicklungsgesch 441:1–89

Bridges PS (1989) Changes in activities with the shift to agriculture in the southeastern United States. Curr Anthropol 30:385–394

Burr DB, Milgrom C, Fyhrie D, Forwood M, Nyska M, Finestone A, Hoshaw S, Saiag E, Simkin A (1996) In vivo measurement of human tibial strains during vigorous activity. Bone 18:405–410

Carlson KJ, Grine FE, Pearson OM (2007) Robusticity and sexual dimorphism in the postcranium of modern hunter-gatherers from Australia. Am J Phys Anthropol 134:9–23

Cowgill LW (2014) Femoral diaphyseal shape and mobility: an ontogenetic perspective. In: Carlson KJ, Marchi D (eds) Reconstructing mobility: environmental, behavioral, and morphological determinants. Springer, New York

Daneshvari SR (2011) Effects of body mass on the skeleton. Ph.D. Dissertation, University of New Mexico, Albuquerque

Demes B (2007) In vivo bone strain and bone functional adaptation. Am J Phys Anthropol 133:717–722

Demes B, Carlson KJ (2009) Locomotor variation and bending regimes of capuchin limb bones. Am J Phys Anthropol 139:558–571

Demes B, Stern JTJ, Hausman MR, Larson SG, McLeod KJ, Rubin CT (1998) Patterns of strain in the macaque ulna during functional activity. Am J Phys Anthropol 106:87–100

Demes B, Qin Y-X, Stern JTJ, Larson SG, Rubin CT (2001) Patterns of strain in the macaque tibia during functional activity. Am J Phys Anthropol 116:257–265

Endo B, Kimura T (1970) Postcranial skeleton of the Amud man. In: Suzuki H, Takai F (eds) The Amud man and his cave site. Academic Press of Japan, Tokyo, pp 231–406

Grine FE, Jungers WL, Tobias PV, Pearson OM (1995) Fossil Homo femur from Berg Aukas, northern Namibia. Am J Phys Anthropol 97:151–185

Holt BM (2003) Mobility in Upper Paleolithic and Mesolithic Europe: evidence from the lower limb. Am J Phys Anthropol 122:200–215

Jurmain R (1999) Stories from the skeleton: behavioral reconstruction in human osteology. Gordon and Breach, Amsterdam

Jurmain R, Alves Cardoso F, Henderson C, Villotte S (2012) Bioarchaeology's holy grail: the reconstruction of activity. In: Grauer AL (ed) A companion to paleopathology. Wiley-Blackwell, Chichester, West Sussex, pp 531–552

Kimura T (1971) Cross-section of human lower leg bones viewed from the strength of materials. J Anthropol Soc Nippon 79:323–336

Kimura T (1974) Mechanical characteristics of human lower leg bones. J Fac Sci Univ Tokyo V Anthropol 4:319–393

Kimura T, Amtmann E (1984) Distribution of mechanical robustness in the human femoral shaft. J Biomech 17:41–46

Kimura T, Takahashi H (1982) Mechanical properties of cross section of lower limb bones in Jomon man. J Anthropol Soc Nippon 90:105–118

Larsen CS, Ruff C (2011) 'An external agency of considerable importance': the stresses of agriculture in the foraging-to-farming transition in eastern North America. In: Pinhasi R, Stock JT (eds) Human bioarchaeology of the transition to agriculture. Wiley-Blackwell, Chichester, West Sussex, pp 293–316

Larsen CS, Crosby AW, Griffin MC, Hutchinson DL, Ruff CB, Russell KF, Schoeninger MJ, Sering LE, Simpson SW, Takács JL, Teaford MF (2002) A biohistory of health and behavior in the Georgia Bight: the agricultural transition and the impact of European contact. In: Steckel RS, Rose JC (eds) The backbone of history: health and nutrition in the western hemisphere. Cambridge University Press, Cambridge, pp 406–439

Lieberman DE, Pearson OM, Polk JD, Demes B, Crompton AW (2003) Optimization of bone growth and remodeling in response to loading in tapered mammalian limbs. J Exp Biol 206:3125–3138
Lieberman DE, Polk JD, Demes B (2004) Prediction long bone loading from cross-sectional geometry. Am J Phys Anthropol 123:156–171
Lovejoy CO, Trinkaus E (1980) Strength and robusticity of the Neandertal tibia. Am J Phys Anthropol 53:465–470
Lovejoy CO, Burstein AH, Heiple KG (1976) The biomechanical analysis of bone strength: a method and its application to platycnemia. Am J Phys Anthropol 44:489–505
Marchi D, Shaw CN (2011) Variation in fibular robusticity reflects variation in mobility patterns. J Hum Evol 61:609–616
Marchi D, Sparacello V, Shaw C (2011) Mobility and lower limb robusticity of a pastoralist Neolithic population from north-western Italy. In: Pinhasi R, Stock JT (eds) Human bioarchaeology of the transition to agriculture. Wiley-Blackwell, Chichester, West Sussex, pp 317–346
Martin R, Saller K (1956) Lehrbuch der Anthropologie. Gustav Fischer, Stuttgart
Martin BD, Burr DB, Sharkey NA (1998) Skeletal tissue mechanics. Springer, New York
Pauwels F (1980) Biomechanics of the locomotor apparatus. Springer, Berlin
Pearson OM (1997) Postcranial morphology and the origin of modern humans. Ph.D. thesis, Stony Brook University, Stony Brook
Pearson OM (2000) Activity, climate, and postcranial robusticity: implications for modern human origins and scenarios of adaptive change. Curr Anthropol 41:569–607
Pearson OM, Grine FE (1996) Cortical thickness and relative bending moments in human long bones: correlations among elements [abstract]. Am J Phys Anthropol Suppl 22:S183
Pearson OM, Lieberman DE (2004) The aging of Wolff's "Law:" ontogeny and responses to mechanical loading in cortical bone. Yearb Phys Anthropol 47:63–99
Pearson OM, Millones M (2005) Rasgos esqueléticos de adaptación al clima y a la actividad entre los habitantes aborígenes de Tierra del Fuego. An Inst Patagon Ser Cienc Hum 33:37–51
Pearson OM, Cordero RM, Busby AM (2006) How different were Neanderthals' habitual activities? A comparative analysis with diverse groups of recent humans. In: Harvati K, Harrison T (eds) Neandertals revisited: new approaches and perspectives. Springer, New York, pp 89–112
Rantalainen T, Nikander R, Heinonen A, Suominen H, Sievänen H (2010) Direction-specific diaphyseal geometry and mineral mass distribution of tibia and fibula: a pQCT study of female athletes representing different exercise loading types. Calcif Tissue Int 86:447–454
Ruff C (1987) Sexual dimorphism in human lower limb bone structure: relationship to subsistence strategy and sexual division of labor. J Hum Evol 16:391–416
Ruff CB (1991) Climate and body shape in hominid evolution. J Hum Evol 21:81–105
Ruff CB (1994a) Morphological adaptation to climate in modern and fossil hominids. Yearb Phys Anthropol 37:65–107
Ruff CB (1994b) Biomechanical analysis of northern and southern plains femora: behavioral implications. In: Owsley DW, Jantz RL (eds) Skeletal biology in the Great Plains: migration, warfare, health, and subsistence. Smithsonian Institution Press, Washington, pp 235–245
Ruff CB (1995) Biomechanics of the hip and birth in early Homo. Am J Phys Anthropol 98:527–574
Ruff CB (1999) Skeletal structure and behavioral patterns of prehistoric Great Basin populations. In: Hemphill BE, Larsen CS (eds) Prehistoric lifeways in the Great Basin wetlands: bioarchaeological reconstruction and interpretation. University of Utah Press, Salt Lake City, pp 290–320
Ruff CB (2002) Long bone articular and diaphyseal structure in Old World monkeys and apes. I: locomotor effects. Am J Phys Anthropol 119:305–342
Ruff CB (2008) Biomechanical analyses of archaeological human skeletons. In: Katzenberg MA, Saunders SR (eds) Biological anthropology of the human skeleton, 2nd edn. Wiley-Liss, New York, pp 183–206
Ruff C (2010) Structural analysis of postcranial skeletal remains. In: Morgan M (ed) Pecos Pueblo revisited: the biological and social context (Papers of the Peabody Museum of Peabody

Museum of Archaeology and Ethnology, Harvard University), vol 85. Peabody Museum of Peabody Museum of Archaeology and Ethnology, Harvard University, Cambridge, MA, pp 93–108

Ruff CB, Hayes WC (1983) Cross-sectional geometry of Pecos Pueblo femora and tibiae—a biomechanical investigation: I. Method and general patterns of variation. Am J Phys Anthropol 60:359–381

Ruff CB, Larsen CS, Hayes WC (1984) Structural changes with the transition to agriculture on the Georgia Coast. Am J Phys Anthropol 64:125–136

Ruff CB, Holt BM, Sladek V, Berner M, Murphy WA Jr, zur Nedden D, Seidler H, Recheis W (2006) Body size, body proportions, and mobility in the Tyrolean "Iceman". J Hum Evol 51:91–101

Shackleford LL (2007) Regional variation in the postcranial robusticity of Late Upper Paleolithic humans. Am J Phys Anthropol 133:655–668

Shang H, Tong H, Zhang S, Chen F, Trinkaus E (2007) An early modern human from Tianyuan Cave, Zhoukoudian, China. Proc Natl Acad Sci 104:6573–6578

Shaw CN, Stock JT (2009a) Intensity, repetitiveness, and directionality of habitual adolescent mobility patterns influence the tibial diaphysis morphology of athletes. Am J Phys Anthropol 140:149–159

Shaw CN, Stock JT (2009b) Habitual throwing and swimming correspond with upper limb diaphyseal strength and shape in modern human athletes. Am J Phys Anthropol 140:160–172

Shaw CN, Stock JT (2011) The influence of body proportions on femoral and tibial midshaft shape in hunter-gatherers. Am J Phys Anthropol 144:22–29

Shaw CN, Hofmann CL, Petraglia MD, Stock JT, Gottschall JS (2012) Neandertal humeri may reflect adaptation to scraping tasks, but not spear thrusting. PLoS One 7:e40349

Sparacello VS (2013) The bioarchaeology of changes in social stratification, warfare, and habitual activities in the Iron Age Samnites of Central Italy. Ph.D. thesis, University of New Mexico, Albuquerque

Sparacello V, Marchi D (2008) Mobility and subsistence economy: a diachronic comparison between two groups settled in the same geographical area (Liguria, Italy). Am J Phys Anthropol 136:485–495

Sparacello VS, Pearson OM (2010) The importance of accounting for the area of the medullary cavity in cross-sectional geometry: a test based on the femoral midshaft. Am J Phys Anthropol 143:612–624

Sparacello VS, Pearson OM, Cowgill L (2010) Growing up in the Gravettian: ontogeny of cross-sectional geometry in the lower limb [abstract]. Am J Phys Anthropol Suppl 50:S221

Stock JT (2006) Hunter-gatherer postcranial robusticity relative to patterns of mobility, climatic adaptation, and selection for tissue economy. Am J Phys Anthropol 131:194–204

Stock J, Pfeiffer S (2001) Linking structural variability in long bone diaphyses to habitual behaviors: foragers from the southern African Later Stone Age and the Andaman Islands. Am J Phys Anthropol 115:337–348

Stock JT, Pfeiffer SK (2004) Long bone robusticity and subsistence behaviour among Later Stone Age foragers of the forest and fynbos biomes of South Africa. J Archaeol Sci 31:999–1013

Trinkaus E (2009) The human tibia from Broken Hill, Kabwe, Zambia. PaleoAnthropology 2009:145–165

Trinkaus E (2011) The postcranial dimensions of the La Chapelle-aux-Saints 1 Neandertal. Am J Phys Anthropol 145:461–468

Trinkaus E, Ruff CB (1999a) Diaphyseal cross-sectional geometry of Near Eastern Middle Paleolithic humans: the femur. J Archaeol Sci 26:409–424

Trinkaus E, Ruff CB (1999b) Diaphyseal cross-sectional geometry of Near Eastern Middle Palaeolithic humans: the tibia. J Archaeol Sci 26:1289–1300

Trinkaus E, Ruff CB (2012) Femoral and tibial diaphyseal cross-sectional geometry in Pleistocene Homo. PaleoAnthropology 2012:13–62

Wall-Scheffler C (2014) The balance between burden carrying, variable terrain and thermoregulatory pressures in assessing morphological variation. In: Carlson KJ, Marchi D (eds)

Reconstructing mobility: environmental, behavioral, and morphological determinants. Springer, New York

Wehner T, Lutz C, Ulrich S (2009) Internal loads in the human tibia during gait. Clin Biomech 24:299–302

Wescott DJ (2014) The relationship between femur shape and terrestrial mobility patterns. In: Carlson KJ, Marchi D (eds) Reconstructing mobility: environmental, behavioral, and morphological determinants. Springer, New York

Chapter 9
Variation in Mobility and Anatomical Responses in the Late Pleistocene

Laura L. Shackelford

Abstract Diachronic changes in European Late Pleistocene humans indicate anatomical changes in the lower limb associated with decreased mobility. A more global perspective suggests that similar trends toward reduced mobility occurred simultaneously in other parts of the Old World, but Late Pleistocene populations in non-European regions demonstrate significant variation in femoral and tibial cross-sectional geometric properties that are inconsistent with behavioral interpretations.

Samples of Late Pleistocene early modern humans from Europe, Northern Africa, and Asia are analyzed to assess regional variation in postcranial trends. Cross-sectional geometric properties for midshaft femora and tibiae and measures of articular surface areas and mechanical efficiency are evaluated between samples.

Regional Late Pleistocene samples are differentiated by measures of diaphyseal robusticity. Northern African samples are uniquely robust, particularly at the level of the midshaft tibia. Relative to other regional samples, the Asian sample has distinctly gracile femoral and tibial diaphyses. Although this may indicate reduced mobility, this sample also demonstrates relatively high mechanical efficiency at the knees and hips, which may point to an alternative mechanism for counteracting loading on the lower limbs from high mobility or terrain differences.

Keywords Cross-sectional geometry • Postcrania • Upper Paleolithic • Southeast Asia • North Africa • Robusticity

L.L. Shackelford (✉)
Department of Anthropology, University of Illinois at Urbana-Champaign,
607 S. Mathews Avenue, MC-148, 209L Davenport Hall, Urbana, IL 61801, USA
e-mail: llshacke@illinois.edu

K.J. Carlson and D. Marchi (eds.), *Reconstructing Mobility: Environmental, Behavioral, and Morphological Determinants*, DOI 10.1007/978-1-4899-7460-0_9,

9.1 Introduction

Early modern humans from the European Upper Paleolithic (ca. 35–10 ka) show changes in lower limb robusticity and cross-sectional shape through time (Holt 2003). A decrease in femoral and tibial robusticity, particularly in an anteroposterior (AP) direction, is associated with a decrease in mobility coinciding with the last glacial maximum, and this interpretation is reinforced by archaeological, faunal, and climatological data (Straus 1995; Churchill et al. 2000; van Andel et al. 2003; Holt and Formicola 2008). Archaeological data from available Late Pleistocene (ca. 20–10 ka) sites in North Africa and Southeast Asia suggest that similar trends towards reduced mobility and changing subsistence patterns occurred throughout the Old World (Gorman 1970; Hillman et al. 1989; Anderson 1990; Close and Wendorf 1990; Hershkovitz et al. 1995; Bar-Yosef 2002), but populations in these non-European regions demonstrate significant variation in femoral and tibial cross-sectional geometric properties after the glacial maximum (Shackelford 2005, 2007). Fossils from northern Africa have been described as having particularly robust postcranial remains due in part to a nomadic existence that involved large game hunting and a high protein diet (Smith 1979; Lahr and Arensburg 1995). In contrast, Late Pleistocene Southeast Asian fossils are unique within this wider geographic perspective, demonstrating low levels of strength and robusticity. This is particularly unusual since recent studies have found correlations between long bone robusticity at the femoral and tibial midshafts and rugged or mountainous terrain, a geographic feature specific to the majority of the Southeast Asian sample (Marchi et al. 2006, 2011; Marchi 2008). Given this anatomical variation, it is less clear how these comparisons of cross-sectional geometric properties should be interpreted from a behavioral or subsistence strategy perspective.

Since the lower limb is used primarily for locomotion, differences in lower limb robusticity are most commonly explained as differences in the intensity or repetitiveness of terrestrial mobility. In this context, a traditional definition of mobility as the daily movement of an individual across a landscape is used (Kelly 1992, 1995). Decades of experimental research have documented the sensitivity of bones to habitual loading and demonstrated that the skeleton can remodel to provide both peak load resistance and fatigue resistance to structural overloading (Cochran 1972, 1974; Lanyon et al. 1975; Biewener et al. 1981; Biewener and Bertram 1993; Forwood and Turner 1994, 1995; Mosley et al. 1997; Robling et al. 2001; Saxon et al. 2005; Goodship et al. 2009). However, increasingly refined experimental research has demonstrated that there is a subset of strains that has the greatest effect on the adaptive response. While high frequency, low magnitude loadings can create an osteogenic response, bone is highly tuned to its customary loading pattern and the disturbance of this pattern is important for adaptive bone remodeling (Lanyon et al. 1979; O'Connor and Lanyon 1982; Lanyon and Rubin 1984; Rubin and Lanyon 1984; Biewener and Bertram 1993; Judex and Zernicke 2000a; Goodship et al. 2009; Judex and Rubin 2010). As such, a loading environment that produces high magnitude, dynamic strains in unusual distributions is more effective in initiating adaptive responses than is a less active, static, or routine strain environment (Lanyon and O'Connor 1980; Lanyon et al. 1982; Lanyon and Rubin 1984; Rubin and Lanyon 1985; Biewener and Bertram 1993; Turner 1998; Judex and Zernicke 2000b; Ehrlich and Lanyon 2002).

Noninvasive exercise research has demonstrated similar results in humans, with high impact exercises such as squash, cricket, tennis, and gymnastics leading to greater bone remodeling than running, cycling, or swimming (King et al. 1969; Jones et al. 1977; Fehling et al. 1995; Robinson et al. 1995; Burr et al. 1996; Haapasalo et al. 1996, 2000; Frost 1997; Shaw and Stock 2009). Associations between habitual activity and localized osteogenic response in the femora and tibiae of humans have also been documented through comparisons of athletes and control subjects (Shaw and Stock 2009; Marchi and Shaw 2011) and in Holocene populations with known subsistence patterns (Stock and Pfeiffer 2001, 2004; Weiss 2003; Stock 2006). These interpretations are complicated, however, by additional factors. Multiple studies have identified variable effects of terrain on lower limb robusticity (Ruff 1999; Marchi et al. 2006, 2011; Marchi 2008; Marchi and Shaw 2011). While diaphyseal shape is often used to infer behavioral activities in past populations, the type of movement performed may also play a role in bone remodeling (Carlson and Judex 2007; Judex and Carlson 2009; Marchi and Shaw 2011). Furthermore, different diaphyseal levels may be variably influenced by body proportions or may have greater associations with mobility than others (Ruff 2000; Stock 2006; Shaw and Stock 2011).

An additional consideration in the analysis of robusticity is the variation and plasticity of other skeletal regions including articular surface areas (ASAs) and the mechanical advantage of muscles. While subarticular trabecular structure is the primary joint feature that is generally associated with bone adaptation to mechanical loading, articular size and shape are also affected by the mechanical environment. During the active growth phase, size and shape changes occur in articular surfaces to accommodate loading, although the extent to which adult ASAs are affected by mechanical loading is unclear (Carter and Wong 1990; Hamrick 1999; Carter and Beaupré 2001; Plochocki 2004; Plochocki et al. 2006). In a 2001 study, Lieberman et al. found no ASA response to mechanical loading in juvenile, subadult or adult sheep even though changes in diaphyseal cross sections were observed. This and other research has concluded that ASAs are highly constrained due to the necessity of maintaining joint congruence (Ruff et al. 1991; Lieberman et al. 2001). However, analyses of upper (proximal and distal humerus, proximal and distal radius, metacarpal II) and lower (proximal and distal femur, proximal and distal tibia) limb ASAs have found directional asymmetry in joint dimensions, suggesting phenotypic plasticity exists even after the active growth phase ends (Plochocki 2004; Lazenby et al. 2008). A lack of total constraint is further indicated by studies of malalignment at the knee joint, which indicate that mechanical loading leads to functional adaptation in ASAs even into advanced age (Eckstein et al. 2009).

Finally, skeletal features have been interpreted with respect to their mechanical advantage acting around a joint. In response to the observation that Neandertals had posteriorly displaced tibial condyles as well as AP-thick patellae, Miller and Gross (1998) proposed that they had exceptionally efficient quadriceps muscles during knee extension because these skeletal features functioned to increase the *M. quadriceps femoris* moment arm. Other researchers have countered this argument based on body proportions (Trinkaus and Rhoads 1999), but similar skeletal dimensions have been used to approximate the efficiency of the gluteal abductors at the hip joint (Ruff 1995), the Achilles tendon at the ankle joint (Miller and Gross 1998; Raichlen et al. 2011), as well as multiple upper limb joints (Trinkaus and Churchill 1988; Trinkaus and Villemeur 1991; Churchill and Rhodes 2006).

Variation in levels of robusticity of Late Pleistocene Old World fossils implies varied levels of mechanical loading between them. However, the small sample from Southeast Asia, the majority of which comes from Tam Hang in northern Laos ($N=6$), appears unexpectedly gracile when measured by one common standard—that of cross-sectional geometry—despite originating from a rugged, mountainous environment. This study aims to consider robusticity in a larger context in order to get a better picture of overall strength of the lower limbs of Late Pleistocene populations by quantifying and comparing measures of cross-sectional geometry, ASAs and muscular efficiency.

9.2 Materials

The Late Pleistocene (LP) sample comprises fossils from Europe, Africa, and Asia dated from ca. 20–10 ka and is divided regionally (Table 9.1). Late Pleistocene rather than Late Upper Paleolithic is used to describe the samples temporally because Upper Paleolithic is a European designation that does not aptly describe the cultural traditions of other geographic regions. The European LP sample includes fossils from France, Germany, Italy, England, and Switzerland, identified as Epipaleolithic and occasionally early Mesolithic. The Asian LP sample includes fossils from sites in Japan (Minatogawa) and Laos (Tam Hang).

The African LP sample is divided into two regional samples for analysis: Mediterranean and Nile Valley. The Mediterranean sample includes fossils from two northern African sites—Afalou-Bou-Rhummel, Algeria and the ossuary site from Taforalt, Morocco—and several individuals from the Levant (El Wad, Kebara, Neve David, Ein Gev I). The Nile Valley sample includes fossils from the Epipaleolithic sites of Jebel Sahaba and Wadi Halfa in Sudan. Initially, fossils from the Mediterranean and Nile Valley were treated as a single sample representing North Africa. Paired *t*-tests on features of body size, proportions, and robusticity (results not reported), however, indicated that differences between them warranted separation into two distinct groups in further analyses.

All specimens were determined to be adults based on epiphyseal fusion and verified by dental eruption, when associated crania were available. Data were collected on one femur and one tibia from each individual, as available. Males and females were analyzed separately, but due to very small sample sizes (particularly for females), statistical power of single-sex analyses rarely permitted significance. Thus, males and females were combined for statistical analyses, and single-sex and combined-sex results were found to be similar. As such, only combined-sex results are presented.

Table 9.1 Fossils included in analyses

European LP[a]	Asian LP	Mediterranean LP	Nile Valley LP
$N=27$ ($m=16$; $f=8$; ind=3)	$N=10$ ($m=3$; $f=7$)	$N=39$ ($m=11$; $f=5$; ind=23)[b]	$N=39$ ($m=19$, $f=18$, ind=2)
Arene Candide 2, 4[c], 5, 10[c],12[c]	Minatogawa[d] 1,2,3,4	Afalou 1, 2, 3, 11, 13, 25, 27, 28	Jebel Sahaba 4, 5, 6, 10, 15, 16, 17, 18, 19, 20, 21, 22, 25, 26, 28, 29, 31, 33, 38, 39, 40, 41, 42, 102
Le Bichon 1	Tam Hang 20534, 20535, 20536, 20537, 20538, 20540	El Wad 10252, 10256, 10259, 10260, 10263, 10269, 10290	Wadi Halfa 1, 3, 9, 11, 14, 24, 25, 26, 28, 31, 32, 34, 36, 37, 317
Bruniquel 24[c]		Kebara 10352	
Cap Blanc 1[c]		Taforalt XV8, 9, 15, 17, 26, 28; XVI4, 5, 9, 15, 21; XXVII; XXV4, 5, 13, 22, 26; XIX4, 17; XX1	
Chancelade 1[c]		Ein Gev 1	
Farincourt 1		Neve David[e]	
Gough's Cave 1		Ohalo 2[e]	
Grotte-des-Enfants 3[c]			
Laugerie-Basse 9[e], 51928[c]			
Neussing 2[c]			
Oberkassel 1[c,e], 2[c,e]			
Le Peyret 5, 6			
Le Placard 15			
Riparo Continenza 1[c]			
Riparo Tagliente 1[c]			
Rochereil[e]			
Romanelli 1[c]			
St. Germaine-La-Rivière 4[c]			
San Teodoro 4[c]			
Veyrier 1[c]			

[a]Late Pleistocene
[b]Individuals of indeterminate sex include isolated bones from the ossuary site of Taforalt
[c]Data from Holt (1999)
[d]Data from Baba and Endo (1982)
[e]Data from Erik Trinkaus

9.3 Methods

Comparative data included cross-sectional geometric properties of femoral and tibial diaphyses, ASAs for the proximal femur and distal tibia, and approximations of muscle moment arms for the *M. quadriceps femoris* at the knee and the gluteal abductors at the hip.

9.3.1 Cross-Sectional Geometric Properties

Long bone robusticity was quantified by the application of beam theory, which permits using cross-sectional geometric properties (Ruff and Hayes 1983; Ruff et al. 1993; Shackelford 2007; Stock and Shaw 2007). Cortical bone robusticity was analyzed at the midshaft (50 %) femur and midshaft (50 %) tibia. Cross sections were reconstructed by combining external contour molds with biplanar radiography (Ruff et al. 1993; a detailed description of methods can be found in Shackelford 2005). Reconstructed cross sections were evaluated using a PC version of SLICE (Eschman 1992). When available, data from published sources were used to prevent excessive wear and tear on fossils (Table 9.1).

Cross-sectional properties used for comparing samples included cortical area, second moments of area, and polar moment of area. Cortical area (CA) approximates the axial compressive strength of a bone. Second moments of area, or area moments of inertia (I), were used to determine the bending rigidity of a bone in a single plane. The sum of any two second moments of area calculated about orthogonal axes is the polar moment of area (J), a measure of torsional strength and twice average bending rigidity (Nordin and Frankel 2000; Daegling 2002), which is considered to be the most relevant indicator of a bone's performance under a variety of loads (Lieberman et al. 2004).

For the femur, a ratio of second moments of area in AP and mediolateral (ML) planes (I_x and I_y, respectively) was calculated to assess the distribution of bone at the midshaft. For the tibia, diaphyseal "shape" was calculated as a ratio of the maximum second moment of area (I_{max}) to the minimum second moment of area (I_{min}). Unlike I_x and I_y, I_{max} and I_{min} are independent of orientation, although the shape of the tibia tends to orient I_{max} in a more AP than ML direction (Ruff and Hayes 1983). These measures were used for the tibia rather than I_x and I_y because the former are less prone to error due to slight variations in positioning than the latter (Shaw and Stock 2009).

9.3.2 Articular Surface Dimensions

Articular surface dimensions of the proximal femur were approximated using a linear measurement, the superior-inferior (SI) femoral head diameter (M-18, Brauer 1988). An estimation of the femoral head ASA from linear dimensions that modeled the joint as a partial sphere would provide a more accurate estimate of surface area (Ruff 2002). Unfortunately, these measurements were not available for the majority of specimens in the current analysis. However, Ruff (2002) compared the results of reconstructions of joint surface areas and simpler, single linear measurements. For some joints and some questions, linear measurements were more useful than joint surface areas. At the proximal femur, however, surface area and the femoral head SI diameter gave slightly different information, primarily with respect to potential for hip abduction, but this finding generally applied to between-species comparisons (Ruff 2002).

The ASA of the ankle was estimated by modeling the distal tibial articulation as a rectangle and multiplying the talar articular breadth by the average of the medial and lateral talar articular depths.

9.3.3 Muscle Moment Arms

The quadriceps moment arm was estimated following Trinkaus and Rhoads (1999) as the AP thickness of the patella (M-3, Brauer 1988), which functions as a proxy for the displacement of the patellar tendon. Alternatively, the quadriceps moment arm was approximated by a measure of tibial condylar displacement, defined as "the anteroposterior distance, perpendicular to the diaphyseal axis, from the anterior surface of the tibial tuberosity to the line between the anteroposterior middles of the tibial condyles" (Trinkaus and Rhoads 1999, p. 836).

A modified measure of the femoral neck length has been used to approximate the moment arm for the gluteal abductors about the hip joint in fossil and modern humans (Ruff 1995). Femoral biomechanical neck length was measured as the distance from the femoral diaphyseal axis to the most proximal point on the femoral head (Ruff 1995, Trinkaus personal communication).

9.3.4 Body Mass Estimation

Body mass was estimated for each individual in order to standardize measurements of interest for comparisons across samples. Body mass was estimated using stature and bi-iliac breadth, given that this is a nonmechanical assessment (Ruff 1991, 1994, 2000). Using this method resulted in a slightly smaller sample size than would have been possible if body mass was estimated from femoral head diameter, but since ASAs were of interest for study, this was avoided.

9.3.5 Statistical Standardization

Logged cortical areas were regressed on logged body mass and logged second moments of area and polar moments of area were regressed on logged body mass multiplied by bone length^2 using OLS regression (Ruff 2000; Ruff et al. 2006). Standardized residuals of the samples were compared to determine if population differences were present. Since sample sizes were small and did not meet all normality assumptions, differences between samples in cortical properties were evaluated using nonparametric Kruskal–Wallis tests, followed by Mann–Whitney U-tests to determine where sample differences occurred. Ratios of second moments of area were evaluated using nonparametric Kruskal–Wallis tests, followed by Mann–Whitney U-tests to determine where sample differences occurred.

Similar methods were applied to ASAs and moment arm lengths. To remove the effects of body mass on ASAs, logged ASAs were regressed on logged body mass, following the expectation that compressive stresses on ASAs are proportional to body mass (Lieberman et al. 2001). Logged femoral neck length (gluteal abductor moment arm proxy) was regressed on logged femoral length (Ruff 1995). For the quadriceps moment arm proxies (patellar thickness and tibial condylar displacement), the logged skeletal measurement was regressed on logged body mass multiplied by the body weight moment arm (Trinkaus and Rhoads 1999). Standardized residuals from each regression were analyzed as described above.

9.4 Results

Summary statistics for variables used in the current analyses are provided in Table 9.2. Results of pair-wise comparisons of standardized residuals are provided in Table 9.3 and 9.4.

Comparisons of femoral cross-sectional properties between samples show that the African samples have the highest levels of strength and rigidity, the European sample is intermediate and the Asian sample is the most gracile (Fig. 9.1 and Table 9.3). Axial strength (CA) in both the Mediterranean and Nile Valley samples is significantly greater than in the Asian sample, but there are no other significant differences between samples. In torsional rigidity (*J*), significant differences are found only between the Mediterranean and Asian samples. There are no shape differences at the femoral midshaft ($p = 0.2335$), indicating no significant variation in bending strength and torsional rigidity in an AP versus ML direction between samples (Table 9.3).

Mean axial strength (CA) at the tibial midshaft in the Asian sample is significantly lower than that of all other regional samples (Fig. 9.2 and Table 9.3), with no other significant between-sample differences identified. In average torsional rigidity (*J*), there is a significant difference only between the European and Asian samples, with the European sample again having a greater average value. The Mediterranean sample has the greatest value for the tibial shape ratio (I_{max}/I_{min}) (Table 9.2), and this value is significantly greater than that of the Asian or Nile Valley samples (Table 9.3).

Average ASA of the proximal femur distinguishes the Mediterranean sample from all other regional groups, as well as the European and Nile Valley samples (with the European sample having relatively greater ASA) (Fig. 9.3 and Table 9.4). There are few between-sample differences at the distal tibia, with only the European sample having a distal tibial ASA that is significantly smaller than the African samples (Fig. 9.3 and Table 9.4).

The quadriceps moment arm was estimated using two parameters, patellar thickness and tibial condylar displacement. When using patellar thickness as an estimate, the Asian sample has a significantly smaller average moment arm length than the European or Nile Valley samples (Fig. 9.4 and Table 9.4). When the moment arm

Table 9.2 Summary statistics for comparative samples

		European LP[a]	Asian LP	Mediterranean LP	Nile Valley LP
Femoral 50 % CA[b]	Mean	680.0	604.3	806.3	752.9
	s.d.	100.6	70.8	103.3	86.6
	N	21	10	38	30
Femoral 50 % *J*[c]	Mean	450.9	358.2	501.0	476.4
	s.d.	101.8	62.2	93.0	85.5
	N	20	9	37	27
Femoral 50 % I_x/I_y	Mean	1.34	1.14	1.30	1.24
	s.d.	0.26	0.20	0.31	0.29
	N	21	10	38	30
Tibial 50 % CA[b]	Mean	612.1	505.7	632.1	640.6
	s.d.	85.8	57.1	117.2	107.5
	N	17	8	17	23
Tibial 50 % *J*[c]	Mean	524.9	370.4	532.6	467.3
	s.d.	86.3	73.9	145.7	124.2
	N	16	8	17	21
Tibial 50 % I_{max}/I_{min}	Mean	2.60	2.28	2.89	2.36
	s.d.	0.53	0.41	0.59	0.63
	N	17	9	17	23
Femoral head surface area[b]	Mean	7.05	7.48	7.67	7.03
	s.d.	0.40	0.74	0.74	0.51
	N	19	7	20	24
Distal tibial articular area[b]	Mean	11.28	12.09	14.35	12.59
	s.d.	1.69	1.47	2.90	1.27
	N	21	5	12	20
Patellar thickness[d]	Mean	0.383	0.369	0.367	0.380
	s.d.	0.043	0.038	0.045	0.031
	N	7	6	9	14
TCD[d,e]	Mean	0.649	0.790	0.632	0.625
	s.d.	0.015	0.075	0.0120	0.085
	N	8	8	9	12
Femoral neck length[f]	Mean	8.9	11.9	8.4	8.0
	s.d.	–	1.4	0.6	0.8
	N	1	4	10	20

[a]Late Pleistocene
[b]Standardized by body mass
[c]Standardized by body mass × bone length^2
[d]Standardized by body mass × body weight moment arm (Trinkaus and Rhoads 1999)
[e]*TCD* tibial condylar displacement
[f]Standardized by femoral length (Ruff 2002)

was estimated using tibial condylar displacement, the Asian sample has the greatest average value and is significantly greater than the Nile Valley sample (Fig. 9.2 and Table 9.4). Similarly, in comparisons of the gluteal abductor moment arm at the hip, the Asian sample is unique in its relatively large moment arm, with significant differences between it and both the Mediterranean and Nile Valley samples.

Table 9.3 Results of Mann–Whitney pair-wise comparisons of standardized residuals of cross-sectional geometric properties. *Upper right*: femoral midshaft cross-sectional properties. *Lower left*: tibial midshaft cross-sectional properties

		European LP[1]	Asian LP	Mediterranean LP	Nile Valley LP	
				Femoral midshaft cross-sectional properties		
European LP			0.2628 0.5557 n.s.	<0.0001* 0.1481 n.s.	0.0095* 0.2324 n.s.	CA J Ix/Iy
Asian LP	CA J I_{max}/I_{min}	0.0445* 0.0297* 0.0846		<0.0001* 0.0378* n.s.	0.0004* 0.0860 n.s.	CA J Ix/Iy
Mediterr. LP	CA J I_{max}/I_{min}	0.4084 0.8712 0.2856	0.0133* 0.1374 0.0177*		0.0474* 0.5960 n.s.	CA J Ix/Iy
Nile Valley LP	CA J I_{max}/I_{min}	0.1629 0.1371 0.0710	0.0122* 0.4792 0.8999	0.4275 0.2522 0.0020*		
		Tibial midshaft cross-sectional properties				

[1] Late Pleistocene.
* Significant with multiple comparison corrections.

Table 9.4 Results of Mann–Whitney pair-wise comparisons of standardized residuals of ASAs and muscle moment arms. *Upper right*: articular surface areas. *Lower left*: muscle moment arms

		European LP[1]	Asian LP	Mediterr. LP	Nile Valley LP	
				Articular surface areas		
European LP			0.1047 0.3291	0.8153 0.0006*	0.0002* 0.0110*	Fem. head Distal tibia
Asian LP	Patella TCD[2] Fem neck	0.0383* 0.1893 0.2888		0.1348 0.0820	0.1907 0.4756	Fem. head Distal tibia
Mediterr. LP	Patella TCD Fem neck	0.7508 0.8852 0.5623	0.0875 0.0922 0.0074*		0.0004* 0.0450*	Fem. head Distal tibia
Nile Valley LP	Patella TCD Fem neck	0.3909 1.0000 0.3506	0.0288* 0.0018* 0.0040*	0.1966 0.9151 0.3538		
		Muscle moment arms				

[1] Late Pleistocene.
[2] Tibial condylar displacement.
* Indicates significant with multiple comparison corrections.

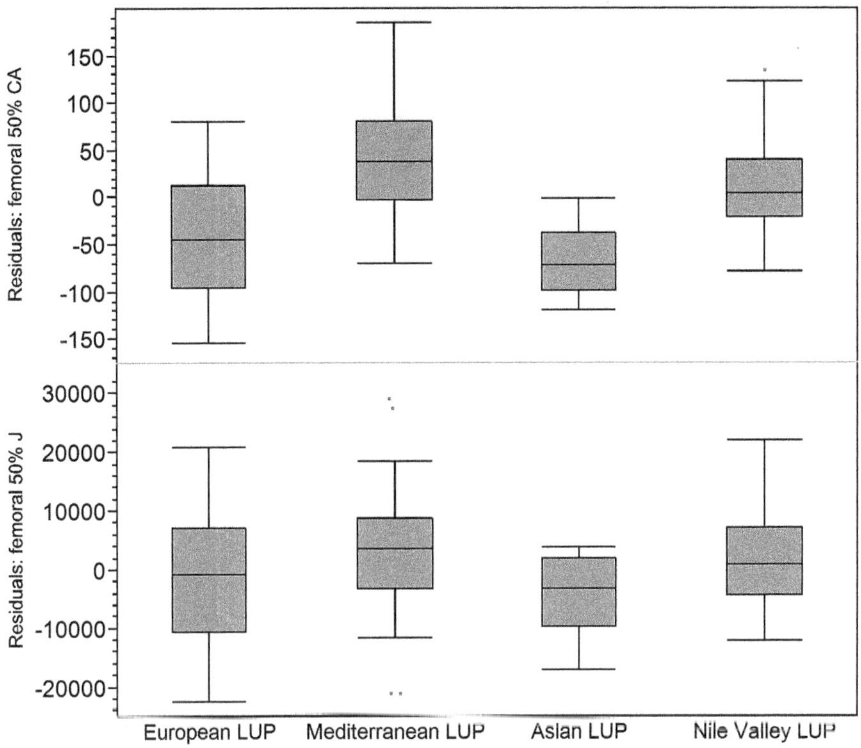

Fig. 9.1 Boxplots of standardized residuals for midshaft femoral cross-sectional properties. *Top*: cortical area; *bottom*: polar moment of area

9.5 Discussion

In these comparisons of LP samples across geographic regions, there are significant differences in diaphyseal strength and rigidity between samples from Europe, Africa (Nile Valley and circum-Mediterranean region) and Southeast Asia. In measures of polar moment of area, which is considered to be the most relevant indicator of a bone's performance under a variety of loads (Lieberman et al. 2004), SE Asians have lower levels of torsional rigidity and axial compressive strength at femoral and tibial midshafts than all other samples (Figs. 9.1, 9.2, Table 9.2). The Mediterranean sample has relatively high measures of bending and torsional strength relative to other regional samples, particularly at the tibial midshaft. Although AP strength at femoral and tibial midshaft sections has been associated with rugged or mountainous terrain (Marchi et al. 2006; Marchi 2008; Marchi and Shaw 2011), these measures are relatively low in the Southeast Asian sample, despite their environment.

In contrast, other measures of strength and robusticity in the Southeast Asian sample are approximately equal or relatively larger than in other regional samples.

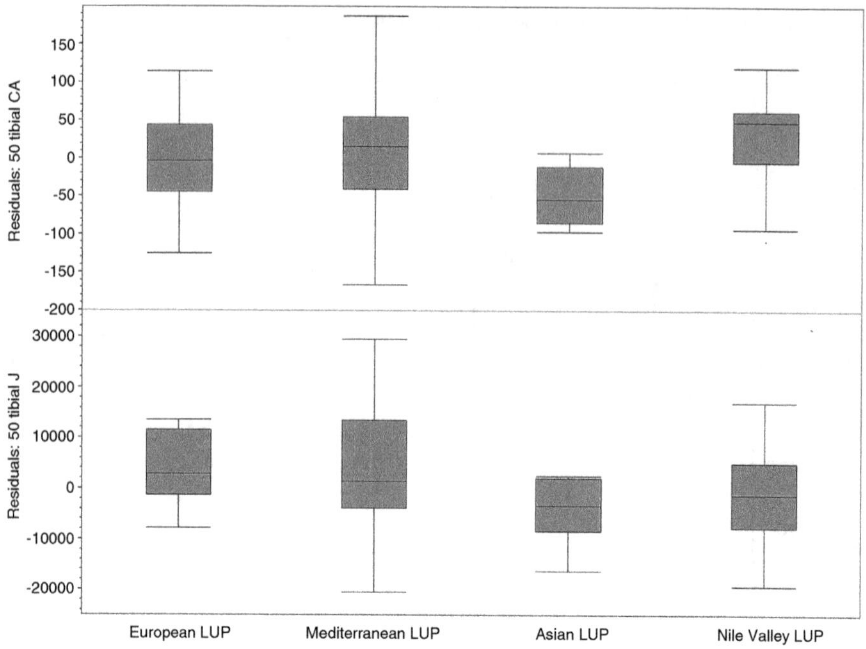

Fig. 9.2 Boxplots of standardized residuals for midshaft tibial cross-sectional properties. *Top*: cortical area; *bottom*: polar moment of area

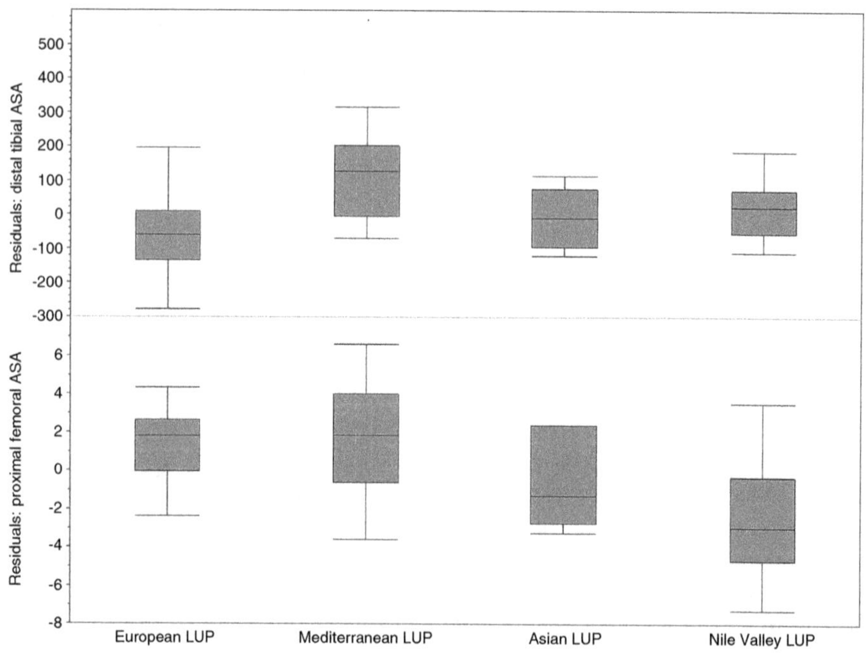

Fig. 9.3 Boxplots of standardized residuals for articular surface areas (ASA). *Top*: distal tibial ASA; *bottom*: proximal femoral ASA

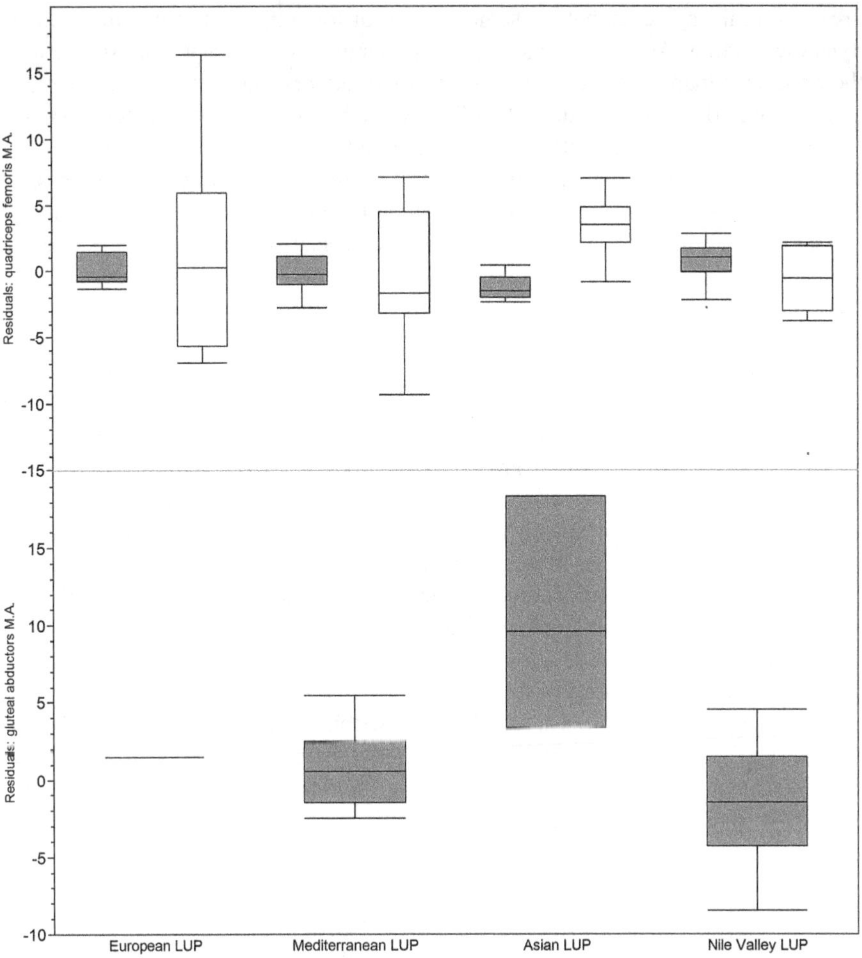

Fig. 9.4 Boxplots of standardized residuals for muscle moment arms (MA). *Top*: quadriceps MA estimated by patellar thickness (*gray boxes*) and tibial condylar displacement (*white boxes*); *bottom*: gluteal abductors MA estimated by femoral neck length

ASAs, for example, follow a different pattern than that seen in cross-sectional properties. At the proximal femur, the Mediterranean sample has the relatively largest ASA, the Asian and European samples are intermediate and the Nile sample has the relatively smallest ASA. A regression of the residuals from this analysis on body mass demonstrates this pattern even though an allometric effect is maintained ($p=0.007$).

In distal tibial ASA, the Mediterranean sample is again relatively large, the Asian and Nile Valley samples are intermediate and the European sample is relatively small, with significant differences between the European and each African sample. In humans, as in most mammals, distal limb segments are smaller than more proximal limb segments, but the same amount of force travels through them under dynamic loading. As a result, more distal joints experience greater stresses across their surfaces than do more proximal joints (Myers and Steudel 1985; Lieberman and Pearson 2001; Lieberman et al. 2001). As such, it follows that greater plasticity would be found in distal joints like the ankle relative to proximal joints like the hip. This pattern of regional variation suggests that the compressive stresses passing through the distal tibial articular surfaces are not mere reflections of body mass.

Despite the outcomes of comparisons of diaphyseal strength and ASAs, the SE Asian sample stands out in measures of mechanical efficiency, particularly for estimations of the gluteal abductors moment arm at the hip. Likewise, when estimated by tibial condylar displacement, the quadriceps moment arm is greatest in the Asian sample although this result disappears when the estimate is made using patellar thickness. This pattern in the SE Asian sample is consistent with a scenario of mechanical efficiency in the lower limbs that would result in lower external muscular forces and diaphyseal loading and consequently lower levels of cortical robusticity. While confirming this scenario requires a much more comprehensive and sophisticated evaluation of lower limb strength, initial data suggest that diaphyseal robusticity may be only one aspect of the anatomical response to loading, and that LP regional samples maintained similar levels of overall lower limb strength, but did so through different anatomical means.

One concern with these results—particularly the comparisons of cross-sectional geometric properties—stems from the combined-sex samples used for analysis and the unequal numbers of males and females within each regional sample. Many modern and prehistoric hunter-gatherer societies have a well-defined division of labor, and analyses of past populations have identified differences in long bone robusticity that are consistent with sex-specific activities (Ruff 1987; Holt et al. 2000). More sedentary, less mobile populations demonstrate reduced sexual dimorphism in the lower limbs, and this difference in lower limb dimorphism is less apparent in Late Pleistocene samples than in preglacial maximum or Holocene samples (Holt 2003). This combined-sex analysis, however, potentially confounds any such behavioral differences, particularly since the Southeast Asian sample is the only female-dominated sample.

The extent of variation in LP samples across the Old World is virtually unexplored for many aspects of the postcrania. As a first approach to a more comprehensive look at robusticity, this study takes an expanded view to include skeletal features that may be associated with alternative ways to generate efficient muscular activity and skeletal features that reflect greater constraint on the skeleton under mechanical loading. While this analysis is limited in its ability to estimate joint sizes and mechanical efficiency, it suggests that a more integrated and sophisticated method for evaluating these variables is necessary in order to capture relevant functional information. Just as important, however, is exploring the relationship between these aspects of the skeleton. Lieberman et al. (2001) found no changes in ASAs as a result of mechanical loading even when diaphyseal properties showed a significant response. As such, it was concluded that cross-sectional geometry may be appropriate for making behavioral inferences, while ASAs were better for estimating body mass at a species level because they are under tighter genetic or functional constraints (Lieberman et al. 2001). These different variables could, therefore, give complementary information. More recent research has demonstrated that aspects of ASAs maintain significant plasticity even into adulthood, calling into question these dichotomous interpretations (Plochocki 2004; Plochocki et al. 2006; Lazenby et al. 2008; Eckstein et al. 2009). Even with a reconsideration of how various skeletal features should be interpreted, this analysis is short-sighted in merely making associations between various methods available to promote efficient mobility. An understanding of the full range of mobility in LP populations will require an investigation of how muscular efficiency affects cortical bone robusticity, and whether these different mechanisms function to complement one another.

Acknowledgments I am grateful to Kris J. Carlson and Damiano Marchi for initiating the symposium leading to this collection of ideas on mobility. Thank you to two anonymous reviewers whose comments greatly improved this manuscript.

References

Anderson DD (1990) Lang Rongrien Rockshelter: a Pleistocene-early Holocene archaeological site from Krabi, Southwestern Thailand. University Museum, Philadelphia

Baba H, Endo B (1982) Postcranial skeleton of the Minatogawa man. In: Suzuki H, Hanihara K (eds) The Minatogawa man. The Upper Pleistocene Man from the Island of Okinawa. Bulletin of the University Museum 19. University of Tokyo, Tokyo, pp 61–195

Bar-Yosef O (2002) The Upper Paleolithic revolution. Annu Rev Anthropol 31:353–393

Biewener AA, Bertram JEA (1993) Skeletal strain patterns in relation to exercise training during growth. J Exp Biol 185:51–69

Biewener AA, Thomason J, Goodship AE, Lanyon LE (1981) The mechanics of horse locomotion-strains developed in the limb bones at different gaits. J Biomech 14:487

Brauer G (1988) Osteometrie. In: Knussman R (ed) Anthropologie I. Fischer, Stuttgart, pp 160–232

Burr DB, Milgrom C, Fyhrie D, Forwood M, Nyska M, Finestone A, Hoshaw S, Saiag E, Simkin A (1996) In vivo measurement of human tibial strains during vigorous activity. Bone 18:405–410

Carlson KJ, Judex S (2007) Increased non-linear locomotion alters diaphyseal bone shape. J Exp Biol 210:3117–3125

Carter DR, Beaupré GS (2001) Skeletal function and form. Cambridge University Press, New York

Carter DR, Wong M (1990) Mechanical stresses in joint morphogenesis and maintenance. In: Mow VC, Radcliffe A, Woo SLY (eds) Biomechanics of diarthrodial joints. Springer, New York, pp 155–174

Churchill SE, Rhodes JA (2006) How strong were the Neandertals? Leverage and muscularity at the shoulder and elbow in Mousterian foragers. Period Biol 4:457–470

Churchill SE, Formicola V, Holliday TW, Holt BW, Schumann BA (2000) The Upper Palaeolithic population of Europe in an evolutionary perspective. In: Roebroeks W, Mussi M, Svoboda J, Fennema K (eds) Hunters of the Golden Age: the mid Upper Paleolithic of Eurasia 30,000-20,000 BP. University of Leiden, Leiden, pp 31–58

Close AE, Wendorf F (1990) North Africa at 18000 BP. In: Gamble C, Soffer O (eds) The world at 18,000 BP, vol 2, Low latitudes. Unwin Hyman, London, pp 41–53

Cochran GVB (1972) Implantation of strain gages on bone in vivo. J Biomech 5:119–123

Cochran GVB (1974) A method for direct recording of electromechanical data from skeletal bone in living animals. J Biomech 7:563–565
Daegling D (2002) Estimation of torsional rigidity in primate long bones. J Hum Evol 43:229–239
Eckstein F, Hudelmaier M, Cahue S, Marshall M, Sharma L (2009) Medial-to-lateral ratio of tibiofemoral subchondral bone area is adapted to alignment and mechanical load. Calcif Tissue Int 84:186–194
Ehrlich PJ, Lanyon LE (2002) Mechanical strain and bone cell function: a review. Osteoporos Int 13:688–700
Eschman PN (1992) SLCOMM version 1.6. Eschman Archeological Services, Albuquerque
Fehling PC, Alekel L, Clasey J, Rector A, Stillman RJ (1995) A comparison of bone mineral densities among female athletes in impact loading and active loading sports. Bone 17:205–210
Forwood MR, Turner CH (1994) The response of rat tibiae to incremental bouts of mechanical loading—a quantum concept for bone formation. Bone 15:603–609
Forwood MR, Turner CH (1995) Skeletal adaptations to mechanical usage—results from tibial loading studies in rats. Bone 17:S197–S205
Frost HM (1997) Why do marathon runners have less bone than weight lifters? A vital biomechanical view and explanation. Bone 20:183–189
Goodship AE, Lawes TJ, Rubin CT (2009) Low-magnitude high-frequency mechanical signals accelerate and augment endochondral bone repair: preliminary evidence of efficacy. J Orthop Res 27:922–930
Gorman CF (1970) Excavations at Spirit Cave, North Thailand: some interim interpretations. Asian Perspect 13:70–107
Haapasalo H, Sievanen H, Kannus P, Oja P, Vuori I (1996) Humeral dimensions after long term unilateral loading. A DXA study of Finnish competitive tennis players. Bone 18:106S
Haapasalo H, Kontulainen S, Sievanen H, Kannus P, Jarvinen M, Vuori I (2000) Exercise-induced bone gain is due to enlargement in bone size without a change in volumetric bone density: a peripheral quantitative computed tomography study of the upper arms of male tennis players. Bone 27:351–357
Hamrick MW (1999) A chondral modeling theory revisited. J Theor Biol 201:201–208
Hershkovitz I, Speirs MS, Frayer D, Nadel D, Wish-Baratz S, Arensburg B (1995) Ohalo II H2: A 19,000-year-old skeleton from a water-logged site at the Sea of Galilee, Israel. Am J Phys Anthropol 96:215–234
Hillman G, Madeys AE, Hather J (1989) Wild plant foods and diet at Late Paleolithic Wadi Kubbaniya: the evidence from charred remains. In: Wendorf F, Schild R, Close AE (eds) The prehistory of Wadi Kubbaniya, vol 2, Stratigraphy, paleoeconomy and environment. SMU Press, Dallas, pp 15–100
Holt BM (1999) Biomechanical evidence of decreased mobility in Upper Paleolithic and Mesolithic Europe. Ph.D. Dissertation, University of Missouri, Columbia
Holt BM (2003) Mobility in Upper Paleolithic and Mesolithic Europe: evidence from the lower limb. Am J Phys Anthropol 122:200–215
Holt BM, Formicola V (2008) Hunters of the Ice Age: the biology of Upper Paleolithic people. Yearb Phys Anthropol 51:70–99
Holt BM, Mussi M, Churchill SE, Formicola V (2000) Biological and cultural trends in Upper Paleolithic Europe. Riv Antropol 78:179–192
Jones HN, Priest JD, Hayes WC, Tichenor CC, Nagel DA (1977) Humeral hypertrophy in response to exercise. J Bone Joint Surg 59A:204–208
Judex S, Carlson K (2009) Is bone's response to mechanical signals dominated by gravitational loading? Med Sci Sports Exerc 41:2037–2043
Judex S, Rubin CT (2010) Is bone formation induced by high-frequency mechanical signals modulated by muscle activity? J Musculoskelet Neuronal Interact 10:3–11
Judex S, Zernicke RF (2000a) High-impact exercise and growing bone: relation between high strain rates and enhanced bone formation. J Appl Physiol 88:2183–2191

Judex S, Zernicke RF (2000b) Does the mechanical milieu associated with high-speed running lead to adaptive changes in diaphyseal growing bone? Bone 26:153–159
Kelly RL (1992) Mobility/sedentism: concepts, archaeological measures and effects. Annu Rev Anthropol 21:43–66
Kelly RL (1995) The foraging spectrum: diversity in hunter-gatherer lifeways. Smithsonian Institution, Washington, DC
King JW, Brelsford HJ, Tullow HS (1969) Analysis of the pitching arm of the professional baseball pitcher. Clin Orthop Relat Res 67:116–123
Lahr MM, Arensburg B (1995) Skeletal robusticity in the Epipaleolithic of North Africa and the Levant. Paléoorient 21:87–96
Lanyon LE, O'Connor JA (1980) Adaptation of bone artificially loaded at high and low physiological strain rates. J Physiol (Lond) 303:P36
Lanyon LE, Rubin CT (1984) Static vs. dynamic loads as an influence on bone remodeling. J Biomech 17:897–905
Lanyon LE, Hampson WGJ, Goodship AE, Shah JS (1975) Bone deformation recorded in vivo from strain gauges attached to human tibial shaft. Acta Orthop Scand 46:256–268
Lanyon LE, Magee PT, Baggott DG (1979) Relationship of functional stress and strain to the processes of bone remodeling—experimental study on the sheep radius. J Biomech 12:593–600
Lanyon LE, Goodship AE, Pye CJ, Macfie JH (1982) Mechanically adaptive bone remodeling. J Biomech 15:141–154
Lazenby RA, Cooper DML, Angus A, Hallgrímsson B (2008) Articular constraint, handedness, and directional asymmetry in the human second metacarpal. J Hum Evol 54:875–885
Lieberman DE, Pearson OM (2001) Trade-off between modeling and remodeling responses to loading in the mammalian limb. Bull Mus Comp Zool 156:269–282
Lieberman DE, Devlin MJ, Pearson OM (2001) Articular area responses to mechanical loading: effects of exercise, age and skeletal location. Am J Phys Anthropol 116:266–277
Lieberman DE, Polk JD, Demes B (2004) Predicting long bone loading from cross-sectional geometry. Am J Phys Anthropol 123:156–171
Marchi D (2008) Relationships between lower limb cross-sectional geometry and mobility: the case of a Neolithic sample from Italy. Am J Phys Anthropol 137:188–200
Marchi D, Shaw C (2011) Variation in fibular robusticity reflects variation in mobility patterns. J Hum Evol 61:609–616
Marchi D, Sparacello VS, Holt BM, Formicola V (2006) Biomechanical approach to the reconstruction of activity patterns in Neolithic western Liguria, Italy. Am J Phys Anthropol 131: 447–455
Marchi D, Sparacello V, Shaw C (2011) Mobility and lower limb robusticity of a pastoralist Neolithic population from North-western Italy. In: Pinhasi R, Stock JT (eds) Human bioarchaeology of the transition to agriculture. Wiley-Blackwell, West Sussex, pp 317–346
Miller JA, Gross MM (1998) Locomotor advantages of Neandertal skeletal morphology at the knee and ankle. J Biomech 31:355–361
Mosley JR, March BM, Lynch J, Lanyon LE (1997) Strain magnitude related changes in whole bone architecture in growing rats. Bone 20:191–198
Myers MJ, Steudel K (1985) Effect of limb mass and its distribution on the energetic cost of running. J Exp Biol 116:363–373
Nordin M, Frankel VH (2000) Basic biomechanics of the musculoskeletal system, 3rd edn. Lippincott Williams & Wilkins, New York
O'Connor JA, Lanyon LE (1982) The effect of strain rate on mechanically adaptive bone remodeling. Orthop Trans 6:240–241
Plochocki J (2004) Bilateral variation in limb articular surface dimensions. Am J Hum Biol 16:328–333
Plochocki J, Riscigno CJ, Garcia M (2006) Functional adaptation of the femoral head to voluntary exercise. Anat Rec A Discov Mol Cell Evol Biol 288A:776–781

Raichlen DA, Armstrong H, Lieberman DE (2011) Calcaneus length determines running economy: implications for endurance running performance in modern humans and Neandertals. J Hum Evol 60:299–308

Robinson TL, Snow-Harter C, Taaffe DR, Gillis D, Shaw J, Marcus R et al (1995) Gymnasts exhibit higher bone mass than runners despite similar prevalence of amenorrhea and oligomenorrhea. J Bone Miner Res 10:26–35

Robling AG, Duijvelaar KM, Geevers JV, Ohashi N, Turner CH (2001) Modulation of appositional and longitudinal bone growth in the rat ulna by applied static and dynamic force. Bone 29: 105–113

Rubin CT, Lanyon LE (1984) Regulation of bone-formation by applied dynamic loads. J Bone Joint Surg 66A:397–402

Rubin CT, Lanyon LE (1985) Regulation of bone mass by mechanical strain magnitude. Calcif Tissue Int 37:411–417

Ruff CB (1987) Sexual dimorphism in human lower limb bone structure: relationship of subsistence strategy and sexual division of labor. J Hum Evol 16:391–416

Ruff CB (1991) Climate and body shape in hominid evolution. J Hum Evol 21:81–105

Ruff CB (1994) Morphological adaptation to climate in modern and fossil hominids. Yearb Phys Anthropol 37:65–107

Ruff CB (1995) Biomechanics of the hip and birth in early Homo. Am J Phys Anthropol 98:527–574

Ruff CB (1999) Skeletal structure and behavioral patterns of prehistoric Great Basin populations. In: Hemphill BE, Larsen CS (eds) Prehistoric lifeways in the Great Basin Wetlands: bioarchaeological reconstruction and interpretation. University of Utah Press, Salt Lake City, pp 290–320

Ruff CB (2000) Body size, body shape, and long bone strength in modern humans. J Hum Evol 38:269–290

Ruff CB (2002) Long bone articular and diaphyseal structure in Old World monkeys and apes. I: locomotor effects. Am J Phys Anthropol 119:305–342

Ruff CB, Hayes WC (1983) Cross-sectional geometry of Pecos Pueblo femora and tibiae-a biomechanical investigation. 2. Sex, age and side differences. Am J Phys Anthropol 60:383–400

Ruff CB, Scott WW, Liu AYC (1991) Articular and diaphyseal remodeling of the proximal femur with changes in body mass in adults. Am J Phys Anthropol 86:397–413

Ruff CB, Trinkaus E, Walker A, Larsen CS (1993) Postcranial robusticity in *Homo*. I: temporal trends and mechanical interpretation. Am J Phys Anthropol 91:21–53

Ruff CB, Holt BM, Sládek V, Berner M, Murphy WA Jr, zur Nedden D et al (2006) Body size, body proportions, and mobility in the Tyrolean "Iceman". J Hum Evol 51:91–101

Saxon LK, Robling AG, Alam I, Turner CH (2005) Mechanosensitivity of the rat skeleton decreases after a long period of loading, but is improved with time off. Bone 36:454–464

Shackelford LL (2005) Regional variation in the postcranial robusticity of Late Upper Paleolithic humans. Ph.D. Dissertation, Washington University, St. Louis

Shackelford LL (2007) Regional variation in the postcranial robusticity of Late Upper Paleolithic humans. Am J Phys Anthropol 133:655–668

Shaw C, Stock J (2009) Habitual throwing and swimming correspond with upper limb diaphyseal strength and shape in modern human athletes. Am J Phys Anthropol 140:160–172

Shaw C, Stock J (2011) The influence of body proportions on femoral and tibial midshaft shape in hunter-gatherers. Am J Phys Anthropol 144:22–29

Smith P (1979) Regional diversity in Epipaleolithic populations. Ossa 6:243–250

Stock J (2006) Hunter-gatherer postcranial robusticity relative to patterns of mobility, climatic adaptation, and selection for tissue economy. Am J Phys Anthropol 131:194–204

Stock J, Pfeiffer S (2001) Linking structural variability in long bone diaphyses to habitual behaviors: foragers from the southern African Later Stone Age and the Andaman Islands. Am J Phys Anthropol 115:337–348

Stock J, Pfeiffer S (2004) Long bone robusticity and subsistence behaviour among Later Stone Age foragers of the forest and fynbos biomes of South Africa. J Archaeol Sci 31:999–1013

Stock J, Shaw C (2007) Which measures of diaphyseal robusticity are robust? A comparison of external methods of quantifying the strength of long bone diaphyses to cross-sectional geometric properties. Am J Phys Anthropol 134:412–423
Straus LG (1995) The Upper Paleolithic of Europe: an overview. Evol Anthropol 4:4–16
Trinkaus E, Churchill SE (1988) Neandertal radial tuberosity orientation. Am J Phys Anthropol 75:15–21
Trinkaus E, Rhoads ML (1999) Neandertal knees: power lifters in the Pleistocene? J Hum Evol 37:833–859
Trinkaus E, Villemeur I (1991) Mechanical advantages of the Neandertal thumb in flexion: a test of an hypothesis. Am J Phys Anthropol 84:249–260
Turner CH (1998) Three rules for bone adaptation to mechanical stimuli. Bone 23:399–407
van Andel TH, Davies W, Weniger B (2003) The human presence in Europe during the Last Glacial period I: migrations and the changing climate. In: van Andel TH, Davies W (eds) Neanderthals and modern humans in the EUROPEAN Landscape during the Last Glaciation. University of Cambridge, Cambridge, pp 31–56
Weiss E (2003) Effects of rowing on humeral strength. Am J Phys Anthropol 121:293–302

Chapter 10
The Balance Between Burden Carrying, Variable Terrain, and Thermoregulatory Pressures in Assessing Morphological Variation

Cara M. Wall-Scheffler

Abstract A series of studies have lately shown specific morphological correlations with energetic savings during locomotor tasks, including burden carrying and walking up inclines. These energetic savings appear significant and make sense given the interactions between morphological variation and behavior in many populations. For example, a wider pelvis for one's mass has been shown to save people energy while carrying both front and back burdens. Ethnographic reports suggest that females, who maintain a wider pelvis for their mass in any given group, carry a much higher proportion of burdens than males—some have even suggested that carrying is "women's work." Simultaneously, a shorter tibia has been correlated with energy efficiency while walking up inclines, while a longer tibia has been shown to increase speed along flat terrain. All of these traits have also been shown to be correlated with climatic variables, particularly temperature, such that morphology may result from some interaction between selection pressures for thermoregulatory and mobility adaptations. Here I review work on the mobility effects of pelvic width and limb proportions in the context of carrying burdens across variable terrain and while balancing thermoregulatory pressures.

Keywords Walking • Energetics • Pelvis • Limb length • Cost of transport • Thermoregulation • Baby carrying

---

C.M. Wall-Scheffler (✉)
Department of Biology, Seattle Pacific University, Seattle, WA, USA

Department of Anthropology, University of Washington, Seattle, WA, USA
e-mail: cwallsch@spu.edu

K.J. Carlson and D. Marchi (eds.), *Reconstructing Mobility: Environmental, Behavioral, and Morphological Determinants*, DOI 10.1007/978-1-4899-7460-0_10,

10.1 Introduction

The study of mobility and mobility strategies within physical anthropology has burgeoned in the past few years, thanks to studies specifically addressing the effects of a variety of morphological variables on mobility patterns. Mobility is defined in various ways and has been used to define everything from specific subsistence types (e.g., marine hunter-gatherer) to broad scale patterns of movement (e.g., residential mobility). Here, mobility encompasses the manner (e.g., speed, encumbered with burdens, with others) by which people locomote (i.e., walk specific distances) across terrain (e.g., with variable incline) to gain access to necessary resources (e.g., food and water). Relating locomotion (how fast and how far) to subsistence strategies (e.g., marine gatherer) has often involved studies of the cross-sectional measures of limb bones for populations with known mobility patterns in terms of subsistence type and sometimes distances moved (Ruff 1984, 1987; Ruff and Hayes 1983; Stock 2006; Stock et al. 2011; Trinkaus and Ruff 1999a, b) (see also Ruff and Larsen 2014, for a complete review). As this volume attests, these studies have greatly aided our ability to interpret the variation in fossil and historical long bone remains in terms of the subsistence strategies of the populations. Cross-sectional measures are also highly influenced by limb lengths and body proportions since limb lengths influence bending moments and thus joint and diaphysis remodeling (Gruss 2007), though see Pearson et al. (2014) for the low correlations between bi-iliac breadth and limb midshaft shape.

Studies investigating mobility (my definition) from an energetic perspective have consequently specifically focused on the role of body size and proportions (e.g., lower limb length, crural index, pelvis width) in saving the entire physiological system energy (DeJaeger et al. 2001; Kramer 1999, 2004; Kramer and Eck 2000; Kramer and Sylvester 2009; Minetti and Alexander 1997; Minetti et al. 1994; Steudel-Numbers 2006; Steudel-Numbers and Tilkens 2004; Steudel 1994, 1996; Steudel and Beattie 1995; Wall-Scheffler 2012a, b), thus allowing more energy to be available to the individual for reproductive costs. The theory behind an energetic perspective of limb lengths and torso widths suggests that when mobility tasks are accomplished on a daily basis, those individuals who spend less energy on the daily task will thus be able to spend their energy savings on reproductive tasks. If particular limb proportions aid energetic economy, these proportions would be more successful. This is particularly true for females who are able to drive down the cost of walking: those females with low costs of mobility would then have reduced interbirth intervals, and increased lifetime fitness for example. Data supporting the importance of energetic constraints on locomotor morphology come from a variety of sources, particularly reproductive ecology. Studies that integrate mobility and reproductive success clearly show that the energetics of daily walking influence interbirth intervals and offspring survivorship (Blurton Jones et al. 1989; Blurton Jones 1986, 1987; Gibson and Mace 2006). Locomotor stress has also been shown to lead to drops in ovarian hormones (Devlin 2011). Changes in mobility (e.g., amount, intensity) have consistently been shown to influence weight loss/weight

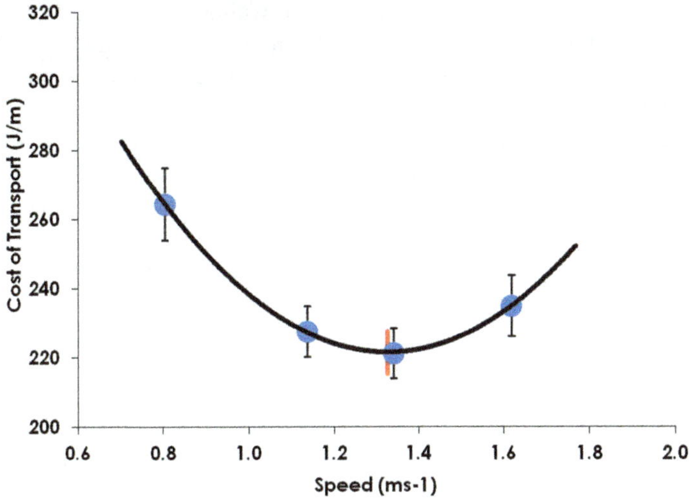

Fig. 10.1 The optimal walking curve. The *vertical red line* demarcates the optimal speed—the speed at which the Cost of Transport is lowest (data from Wall-Scheffler and Myers 2013)

gain, and ovarian function and fecundity have consistently been shown to be particularly sensitive to changes in the metabolic balance (often measured in terms of changes in mass), such that an increased imbalance between energy-in and energy-out decreases fertility (Ellison 2003, 2008; Emaus et al. 2008; Jasienska and Ellison 2004; Pike 2005; Ziomkiewicz et al. 2008). Such sensitivities to changes in metabolic balance (as caused by changes in mobility) have been shown both in high-mass, energy-rich populations and also in lower body mass, less healthy populations.

10.1.1 The Importance of Speed in the Context of Mobility

Mobility strategies are often defined within a particular landscape. Mobility encompasses distances to food patches, water holes, and hunting grounds, as well as the terrain through the landscape. Within this context, the energy used per unit distance (Cost of Transport—CoT) becomes crucial as it helps situate mobility strategies in a particular niche. For humans, regardless of gait choice, the amount of energy used to go a given distance is curvilinear in relation to speed choice (Steudel-Numbers and Wall-Scheffler 2009; Wall-Scheffler 2012b) (Fig. 10.1). That is, people can walk at a variety of speeds, but one speed will cost them the lowest amount of energy when going a particular distance. Studies done on people walking alone have shown that most people choose to walk very close to the speed that minimizes the cost to go that given distance (the "optimal" speed), even when walking for relatively short periods of time (Browning et al. 2006; Browning and Kram 2005; O'Connor and Donelan 2012; Peyrot et al. 2012; Ralston 1958); this further seems to be true for other primates and mammals generally (Langman et al. 2012; O'Neill 2012).

The relationship between speed and daily movement distances (an often used measure of mobility) is complex. Many endotherms show seasonal patterns in daily movement distances but maintain a consistent speed (Jedrzejewski et al. 2001; Kowalzcyk et al. 2006)—not surprising given the pattern of organisms to travel around their optimal speed. Other species, such as the Polish lynx, show an increase in speed during the season with an increase in daily movement distance, suggesting that increased speed can be an important mechanism for animals to increase their movement distances—in the case of the lynx because the search for mates directly depends upon their range (Jedrzejewski et al. 2002). As high resolution studies of human seasonality of speed and movement distances become available (e.g., Pontzer et al. 2012), hopefully some of these relationships will be uncovered for our own species.

10.1.2 The Importance of Body Proportions in the Context of Mobility

Limb proportions have been shown to be crucial for the absolute cost of locomotion (Kramer and Eck 2000) and the relative (for a given mass) cost of locomotion (Pontzer 2007; Steudel-Numbers and Tilkens 2004), but also for the speed of this optimum (Wall-Scheffler 2012b). This means that how much ground an individual can economically cover, and thus how much mobility is energetically feasible, is also dependent on limb lengths and proportions (Wall-Scheffler 2012a). Additionally, limb lengths, limb proportions, and body breadth have also been shown to vary the curvilinear relationship between speed and the Cost of Transport (Wall-Scheffler 2012b), resulting in some body proportions having a wider curve (more speeds from which to choose) or a more narrow curve (fewer speeds could be considered "optimal").

Morphological variation in terms of limb lengths and pelvis breadths has also been shown to have dramatic importance for negotiating variable terrain, with many studies focusing particularly on incline movement (both walking and running) (Block et al. 2009; Higgins and Ruff 2011; Wall-Scheffler et al. 2010). While few studies have focused on energetic contributions of limb proportions (but see Block et al. 2009), and have rather characterized the mechanical work involved in walking or running up hills (Higgins and Ruff 2011; Wall-Scheffler et al. 2010), clear patterns of the importance of limb proportions for reducing work, as well as energy, have surfaced.

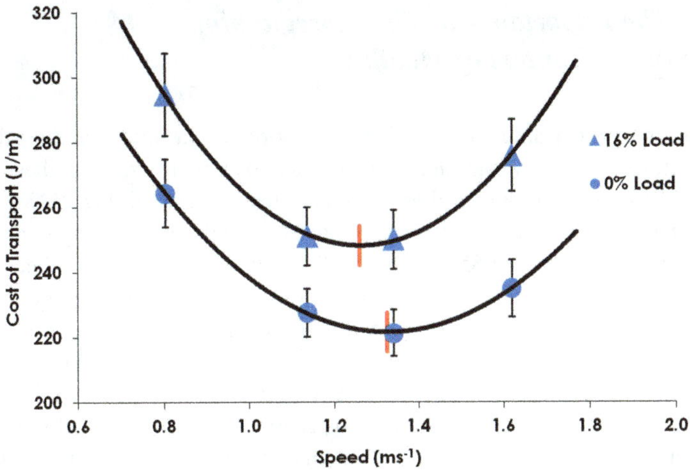

Fig. 10.2 Optimal walking curves for unloaded and front loaded (16 % of body mass) conditions. Of particular note is the different curvature of the two curves; the loaded curve has a more acute curve, and thus the energetic penalty for walking away from the optimal speed is progressively greater (data from Wall-Scheffler and Myers 2013)

10.1.3 The Importance of Tasks in the Context of Mobility

Another key task that has begun to emerge as crucial in order to weigh different selection pressures and outcomes of limb proportions and body shape is load carrying (Gruss et al. 2009; Kramer 2004; Wall-Scheffler et al. 2007; Wall-Scheffler and Myers 2013). As load carrying is a universal task for humans, those individuals who can carry loads for a reduced cost are at a distinct advantage reproductively. Females in particular carry frontal loads continuously for minimally a year (during pregnancy and initial lactation position), at which time the burden (e.g., child) is moved to the side or the back for another few years (before the cycle begins again). Since burden carrying is highly energetically expensive (Kramer 1998; Wall-Scheffler 2007; Watson et al. 2008) and causes statistically significant changes to the CoT curve (Wall-Scheffler and Myers 2013) (Fig. 10.2), understanding morphological possibilities of offsetting these costs is key to elucidating selection pressures on people. The studies looking at this problem find that, for sexually dimorphic traits (such as body size and pelvis width), the female morphology creates energetic savings during load carrying (Gruss et al. 2009; Wall-Scheffler 2012a; Wall-Scheffler et al. 2007) as well as creating important contributions in terms of speed options (Wall-Scheffler and Myers 2013).

10.1.4 The Importance of Thermoregulation in the Context of Mobility

While the energetic framework is of great importance in understanding the wide range of limb and pelvic variation among extant and extinct hominins, these proportions have clear implications for thermoregulation as well (Holliday 1997b; Roberts 1978; Ruff 1994). For example, the importance of distal segment length in heat dissipation (Irving and Krog 1955; Trinkaus 1981) and the importance of heat dissipation for energy expenditure (due to preventing the cardiovascular system from overworking to maintain consistent core and limb temperatures) (Gonzalez-Alonso 2012; Speakman and Król 2010) is a crucial example of the interaction of mobility and thermoregulation. And while one might suppose that thermoregulatory pressures during development might drive limb proportions irrespective of selection for economical locomotion, genetics powerfully impacts lower limb proportions (Schultz 1926), even in the context of short-term environmental factors such as nutrition and hypoxia (Bailey et al. 2007). As such, adult body proportions can be viewed as the outcome of a range of key environmental interactions (Stock et al. 2011; Tilkens et al. 2007) but that are heavily impacted by the cumulative effects of selection, including selection for mobility (Binford 2001).

Thus, limb lengths and body widths prove to be determinants in setting the constraints within which individuals must work, both in terms of gaining access to resources and maintaining physiological homeostasis. A population's set of limb lengths and proportions will be indicative of the selection pressures that population has faced—both in terms of economy (how much ground needs to be covered) and speed (how many other tasks are necessary, how much daylight, how much heat load) within the context of terrain, task (e.g., carrying burdens), and climate (e.g., thermoregulation). Thus, speed, thermoregulation, and energetics must be assessed together in order to understand the selection pressures involved in increasing the energy available for reproduction and accessing resources across any series of niche adaptations which would ultimately improve reproductive fitness. Here I assess current work investigating the interactions among these complex variables within a selectionist framework.

10.2 Lower Limb Length

10.2.1 Energy Savings on the Level

Research on the energetics of walking on level surfaces has consistently shown that longer lower limbs (relative to mass) reduce the cost of walking at a range of comfortable walking speeds (Steudel-Numbers and Tilkens 2004; Wall-Scheffler et al. 2010) (Table 10.1); this is generally considered particularly important for the interpretation of the hominin fossil record. Long lower limbs, relative to stature or mass

Table 10.1 Summary of body proportions' influence on locomotor energetics

	Terrain	Burden	Energetics	CoT curvature	Citation
Long lower limbs	Flat	Front	Reduced	Acute	Steudel-Numbers and Tilkens (2004), Wall-Scheffler and Myers (2013)
Short lower limbs	Incline		Reduced		Block et al. (2009), Higgins and Ruff (2011)
Wide bi-trochanteric	Flat	Back	Reduced	Acute	Wall-Scheffler (2012b), Wall-Scheffler et al. (2007)
Wide bi-iliac	Flat	Front		Obtuse	Data from Block et al. (2009), Wall-Scheffler and Myers (2013)
	Incline		Reduced		

estimations, are interpreted as suggesting modern human-like efficiency, and "striding-bipedality" (Pontzer et al. 2010; Steudel-Numbers 2006). Even more important for ideas about mobility and subsistence is the finding that absolutely longer lower limbs increase the speed at which the minimum CoT occurs, meaning that populations with longer lower limbs have the potential to cover more ground in less time without bearing additional energetic burdens (Wall-Scheffler 2012a, b). The trade-off for this benefit, however, is that those individuals with longer lower limbs also have increased curvature of their CoT curve (Wall-Scheffler 2012b) meaning that when they are not traveling fast, they do pay an energetic penalty, perhaps because of the specific timing of muscle activation patterns (Carrier et al. 2011; Wall-Scheffler et al. 2010). That is, if the muscle activation patterns are not commonly used or involve the coactivation of multiple muscle groups, the cost of locomotion increases (Voloshina et al. 2013). This increase in cost would have dramatic implications if, for example, a large male walks with a small female while she is carrying a load that decreases her optimal speed (see further discussion below). His energetic burden for slowing down to the loaded optimum of a small individual could be nearly 10 % of his daily energy expenditure (DEE) (Wall-Scheffler 2012b). It would thus be useful to test the relationship between sexual dimorphism and absolute lower limb length; if females were smaller, but maintained absolutely longer lower limbs, this could predict more mixed-sex walking groups and/or long daily movement distances.

10.2.2 Lower Limb Length at an Incline

Speed itself seems to be positively driven by the absolute length of the tibia (Bereket 2005; Wall-Scheffler 2012b), which is important as tibia length has recently been implicated in being crucial for the negotiation of uneven terrain (Block et al. 2009; Higgins and Ruff 2011). While on the level surface, a longer tibia confers a higher optimal speed (Bereket 2005; Wall-Scheffler 2012b), on an inclined surface, longer tibiae have been shown to be energetically costly. A long tibia makes long stride lengths more challenging on an incline by effectively removing the mechanical advantages of the longer limbs (Higgins and Ruff 2011); the result is that long tibiae become significantly costlier as the incline becomes steeper (Block et al. 2009).

10.2.3 Lower Limb Length Mechanics

Longer limbs offer energetic advantages on level surfaces for two likely reasons. From a mechanical perspective, long limbs lead to longer stride lengths, and thus fewer strides per distance (Grieve and Gear 1966; Wall-Scheffler and Myers 2013); fewer strides should lead to lower costs (Donelan et al. 2002; Kuo et al. 2005; Weyand et al. 2010). Additionally, longer limbs have consistently been shown to reduce heat load, and increase heat loss (Cross et al. 2008; Tilkens et al. 2007). As mentioned above, such an ability to minimize heat gain and overall heat load would allow the individual to maintain higher speeds without overloading homeostatic systems and increasing energetic cost (Gonzalez-Alonso 2012; Speakman and Król 2010). It is possible that this thermoregulatory component is a key aspect of why longer lower limbs are more energetically economical for a given mass—the surface area to volume ratio increases, allowing more heat to be lost during locomotion, allowing higher speeds to be maintained at a lower energetic cost. While the role of long lower limbs in increasing heat loss has been shown during resting (Tilkens et al. 2007) and has long been theorized to increase heat loss during walking (Cross et al. 2008; Wheeler 1993), recent data coming out have now shown that long lower limbs do in fact correlate with a reduction in heat gain during walking (Fig. 10.3).

10.2.4 Lower Limb Length and Burden Carrying

Despite these important benefits, the role of limb length to augment any of the burden—thermoregulatory or energetic—during load carrying has been undetermined at this point. Initial studies on the energetics of burden carrying have not found any particular role for lower limb length one way or another (Kramer 2010; Wall-Scheffler and Myers 2013). This is particularly interesting because increasing stride length does seem to be important for alleviating energetic costs of increasing load; however, as discussed below, the increased stride length during load carrying seems to come from rotating the pelvis and not from using the lower limbs any differently (Gruss et al. 2009; Rak 1991; Wall-Scheffler et al. 2007; Wall-Scheffler and Myers 2013).

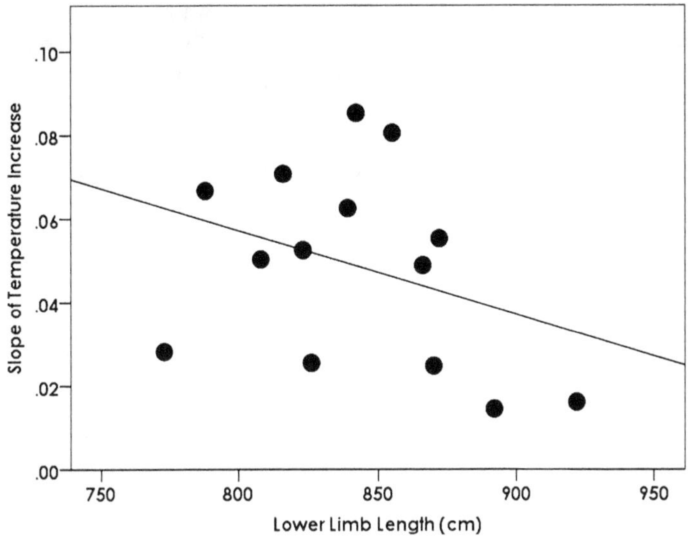

Fig. 10.3 The negative correlation between lower limb length and the slope of the core temperature increase across a bout of stair walking ($N = 14$; $R^2 = 12$ %). Men with longer lower limbs show a slower increase in core temperature over a 15 min stair stepping trial than men with shorter lower limbs. (data are currently being prepared for publication and are available on request)

10.3 Pelvis Width

10.3.1 How Should the Pelvis be Considered?

In the last few years two important ideas relating to the pelvis have emerged. First is the near consensus among paleoanthropologists that a broad/wide pelvis has been the dominant pelvis shape for most of hominin evolution (Bonmati et al. 2010; Kibii et al. 2011; Rosenberg et al. 2006; Simpson et al. 2008, 2010), up to the very late appearance of the narrow pelvis characteristic of anatomically modern human males (Holliday 1997a, 2000; Trinkaus 1983). The second is that a wide pelvis—variably defined by bi-iliac, bi-trochanteric, or bi-acetabular breadth—has some advantages for the individuals who have them (Donelan et al. 2004; Dunsworth et al. 2012; Kuo 2007; Wall-Scheffler 2012a, b; Wall-Scheffler et al. 2007, 2010; Wall-Scheffler and Myers 2013). Both of these are relatively recent ideas as for many years it was thought that the narrow pelvis emerged early [e.g., with *Homo erectus* (Walker and Leakey 1993)] and that a wide pelvis was a (primitive) costly trait due to the creation of large moment arms (Lovejoy 1988; Richmond and Jungers 2008). These shifts in thinking are on the one hand due to numerous important fossils that have been discovered and/or described in recent years, as well as due to datasets featuring other demographics and conditions besides healthy men walking unencumbered on treadmills at their preferred walking speed. In particular, studies assessing people walking at a variety of speeds (e.g., in order to investigate the entire Cost of Transport curve) and with people doing a variety of tasks (e.g., carrying burdens) have been at the forefront of understanding the possible benefits to mobility offered by a broad pelvis.

In order to understand the role the pelvis plays in locomotion, it would be helpful to define what aspects of pelvic shape influence locomotor mechanics, muscle activation, and energetics. Pelvis width is often taken to mean bi-iliac breadth, but important mechanical considerations exist for bi-trochanteric breadth since this encompasses the functional relationship between the femur and the pelvis and influences the angle by which the knee must be adducted under the body. Since the relationship between the lateral margin of the iliac crest and that of the greater trochanters also defines the muscle force that must be generated by the hip abductors (e.g., Richmond and Jungers 2008), the different widths may not necessarily be independently selected (Grabowski et al. 2011), but a comparison between them has not been systematically made (though see Ruff 1995).

Since bi-trochanteric breadth is extremely challenging to get from disarticulated bones, bi-acetabular breadth is another measure that can be used to measure pelvis width as well as the width between the femoral heads (in order to estimate bi-trochanteric width even if femoral neck length and angle is not known); furthermore, bi-acetabular breadth is a measure of reproductive significance. Suffice it to say, we do not yet understand how the three measures are correlated to each other morphologically, much less functionally; selection can clearly act on them independently and we see different patterns of variation in different populations and at different times (Grabowski et al. 2011; Kurki 2008, 2011; Ruff 1995). As such, in the following review, I have tried to be as specific as possible in relating what measures of pelvic width seem to have particular relationships with energetics and cost in terms of mobility. Bi-iliac breadth seems to have more influence on muscle activity as the ilia are the main source of pelvis and thigh muscles, whereas bi-trochanteric breadth seems to have more influence on biomechanical variables such as stride length, likely due to its role in rotating the pelvis in relationship with the femur. Bi-acetabular breadth will be discussed based on its role in reconstructing fossil morphology and variation as any effort to actually measure this in vivo and declare its relevance to mobility studies is in its early stages (e.g., Dunsworth et al. 2012).

Energy Savings on the Level With and Without Burdens

During unloaded walking on a level surface, albeit not a treadmill, pelvis width has been shown to influence both the relative cost (for a given mass) as well as the flexibility of walking speed (Wall-Scheffler 2012b) (Table 10.1). The data show that for a given mass, increased pelvis width (here, as bi-trochanteric) reduces the metabolic cost of walking—this is a similar effect to that of lower limb length discussed above and is additive; you can have both longer limbs and a broader pelvis and have energetic benefits from both—enough to offset the increased metabolic cost that comes with getting bigger (Wall-Scheffler 2012a). However, a wider pelvis (again as bi-trochanteric) also increases the curvature around the minimum speed, meaning that the penalty for traveling away from the optimal speed is absolutely greater, though this may simply be a piece of the series of tradeoffs for increasing size (similar to increased lower limb length). The role of a wide pelvis is exactly the opposite, however, when loads are considered. Data of women walking with frontal loads have shown that a wider pelvis (here as bi-iliac) dramatically (5 %) decreases the curvature of the CoT curve, making women with wider bi-iliac breadths significantly more able to vary their walking speed with no metabolic penalty (Wall-Scheffler and Myers 2013). Since smaller people generally have a wider CoT curve anyway (Wall-Scheffler 2012b), being both small and having a broad pelvis allows women a substantial variety of walking speeds while they are carrying loads all at the same optimum cost. Data on women walking at a single speed with a backload similarly show that women with a wider pelvis (bi-trochanteric) have a reduced cost for carrying the load compared to women with a more narrow pelvis (Wall-Scheffler et al. 2007). This is particularly interesting since much of the ethnographic literature supports a scenario in which women walk together (Binford 2001)—generally with different sorts of food, baby, and household loads (Brightman 1996; Hilton and Greaves 2008). Having a broad pelvis allows all the women to be able to walk together essentially at their optimum speed.

Data on muscle activity during walking has shown that individuals with wider bi-iliac breadths have significantly decreased muscle activity in the hip adductors and hamstrings while walking on inclines (Wall-Scheffler et al. 2010) suggesting energy savings. This is confirmed with unpublished data on the metabolic costs of incline walking (data from Block et al. 2009), which illustrates a cost-saving mechanism of bi-iliac breadth for incline walking. In these data, bi iliac breadth has again a negative relationship with cost even when the relationship between mass and the interaction between tibia and slope are considered.

10.3.2 Pelvis Width Mechanics

How can we compare these two different mechanisms of cost saving, one involving bi-trochanteric breadth and one involving bi-iliac breadth? The mechanisms seem to relate in part to increased stride length, as well as increased stability; the mechanisms do not seem to relate to thermoregulatory advantages contrary to the situation with lower limb length. Clearly, both biogeographical patterns (e.g., Bergmann's rule) as well as data (Wall-Scheffler 2010; Wall-Scheffler and Myers 2012) have shown that wider pelves/torsos lead to increased body temperature both at rest and while walking, so the energetic benefits must come from elsewhere. Studies using multiple methods and looking at both loaded and unloaded walkers on a level surface have shown that increased pelvis width (here as bi-trochanteric breadth) leads to longer stride lengths (Gruss et al. 2009; Rak 1991; Wall-Scheffler et al. 2007; Wall-Scheffler and Myers 2013). As discussed above, increasing stride length has the potential to reduce total strides taken and thus reduce metabolic cost for going a particular distance. It is not clear at this point whether the relationship between bi-trochanteric breadth and stride length exists on an incline.

Having a broader pelvis (and all the tissue and musculature that this entails) particularly in terms of the breadth of the ilia, pulls the center of mass downwards, theoretically increasing stability and partially explaining why females have a lower center of mass than males. It has further been clearly shown that stability—particularly in the mediolateral plane—is important in the energetics of locomotion, and that increased stability reduces the energetic cost of walking (Donelan et al. 2004; Voloshina et al. 2013). Furthermore, increased stability through the actions of the hip abductors specifically [which comes with a broader pelvis (Wall-Scheffler et al. 2010)] may thus decrease the entire system's cost of walking (Kuo 1999, 2007; Kuo et al. 2005).

10.4 Application: Neanderthal Mobility

Clearly not all variations among extinct or extant populations' body proportions can be explained by climatic variation alone (Kurki et al. 2008), and the studies assessed here show a pattern of specific body proportions alleviating the energetic burden of mobility under particular situations. Additionally, some of the thermoregulatory benefits of certain proportions indirectly offer clear advantages for mobility: a reduction in heat load allows people to walk faster for longer for less energetic cost (Wheeler 1993). What then do these energetic, speed, and thermoregulatory findings mean for mobility?

Since the optimal walking speed significantly correlates with tibia length, this relationship illuminates a link between something measurable in the fossil record (i.e., tibia length) and a behavior of energetic importance (i.e., speed) (Bereket 2005; Wall-Scheffler 2012b). Irrespective of the selection pressures on the short tibiae of Neanderthals for example, the result of their shorter limbs will have ramifications for their mobility and their walking speeds. The regression equation developed from the model of tibia length predicting optimal speed [after utilizing Porter's correction of external measures to skeletal measures (Porter 1996)] (Wall-Scheffler 2012b), was applied to fossil hominins for whom tibia length is available; these results are shown in Table 10.2. The range of tibiae lengths in the original study population (Wall-Scheffler 2012b) was 279.2–420.4 mm, so most of the fossil specimens fell within this range. Since the regression equation explains 36 % of the variation, these optimal speeds should be considered preliminary at best; however, they do supply a meaningful context for the morphological variation within and between populations. For both Neanderthals ($n=12$) and Anatomically Modern Humans (AMH) ($n=16$) samples, females show slower optimal walking speeds ($p=0.002$ between tibia length each group) than males (1.31 and 1.22 ms^{-1} for Neanderthal males and females, respectively, and 1.43 and 1.34 ms^{-1} for AMH males and females). For both species, male optimal travel speeds are roughly 7 %

Table 10.2 Optimal speed predicted for fossil *Homo*, based on the equation: Optimal speed (ms^{-1}) = 0.00205 × Tibia length (mm) + 0.5811 [Wall-Scheffler (2012b)]

Species	Specimen	Sex	Tibia length (mm)	Speed (ms^{-1})	Source
H. erectus	KNM-ER 15000	M?	380-dead	1.36	Antón (2003)
	KNM-ER 15000		*(403.4)-adult*	1.41	Graves et al. (2010)
	Dmanisi D3901	M?	306	1.21	Lordkipanidze et al. (2007)
	Ngandong B		*(360)*	1.32	Antón (2003)
H. heidelbergensis	Kabwe	M	416	1.43	Trinkaus (2009)
	Boxgrove	M?	*(396.8)*	1.39	Stringer et al. (1998)
European Neanderthals	La Chapelle 1	M	*(340)*	1.28	Holliday (1995), Trinkaus (1980)
	La Ferrassie 1	M	*(370)*	1.34	Holliday (1995), Trinkaus (1980)
	Spy 2	M	331	1.26	Holliday (1995), Trinkaus (1980)
	La Ferrassie 2	F	311	1.22	Holliday (1995), Trinkaus (1980)
	Palomas 96	F	304	1.20	Walker et al. (2011)
Crimea Neanderthal	Kiik-Koba 1	M	(346)	1.29	Trinkaus (1980)
West Asian Neanderthals	Amud 1	M	*(482)**	1.57	Holliday (1995), Trinkaus (1980)
	Shanidar 1	M	*(355)*	1.31	Holliday (1995), Trinkaus (1980)
	Shanidar 2	M	337.5	1.27	Holliday (1995), Trinkaus (1980)
	Shanidar 5	M	*(355)*	1.31	Holliday (1995), Trinkaus (1980)
	Shanidar 6	F	*(300)*	1.20	Holliday (1995), Trinkaus (1980)
	Tabun C1	F	319	1.24	Holliday (1995), Trinkaus (1980)
Middle Paleolithic *H. sapiens*	Skhul 4	M	434*	1.47	Holliday (1995)
	Skhul 5	M	*(443)**	1.49	Holliday (1995)
	Skhul 6	M	*(405)*	1.41	Holliday (1995)
	Qafzeh 8	M	*(436)**	1.48	Holliday (1995)
	Qafzeh 9	F	*(400)*	1.40	Holliday (1995)
	Qafzeh 3	F	*(357)*	1.31	Holliday (1995)
Upper Paleolithic *H. sapiens*	Dolni Vestonice 14	M	418	1.44	Holliday (1995)
	Grotte des Enfants 4	M	455*	1.51	Holliday (1995)
	Paviland	M	398	1.40	Holliday (1995)
	Predmosti 3	M	423*	1.45	Holliday (1995)
	Predmosti 9	M	351	1.30	Holliday (1995)
	Predmosti 14	M	396	1.39	Holliday (1995)
	Predmosti 4	F	364	1.33	Holliday (1995)
	Predmosti 10	F	359	1.32	Holliday (1995)
	Cro-Magnon 2	F	384	1.37	Holliday (1995)
	Grotte des Enfants 5	F	367	1.33	Holliday (1995)

Sex was determined by the source author. Tibia lengths in parentheses denote estimations. An asterisk denotes a tibia length that falls outside the variation of the regression sample (279.2–420.39 mm) faster than females. There was no difference between the optimal speed of European Neanderthals and West Asian Neanderthals ($p=0.722$ for tibia lengths), nor between those of Middle Paleolithic and Upper Paleolithic AMH ($p=0.251$ for tibia lengths). West Asian AMH ($N=6$) had faster optimal walking speeds than West Asian Neanderthals ($N=6$) (1.43 ms^{-1} for AMH and 1.29 ms^{-1} for Neanderthals); each group had four individuals considered males and two considered females. Despite the small samples from *H. erectus* and *H. heidelbergensis*, Table 10.2 also allows a comparison of the variation in estimated optimal speed by species and suggests that Neanderthals had much slower optimal speeds than other *Homo* species, even than the smaller individuals from Dmanisi, whereas *H. heidelbergensis* had optimal speeds comparable with those of *H. sapiens*.

Such results offer a helpful perspective on the ongoing discussions of Neanderthal mobility patterns, which have generally been interpreted to show a small range of total Neanderthal movement [annual or lifetime of 20–30 km (Barton 2000; Macdonald et al. 2009; Richards et al. 2008)]. Neanderthals' levels of sexual dimorphism of limb lengths are similar to AMH, a fact already much discussed in the literature (Arsuaga et al. 1997; Trinkaus 1980). The consequence of this dimorphism allows for a pattern of speed dimorphism equivalency between Neanderthals and their contemporaneous AMH. This is interesting particularly because Neanderthals are not interpreted to have labor dimorphism similar to AMH (Kuhn and Stiner 2006), and thus may be less likely to have practiced single-sex travel parties (Wall-Scheffler 2012b). Since Neanderthal pelves show wider bi-acetabular breadths (Ponce de Leon et al. 2008), it is possible instead that Neanderthals accepted some small energetic penalty for walking at suboptimal speeds with individuals of different sizes, but minimized this cost with a wider pelvic breadth (in all dimensions).

Neanderthal optimal walking speeds, however, are estimated to be significantly slower than those of AMH to the extent that the mean optimal speed of Neanderthal males is slower than the mean optimal speed of AMH females. Despite the thermoregulatory and incline-walking advantages, short distal segments offered Neanderthals, their short tibiae do offer a potential dilemma during short winter days, particularly if resources are difficult to acquire and/or widespread. With daylight as a likely constraint on foraging, all high latitude populations must negotiate finding enough nutrients and water in highly variable amounts of time; based on the latitude of European Neanderthal sites, this time can decrease to 7–9 h during the winter. The combination of such short days with Neanderthals' slow optimal travel speeds suggests that Neanderthals may have used small daily movement distances. Corroborating evidence is demonstrated by a number of archeological studies showing sites with an accumulation of ungulate remains typical of the winter season near sites with more year-round accumulations of ungulate remains (Daujeard and Moncel 2010; Pike-Tay et al. 1999; Wall 2005), implying shorter daily movement distances during the shortened winter days. Furthermore, Macdonald et al. (2009) also suggest that Neanderthals were moving shorter distances with each foraging excursion, but that Neanderthals may have moved more frequently to exploit an increased amount of resources; continued assessment of the location of season-specific sites will help further demonstrate the application of this idea. Additionally, if Neanderthals' slower speeds promoted time stress during the short winter days, this provides a selection pressure for the smaller group sizes suggested for Neanderthals (Burke 2006; Pettitt 1997) due to reduced time for sociality (Dunbar 1992). It may further be interesting to predict whether Neanderthal regionally intense behavior allowed them to locate resources quickly and whether their material culture (and lack of sexual division of labor?) allowed them to process material very quickly.

Work on the Pumé of Venezuela offers another interesting layer onto our assessment of Neanderthal morphology. Hilton and Greaves (2008) found that female decision-making regarding how much they would load themselves varied with the season and the distance to the food patch: the further the distance, the more they would load themselves rather than make multiple trips. Thus, the intersection of time and cost becomes of great importance. If time is a constraint, then increasing loading may be vital to a successful foraging (or hunting) attempt. Neanderthal morphology—with its broad pelvis—is particularly useful for walking with heavy burdens, a practice which could have been crucial under time-constrained situations. Since burdened people already walk quite a bit more slowly than unburdened people, the short tibiae and slower optimal speeds of Neanderthals may in fact be a function of regular burdened locomotion over inclined and variable terrain. A crucial piece of future research thus must focus on the intersection of travel time and cost (including pressures of terrain and burden) while walking, in order to better understand how these morphological factors act under more complex mobility regimes.

Acknowledgements Thank you very much to K.J. Carlson and D. Marchi for their invitation to write this chapter and to all the symposium participants for the interesting discussions. Thanks also to C. Wall, K. Steudel, M.J. Myers, and two anonymous reviewers for helpful comments that greatly improved the manuscript.

References

Antón SC (2003) Natural history of *Homo erectus*. Yearbook Phys Anthropol 46:126–169

Arsuaga JL, Carretero JM, Lorenzo C, Gracia A, Martínez I, Bermúdez de Castro JM, Carbonell E (1997) Size variation in Middle Pleistocene humans. Science 277:1086–1088

Bailey SM, Xu J, Feng JH, Hu X, Zhang C, Qui S (2007) Tradeoffs between oxygen and energy in tibial growth at high altitude. Am J Hum Biol 19:662–668

Barton RNE (2000) Mousterian hearths and shellfish: late Neanderthal activities on Gibraltar. In: Stringer CB, Barton RNE, Finlayson C (eds) Neanderthals on the edge. Oxbow Books, Oxford, pp 211–220

Bereket S (2005) Effects of anthropometric parameters and stride frequency on estimation of energy cost of walking. J Sports Med Phys Fitness 42:152–161

Binford LR (2001) Constructing frames of reference: an analytical method for archaeological theory building using Hunter-Gatherer and environmental data sets. University of California Press, Berkeley

Block LM, Chumanov ES, Steudel-Numbers KL, Wall-Scheffler CM (2009) How sexual dimorphism and kinematics interact to impact cost over variable terrain. Am J Phys Anthropol S48:94

Blurton Jones N, Hawkes K, O'Connell JF (1989) Modelling and measuring costs of children in two foraging societies. In: Standen V, Foley RA (eds) The behavioural ecology of humans and other mammals. Blackwell Scientific Publications, Oxford, pp 367–390

Blurton Jones NG (1986) Bushman birth spacing: a test of optimal interbirth intervals. Ethol Sociobiol 7:91–105

Blurton Jones NG (1987) Bushman birth spacing: direct tests of some simple predictions. Ethol Sociobiol 8:183–203

Bonmati A, Gomez-Olivencia A, Arsuaga JL, Carretero JM, Gracia A, Martínez I, Lorenzo C, Bermúdez de Castro JM, Carbonell E (2010) Middle Pleistocene lower back and pelvis from an aged human individual from teh Sima de los Huesos site, Spain. Proc Natl Acad Sci U S A 107:18386–18391

Brightman R (1996) The sexual division of foraging labor: biology, taboo, and gender politics. Comp Study Soc History 38:687–729

Browning RC, Baker EA, Herron JA, Kram R (2006) Effects of obesity and sex on the energetic cost and preferred speed of walking. J Appl Physiol 100:390–398

Browning RC, Kram R (2005) Energetic cost and preferred speed of walking in obese versus normal weight women. Obes Res 13:891–899

Burke A (2006) Neanderthal settlement patterns in Crimea: a landscape approach. J Anthropol Archaeol 25:510–523

Carrier DR, Adners C, Schilling N (2011) Musculoskeletal system of humans is not tuned to maximize the economy of locomotion. Proc Natl Acad Sci U S A 108:18631–18636

Cross A, Collard M, Nelson A (2008) Body segment differences in surface area, skin temperature and 3D displacement and the estimation of heat balance during locomotion in hominins. PLoS One 3:e2464

Daujeard C, Moncel M-H (2010) On Neanderthal subsistence strategies and land use: a regional focus on the Rhone Valley area in southeastern France. J Anthropol Archaeol 29:368–391

DeJaeger D, Willems PA, Heglund NC (2001) The energy cost of walking in children. Pflugers Arch 441:538–543

Devlin MJ (2011) Estrogen, exercise and the skeleton. Evol Anthropol 20:54–61

Donelan JM, Kram R, Kuo AD (2002) Mechanical work for step-to-step transitions is a major determinant of the metabolic cost of human walking. J Exp Biol 205:3717–3727

Donelan JM, Shipman DW, Kram R, Kuo AD (2004) Mechanical and metabolic requirements for active lateral stabilization in human walking. J Biomech 37:827–835

Dunbar RIM (1992) Time: a hidden constraint on the behavioural ecology of baboons. Behav Ecol Sociobiol 31:35–49

Dunsworth HM, Warrener AG, Deacon T, Ellison PT, Pontzer H (2012) Metabolic hypothesis for human altriciality. Proc Natl Acad Sci U S A 109:15212–15216

Ellison PT (2003) Energetics and reproductive effort. Am J Hum Biol 15:342–351

Ellison PT (2008) Energetics, reproductive ecology and human evolution. PaleoAnthropol 2008:172–200

Emaus A, Veierod MB, Furberg A-S, Espetvedt S, Friedenreigh C, Ellison PT, Jasienska G, Andersen LB, Thune I (2008) Physical activity, heart rate, metabolic profile and estradiol in premenopausal women. Med Sci Sports Exerc 40:1022–1030

Gibson MA, Mace R (2006) An energy-saving development initiative increases birth rate and childhood malnutrition in rural Ethiopia. PLoS Med 3:476–484

Gonzalez-Alonso J (2012) Human thermoregulation and the cardiovascular system. Exp Physiol 97(3):340–6. doi:10.1113/expphysiol.2011.058701

Grabowski MW, Polk JD, Roseman CC (2011) Divergent patterns of integration and reduced constraint in the human hip and the origins of bipedalism. Evolution 65:1336–1356

Graves RR, Lupo AC, McCarthy RC, Wescott DJ, Cunningham DL (2010) Just how strapping was KNM-WT 15000? J Hum Evol 59:542–554

Grieve DW, Gear RJ (1966) The relationships between length of stride, step frequency, time of swing and speed of walking for children and adults. Ergonomics 9:679–399

Gruss LT (2007) Limb length and locomotor biomechanics in the genus *Homo*: an experimental study. Am J Phys Anthropol 134:106–116

Gruss LT, Wall-Scheffler CM, Malik N (2009) Infant carrying in humans: interactions between morphometric and gait parameters. Am J Phys Anthropol S48:182–183

Higgins RW, Ruff CB (2011) The effects of distal limb segments shortening on locomotor efficiency in sloped terrain: implications for Neandertal locomotor behavior. Am J Phys Anthropol 146:336–345

Hilton CE, Greaves RD (2008) Seasonality and sex differences in travel distance and resource transport in Venezuelan foragers. Curr Anthropol 49:144–153

Holliday TW (1995) Body size and proportions in the late Pleistocene western old world and the origins of modern humans. University of New Mexico, Albuquerque

Holliday TW (1997a) Body proportions in Late Pleistocene Europe and modern human origins. J Hum Evol 32:423–454

Holliday TW (1997b) Postcranial evidence of cold adaptation in European Neanderthals. Am J Phys Anthropol 104:245–258

Holliday TW (2000) Evolution at the crossroads: modern human emergence in Western Asia. Am Anthropol 102:54–68

Irving L, Krog J (1955) Skin temperature in the arctic as a regulator of heat. J Appl Physiol 7:354–363

Jasienska G, Ellison PT (2004) Energetic factors and seasonal changes in ovarian function in women from rural Poland. Am J Hum Biol 16:563–580

Jedrzejewski W, Schmidt K, Okarma H, Kowalczyk R (2002) Movement pattern and home range use by the Eurasian lynx in Bialowieza Primeval forest (Poland). Annales Zoologici Fennici 39:29–41

Jedrzejewski W, Schmidt K, Theuerkauf J, Jedrzejewska B, Okarma H (2001) Daily movements and territory use by radio-collared wolves (*Canis lupus*) in Bialowieza Primeval Forest in Poland. Can J Zool 79:1993–2004

Kibii JM, Churchill SE, Schmid P, Carlson KJ, Reed ND, de Ruiter DJ, Berger LR (2011) A partial pelvis of *Australopithecus sediba*. Science 333:1407–1411

Kowalzcyk R, Zalewski A, Jedrzejewska B (2006) Daily movement and territory use by badgers *Meles meles* in Bialowieza Primeval Forest, Poland. Wildlife Biol 12:385–391

Kramer PA (1998) The costs of human locomotion: maternal investment in child transport. Am J Phys Anthropol 107:71–85

Kramer PA (1999) Modeling the locomotor energetics of extinct hominids. J Exp Biol 202:2807–2818

Kramer PA (2004) The behavioral ecology of locomotion. In: Meldrum DJ, Hilton CE (eds) From Biped to Strider: the emergence of modern human walking, running and resource transport. Plenum, New York, pp 101–115

Kramer PA (2010) The effect of energy expenditure of walking on gradients or carrying burdens. Am J Hum Biol 22:497–507

Kramer PA, Eck GG (2000) Locomotor energetics and leg length in hominid bipedality. J Hum Evol 38:651–666

Kramer PA, Sylvester AD (2009) Bipedal form and locomotor function: understanding the affects of size and shape on velocity and energetics. PaleoAnthropol 238–251

Kuhn SL, Stiner MC (2006) What's a mother to do? the division of labor among Neandertals and modern humans in Eurasia. Curr Anthropol 47:953–980

Kuo AD (1999) Stabilization of lateral motion in passive dynamic walking. Int J Robot Res 18:917–930

Kuo AD (2007) Choosing your steps carefully: trade-offs between economy and versatility in dynamic walking bipedal robots. IEEE Robot Autom Mag 14:18–29

Kuo AD, Donelan JM, Ruina A (2005) Energetic consequences of walking like an inverted pendulum: step-to-step transitions. Exerc Sport Sci Rev 33:88–97

Kurki HK (2011) Pelvic dimorphism in relation to body size and body size dimorphism in humans. J Hum Evol 61:631–643

Kurki HK, Ginter JK, Stock JT, Pfeiffer S (2008) Adult proportionality in small-bodied foragers: a test of ecogeographic expectations. Am J Phys Anthropol 136:28–38

Langman VA, Rowe MF, Roberts TJ, Langman NV, Taylor CR (2012) Minimum cost of transport in Asian elephants: do we really need a bigger elephant? J Exp Biol 215:1509–1514

Lordkipanidze D, Jashashvili T, Vekua A, Ponce de Leon MS, Zollikofer CPE, Rightmire GP, Pontzer H, Ferring R, Oms O, Tappen M, Bukhsianidze M, Agusti J, Kahlke R, Kiladze G, Martinez-Navarro B, Mouskhelishvili A, Nioradze M, Rook L (2007) Postcranial evidence from early *Homo* from Dmanisi, Georgia. Nature 449:305–310

Lovejoy CO (1988) Evolution of human walking. Sci Am 256:118–125

Macdonald K, Roebroeks W, Verpoorte A (2009) An energetics perspective on the Neanderthal record. In: Hublin J-J, Richards MP (eds) The evolution of hominin diets: integrating approaches to the study of palaeolithic subsistence. Springer Science+Business Media B.V, New York, pp 211–220

Minetti AE, Alexander RM (1997) A theory of metabolic costs for bipedal gaits. J Theor Biol 186:467–476

Minetti AE, Saibene FP, Ardigo LP, Atchou G, Schena F, Ferretti G (1994) Pygmy locomotion. Eur J Appl Physiol 68:285–290

O'Connor SM, Donelan JM (2012) Fast visual prediction and slow optimization of preferred walking speed. J Neurophysiol 107:2549–2559

O'Neill M (2012) Gait-specific metabolic costs and preferred speeds in ringtailed lemurs (*Lemur catta*), with implications for the scaling of locomotor costs in primates. Am J Phys Anthropol 149:356–364

Pearson OM, Petersen TR, Sparacello VS, Daneshvari S, Grine FE (2014) Activity, "body shape", and cross-sectional geometry of the femur and tibia. In: Carlson KJ, Marchi D (eds) Reconstructing mobility: environmental, behavioral, and morphological determinants. Springer, New York

Pettitt PB (1997) High resolution Neanderthals? Interpreting middle Palaeolithic intrasite spatial data. World Archaeol 92:208–224

Peyrot N, Thivel D, Isacco L, Morin JB, Belli A, Duche P (2012) Why does walking economy improve after weight loss in obese adolescents? Med Sci Sports Exerc 44:659–665

Pike-Tay A, Valdes VC, de Quiros FB (1999) Seasonal variations of the Middle-Upper Paleolithic transition at El Castillo, Cueva Morin and El Pendo (Cantabria, Spain). J Hum Evol 36:283–317

Pike IL (2005) Maternal stress and fetal responses: evolutionary perspectives on preterm delivery. Am J Hum Biol 17:55–65

Ponce de Leon MS, Golovanova L, Doronichev V, Romanova G, Akazawa T, Kondo O, Ishida H, Zollikofer CPE (2008) Neanderthal brain size at birth provides insights into the evolution of human life history. Proc Natl Acad Sci U S A 105:13764–13768

Pontzer H (2007) Limb length and the scaling of locomotor cost in terrestrial animals. J Exp Biol 210:1752–1761

Pontzer H, Raichlen DA, Wood BM, Mabulla AZP, Racette SB, Marlowe FW (2012) Hunter-gatherer energetics and human obesity. PLoS One 7:e40503

Pontzer H, Rolian C, Rightmire GP, Jashashvili T, Ponce de León MS, Lordkipanidze D, Zollikofer CPE (2010) Locomotor anatomy and biomechanics of the Dmanisi hominins. J Hum Evol 58:492–504

Porter AMW (1996) Physique and the skeleton. PhD Dissertation, University of London, London

Rak Y (1991) Lucy's pelvic anatomy: its role in bipedal gait. J Hum Evol 20:283–290

Ralston HJ (1958) Energy-speed relation and optimal speed during level walking. Eur J Appl Physiol 17:277–283

Richards M, Harvati K, Grimes V, Smith C, Smith T, Hublin J-J, Karkanas P, Panagopoulou E (2008) Strontium isotope evidence of Neanderthal mobility at the site of Lakonis, Greece using laser-ablation PIMMS. J Archaeol Sci 35:1251–1256

Richmond BG, Jungers WL (2008) *Orrorin tugenensis* femoral morphology and the evolution of hominin bipedalism. Science 319

Roberts DF (1978) Climate and human variability. Cummings, Philippines

Rosenberg KR, Zuné L, Ruff CB (2006) Body size, body proportions, and encephalization in a Middle Pleistocene archaic human from northern China. Proc Natl Acad Sci U S A 103:3552–3556

Ruff CB (1984) Allometry between length and cross-sectional dimensions of the femur and tibia in *Homo sapiens sapiens*. Am J Phys Anthropol 65:347–358

Ruff CB (1987) Sexual dimorphism in human lower limb bone structure: relationship to subsistence strategy and sexual division of labor. J Hum Evol 16:391–416

Ruff CB (1994) Morphological adaptation to climate in modern and fossil hominids. Yearbook Phys Anthropol 37:65–107

Ruff CB (1995) Biomechanics of the hip and birth in early *Homo*. Am J Phys Anthropol 98:527–574

Ruff CB, Hayes WC (1983) Cross-sectional geometry of Pecos Pueblo femora and tibiae-a biomechanical investigation. I. Method and general patterns of variation. Am J Phys Anthropol 60:359–381

Ruff CB, Larsen CS (2014) Long bone structural analyses and the reconstruction of past mobility—a historical review. In: Carlson KJ, Marchi D (eds) Reconstructing mobility: environmental, behavioral, and morphological determinants. Springer, New York

Schultz AH (1926) Fetal growth of man and other primates. Quart Rev Biol 1:465–521

Simpson SW, Quade J, Levin NE, Butler R, Dupont-Nivet G, Everett M, Semaw S (2008) A female *Homo erectus* pelvis from Gona, Ethiopia. Science 322:1089–1092

Simpson SW, Spurlock LB, Lovejoy CO, Latimer B (2010) A new reconstruction of the KNM-WT 15000 juvenile male pelvis. Am J Phys Anthropol 50:217

Speakman JR, Król E (2010) Maximal heat dissipation capacity and hyperthermia risk: neglected key factors in the ecology of endotherms. J Animal Ecol 79:726–746

Steudel-Numbers K (2006) Energetics in *Homo erectus* and other early hominins: the consequences of increased lower limb length. J Hum Evol 51:445–453

Steudel-Numbers K, Tilkens M (2004) The effect of lower limb length on the energetic cost of locomotion: implications for fossil hominins. J Hum Evol 47:95–109

Steudel-Numbers K, Wall-Scheffler CM (2009) Optimal running speed and the evolution of hominin hunting strategies. J Hum Evol 56:355–360

Steudel K (1994) Locomotor energetics and hominid evolution. Evol Anthropol 3:42–48

Steudel K (1996) Limb morphology, bipedal gait and the energetics of hominid locomotion. Am J Phys Anthropol 99:345–355

Steudel K, Beattie J (1995) Does limb length predict the energetic cost of locomotion in mammals? J Zool 235

Stock JT (2006) Hunter-gatherer postcranial robusticity relative to patterns of mobility, climatic adaptation, and selection for tissue economy. Am J Phys Anthropol 131:194–204

Stock JT, O'Neill M, Ruff CB, Zabecki M, Shackelford L, Rose J (2011) Body size, skeletal biomechanics, mobility and habitual activity from the Late Palaeolithic to mid-Dynastic Nile Valley. In: Pinhasi R, Stock JT (eds) Human bioarchaeology of the transition to agriculture. Wiley-Blackwell, Oxford, pp 347–370

Stringer CB, Trinkaus E, Roberts MB, Parfitt SA, Macphail RI (1998) The middle Pleistocene human tibia from Boxgrove. J Hum Evol 34:509–547

Tilkens M, Wall-Scheffler CM, Weaver T, Steudel-Numbers K (2007) The effect of body proportions on thermoregulation: an experimental assessment of Allen's rule. J Hum Evol 53:286–291

Trinkaus E (1980) Sexual differences in Neanderthal limb bones. J Hum Evol 9:377–397

Trinkaus E (1981) Neanderthal limb proportions and cold adaptation. In: Stringer CB (ed) Aspects of human evolution. Taylor & Francis, London, pp 187–224
Trinkaus E (1983) Neandertal postcrania and the adaptive shift to modern humans. In: Trinkaus E (ed) The Mousterian legacy: human biocultural change in the upper Pleistocene. Archaeopress, Oxford, pp 165–200
Trinkaus E (2009) The human tibia from Broken Hill, Kabwe, Zambia. PaleoAnthropol 145–165
Trinkaus E, Ruff CB (1999a) Diaphyseal cross-sectional geometry of Near Eastern Middle Palaeolithic humans: the femur. J Archaeol Sci 26:409–424
Trinkaus E, Ruff CB (1999b) Diaphyseal cross-sectional geometry of Near Eastern Middle Palaeolithic humans: the tibia. J Archaeol Sci 26:1289–1300
Voloshina AS, Kuo AD, Daley MA, Ferris DP (2013) Biomechanics and energetics of walking on uneven terrain. J Exp Biol 216(Pt 21):3963–70. doi:10.1242/jeb.081711
Walker A, Leakey R (1993) The postcranial bones. In: Walker A, Leakey R (eds) The Nariokotome *Homo erectus* Skeleton. Harvard University Press, Cambridge, pp 95–160
Walker MJ, Ortega J, Parmova K, Lopez MV, Trinkaus E (2011) Morphology, body proportions, and postcranial hypertrophy of a female Neanderthal from the Sima de las Palomas, southeastern Spain. Proc Natl Acad Sci U S A 108:10087–10091
Wall-Scheffler CM (2007) Digital cementum luminance analysis and the Haua Fteah hominins: how seasonality and season of use changed through time. Archaeometry 49:815–826
Wall-Scheffler CM (2010) Evolutionary tradeoffs: how thermoregulation and the cost of mobility impact morphology. Am J Phys Anthropol 141:238
Wall-Scheffler CM (2012a) Energetics, locomotion and female reproduction: implications for human evolution. Ann Rev Anthropol 41:71–85
Wall-Scheffler CM (2012b) Size and shape: morphology's impact on human speed and mobility. J Anthropol 2012:1–9
Wall-Scheffler CM, Chumanov ES, Steudel-Numbers K, Heiderscheit BC (2010) EMG activity across gait and incline: the impact of muscular activity on human morphology. Am J Phys Anthropol 143:601–611
Wall-Scheffler CM, Geiger K, Steudel-Numbers K (2007) Infant carrying: the role of increased locomotory costs in early tool development. Am J Phys Anthropol 133:841–846
Wall-Scheffler CM, Myers MJ (2012) Is female morphology selected for economy and male morphology selected for efficiency? evidence from studies on humans. Am J Phys Anthropol 54:296
Wall-Scheffler CM, Myers MJ (2013) Reproductive costs for everyone: how female loads impact human mobility strategies. J Hum Evol 64:448–456
Wall CM (2005) The seasonality of site deposition of Gibraltar Neanderthals: evidence from Gorham's and Vanguard Caves. J Iberian Archaeol 7:9–22
Watson JC, Payne RC, Chamberlain AT, Jones RK, Sellers WI (2008) The energetic costs of load-carrying and the evolution of bipedalism. J Hum Evol 54:675–683
Weyand PG, Smith BR, Puyau MR, Butte NF (2010) The mass-specific energy cost of human walking is set by stature. J Exp Biol 213:3972–3979
Wheeler PE (1993) The influence of stature and body form on hominid energy and water budgets; a comparison of *Australopithecus* and early *Homo* physiques. J Hum Evol 24:13–28
Ziomkiewicz A, Ellison PT, Lipson SF, Thune I, Jasienska G (2008) Body fat, energy balance and estradiol levels: a study based on hormonal profiles from complete menstrual cycles. Hum Reprod 23:2555–2563

Chapter 11
Femoral Diaphyseal Shape and Mobility: An Ontogenetic Perspective

Libby W. Cowgill

Abstract The ratio of midshaft femoral anteroposterior (I_x) to mediolateral (I_y) second moments of area has been suggested as a possible indicator of mobility, and has been applied as an analytic tool to paleoanthropological and archaeological samples with variable success. Under this model, biomechanical loads associated with increased mobility modify the shape of the femoral midshaft from a roughly circular cross section to an anteroposteriorly reinforced one. While previous research indicates that immature femora respond to changes in the manner of locomotor loading (Ruff 2003; Cowgill et al. 2010), relatively fewer studies have examined population-level differences in immature femoral shape as a product of overall group mobility.

This study uses seven immature Holocene human samples ($n = 522$) to explore two questions: (1) When do population differences in midshaft femoral shape emerge during ontogeny? (2) Does the midshaft femoral mobility index correlate with other cross-sectional properties of the femur? The results of this analysis indicate that while population-level shape differences appear relatively early in human ontogeny (>6 years of age), these differences may not correspond to expected levels of group mobility, and may be a product of differences in body proportions. In addition, while many factors are implicated, it remains unclear what intrinsic dynamics directly impact midshaft femoral shape during ontogeny, and, while mobility levels may be a factor, additional influences need to be evaluated.

Keywords Femur • Growth • Body proportions • Robusticity • Mobility

L.W. Cowgill (✉)
Department of Anthropology, University of Missouri, Columbia, Swallow Hall 107, Columbia, MO 63211, USA
e-mail: cowgilll@missouri.edu

K.J. Carlson and D. Marchi (eds.), *Reconstructing Mobility: Environmental, Behavioral, and Morphological Determinants*, DOI 10.1007/978-1-4899-7460-0_11,

In both archaeological and fossil groups, femoral strength ratios have been used as indicators of levels of adult terrestrial "mobility" (Ruff 1987; Larsen et al 1995; Holt 2003; Wescott 2006), which, for the purposes of this analysis, is defined as the sum total of locomotor activities performed using the lower limb, following Pearson and colleagues (2014). Specifically, the ratio of anteroposterior (AP) (I_x) to mediolateral (ML) (I_y) second moments of area has been previously described as a "mobility index," with higher ratios roughly corresponding to more long distance travel (Ruff 1987). This contention is supported by experimental studies demonstrating that hamstring and quadriceps contraction engenders high AP bending loads in the lower limb (Morrison 1968; Wescott 2006). Within anthropology, differences in femoral shape have been documented across temporal periods, subsistence groups, and the sexes (Ruff 1987; Holt 2003).

To date, however, no studies have examined how levels of group mobility affect immature individuals within groups, or how differences in femoral shape manifest over the course of growth. Previous research indicates that immature femora respond to changes in the manner and level of locomotor loading (Ruff 2003; Cowgill et al. 2010). In a longitudinal sample, Ruff (2003) was able to detect the cessation of crawling and the onset of bipedal locomotion via changes in the velocity of bone appositional deposition. This period is typified by a decrease in immature humeral strength and an increase in femoral strength, as the femur assumes a primarily supportive role and the humerus switches to a predominantly manipulative milieu.

Furthermore, in addition to changes in strength related to large transitions in locomotor patterns, femoral shape in immature individuals changes in response to more subtle shifts in locomotor dynamics. Cowgill and colleagues (2010) detected differences in immature femoral shape in response to changes in walking patterns in early childhood. Early walking in children is characterized by differences from mature walking instance, step length, cadence, and walking velocity. These differences are likely a product of a variety of factors including incomplete musculoskeletal and neurological development, a relatively broad pelvis and a proportionately short lower limb, as well as the lack of a valgus angle. The "waddling" gait patterns typical of young children expose the lower limb to high ML loads, which induce bone deposition on the femur and create high values of ML bending strength (I_y).

The studies above suggest that populations do differ in femoral shape and terrestrial mobility levels, and that differences in femoral shape are likely a product, at least in part, of changes that occur as part of overall group mobility during ontogeny. In populations with high mobility, once children exceed a certain age and size, it would no longer have been feasible to carry them, and their general levels of lower limb strength should reflect overall levels of populational mobility. Furthermore, in a large proportion of non-urban societies, children begin to contribute to subsistence between the ages of 6 and 10, and after the age of 10, the labor patterns of children are similar to those of same-sex adults (Moberg 1985; Bradley 1993). Numerous accounts exist of children in hunting and gathering economies undertaking subsistence-related tasks (Laughlin 1968; Blurton Jones et al. 1989, 1994; Hawkes et al. 1995; Bird and Bliege Bird 2000, 2002, 2005; Bird-David 2005; Tucker and Young 2005), and among agricultural populations, children are almost universally valued and important components of the agricultural work force (Minturn and Hitchcock 1963; Romney and Romney 1963; Nydegger and Nydegger 1963; White 1975; Cain 1977; Nag et al. 1978; Bradley 1984, 1993; Wyer 1986; Ratha and Mahakud 1988; Korkiakangas 1990; Porter 1996). Given this, both subsistence strategy and population mobility levels are likely to affect the lower limb loading of immature individuals.

By exploring correlations among shape indices and other measures of long bone strength and shape during development, new insights into the detection of differences in mobility patterns may be attained. Therefore, this research addresses two questions. First, when do population differences in midshaft femoral mobility index (I_x/I_y) emerge during ontogeny? By examining these changes within three age categories, it may be possible to determine at what point during growth differing population levels begin to affect immature individuals. Second, does the midshaft femoral mobility index correlate with other cross-sectional properties and metrics of the femur? In addition to mobility, other intrinsic factors such as overall femoral and tibial strength, angulation of the femoral neck and distal metaphyses, femoral curvature, and body proportions may influence midshaft femoral shape.

11.1 Materials and Methods

The primary data for this analysis consist of femoral cross-sectional properties and metrics from seven Holocene human skeletal samples (Cowgill 2010). Measurements were collected from a total of 522 immature individuals under the age of 18, although actual sample size may vary by analysis. The seven samples were selected to represent the broadest possible range of historical and archaeological time periods, geographic locations, and subsistence strategies. Individuals displaying indicators of obvious developmental pathology were excluded, although observations of nonspecific developmental stress (Harris lines, cribra orbitalia, porotic hyperostosis) were not considered grounds for exclusion.

11.1.1 Samples

While details of the comparative sample have been published elsewhere (Cowgill 2010), and are summarized in Table 11.1, they are discussed at greater length here for additional clarity. The California Amerindian sample used in this analysis is derived from 28 sites in the Alameda, Sacramento, and San Joaquin counties of north-central California, primarily clustered along the San Francisco Bay and the Sacramento and San Joaquin River valleys. California Amerindians of this area are best characterized as precontact, semisedentary foraging populations, reliant on deer, elk, antelope, fishing, and extensive exploitation of acorns. Indian Knoll is an Archaic Period shell-midden site located on the Green River in Kentucky

Table 11.1 Sample description, size, date, latitude, and location

Sample	Original location	Approximate time period	*N*	Sample location
California Amerindian	Northern California	500–4600 BP	91	Phoebe Hearst Museum at the University of California, Berkeley (Berkeley, CA)
Dart	Johannesburg, South Africa	Twentieth century	73	School of Medicine, University of Witwatersrand (Johannesburg, South Africa)
Indian Knoll	Green River, Kentucky	4143–6415 BP	95	University of Kentucky, Lexington (Lexington, KY)
Kulubnarti	Batn el Hajar, Upper Nubia	Medieval (sixth to fourteenth century)	99	University of Colorado, Boulder (Boulder, CO)
Luis Lopes	Lisbon, Portugal	Twentieth century	47	Bocage Museum (Lisbon, Portugal)
Mistihalj	Bosnia-Herzegovina	Medieval (fifteenth century)	52	Peabody Museum at Harvard University (Cambridge, MA)
Point Hope	Point Hope, Alaska	300–2100 BP	65	American Museum of Natural History (New York, NY)

(Webb 1946). Individuals from Indian Knoll were likely semisedentary with prolonged residences at seasonally occupied sites, who experienced relatively high population densities, and relied heavily on a narrow spectrum of essential resources, such as deer, turkey, mussels, nuts, and a variety of locally collected plant materials (Winters 1974). The site of Kulubnarti is located in Upper Nubia in the Batn el Hajar region, approximately 130 km south of Wadi Halfa, where two medieval Christian cemeteries containing 406 burials were excavated in 1979 (van Gerven et al. 1995). With marginal subsistence levels, individuals traditionally lived in small villages, participated in small-scale agriculture, and likely suffered from chronic nutritional difficulty combined with bouts of infectious disease during growth (Van Gerven et al. 1990). Mistihalj is a medieval burial site located on the border between Bosnia-Herzegovina and Montenegro. The remains at Mistihalj are culturally associated with the Vlakhs, an indigenous Balkan ethnic group, who primarily engaged in breeding sheep, horses, mules, and cattle, and who migrated seasonally over varied terrain (Alexeeva et al. 2003). The Dart Collection is an ethnically mixed, native African cadaver sample derived from hospitals in the Transvaal region in South Africa (Saunders and Devito 1991). Approximately 74 % of all individuals died prior to 1950, and approximately 92 % of the individuals within this sample are Bantu-speaking South African Blacks. Due to the diversity of this region, it is difficult to classify this sample area as exclusively rural or urban. The Luis Lopes skeletal collection consists of twentieth century Portuguese from several cemeteries in Lisbon. In general, the sample is best categorized as an urban population of low to middle socioeconomic status (Cardoso 2005). The site of Point Hope, Alaska is situated on a peninsula in the Chuckchi Sea, approximately 200 km north of the Arctic Circle

(Larsen and Rainey 1948). Earlier periods of the Point Hope stratigraphic sequence are characterized by a reliance on caribou hunting, whereas later cultural horizons indicate a more extensive dependence on the exploitation of maritime resources such as walruses, seals, and whales (Larsen and Rainey 1948; Rainey 1971). Immature skeletal remains from the multiple cultural periods excavated at Point Hope were combined into a single sample for this analysis, as previous analyses of Point Hope adults and immature individuals found little biomechanical difference between the early and late periods at this site (Shackelford 2007, 2014; Cowgill 2014).

11.1.2 Aging and Sex

Age was undocumented for six of the seven samples used in this study, and crown and root formation evaluated from lateral mandibular radiographs was used whenever dental and postcranial remains were reliably associated. Crown and root formation was assessed following the developmental standards set by Smith (1991) for permanent dentition and Liversidge and Molleson (2004) for deciduous dentition. Each set of dentition was scored twice on two consecutive days, and individual teeth that produced different formation stage scores were evaluated a third time to resolve inconsistencies. When no dentition was directly associated with the postcranial remains, chronological age was predicted from within-sample least squares regression of femoral, tibial, or humeral length on age for each of the comparative samples in order to maximize sample size (Cowgill 2010). By developing age-prediction equations specific to each sample, difficulties arising from the application of a formula developed on individuals differing in body size or proportions to an archaeological target sample are partially mitigated. In order to evaluate the age at which population differences in femoral shape occur, the sample was divided into three broad age categories: 0.0–5.9, 6.0–11.9, and 12.0–17.9 years. Given the well-known difficulties of determining sex in immature samples (Scheuer and Black 2000), this was not attempted here and both males and females are analyzed together.

11.1.3 Cross-Sectional Geometry

Biomechanical data for this analysis consist of the ratio between midshaft AP and ML second moments of area for immature femora. Biomechanical length for unfused humeri and femora were measured following Trinkaus and colleagues (2002a, b). Cross-sectional levels were chosen to best approximate the 50 % section level in fused elements. In immature femora, however, 50 % of diaphyseal length was calculated as 45.5 % of femoral intermetaphyseal length, as this measurement best corresponds to the location of the 50 % level in individuals with fused distal femoral epiphyses due to the relatively larger contribution of the distal epiphysis to biomechanical length in fused femora (Ruff 2003).

All cross-sectional properties were collected using a method similar to O'Neill and Ruff's (2004) "latex cast method" (LCM) and the method used by Sakaue (1998), which rely on AP and ML radiographs and silicone molding putty. In order to reconstruct femoral and humeral cross-sectional properties, the external surface of the diaphysis was molded with Cuttersil Putty Plus™ silicone molding putty. Anterior, posterior, medial, and lateral cortical bone thicknesses were measured with digital calipers, and measurements were corrected for parallax distortion by comparing external breadths measured on the radiograph with external breadths measured on the element. Once corrected for parallax, the four cortical bone thicknesses were plotted onto the two-dimensional copy of the original mold, and the endosteal contours were interpolated by using the subperiosteal contour as a guide. The resultant sections were enlarged on a digitizing tablet, and the endosteal and periosteal contours digitized. Cross-sectional properties were computed from the sections in a PC-DOS version of SLICE (Nagurka and Hayes 1980; Eschman 1992).

11.1.4 Size Standardization

Size standardization was needed for two separate purposes in this analysis. First, cross-sectional geometric properties must be standardized by some biomechanically relevant proxy for size, such as body mass (for areas) or body mass × beam length (for bending moments). Body mass was predicted based on formulae developed specifically for immature individuals, which predict body mass from femoral distal metaphyseal ML breadth and femoral head size (Ruff 2007). To remove the effect of body mass on femoral cross-sectional properties, cross-sectional properties were regressed on body mass (total and cortical area) or body mass × beam length^2 (polar second moment of area) using Least Squares regression (Ruff 2000; Cowgill 2010). Standardized residuals, which are the raw residuals divided by the standard deviation of residuals, were then used in comparisons of population differences.

Second, body mass was also used to standardize femoral ratios for age and size. Given that femoral ratios are correlated with both age and body mass ($p < 0.001$, see results below), it is necessary to remove this variation prior to making comparison across multiple samples. Without age or body mass correction, random differences in age distribution among samples will strongly bias results. As correlations between age and body mass are high ($r = 0.965$), both are good choices for standardization, and body mass is used here to prevent error due to random age sampling effects.

11.1.5 Long Bone Metrics

In order to address the second research question, femoral I_x/I_y is compared to several femoral metrics. Femoral metrics and properties include: femoral and tibial total area, cortical area, polar second moment of area, tibial I_x/I_y, neck-shaft angle, bicondylar angle, crural index, femoral curvature subtense, and position of maximum curvature (distance between max. curve subtense and proximal terminus of the chord/chord curvature length). Cross-sectional properties were collected using the methodology detailed above. Measurements of femoral curvature were taken following Shackelford and Trinkaus (2002). Maximum diaphyseal lengths were recorded from all immature humeri, radii, femora, and tibiae to the nearest millimeter. Crural indices were calculated as tibial length/femoral length · 100.

11.1.6 Statistical Analysis

In order to assess when population-specific differences in femoral shape manifest during growth (first research question), femoral I_x/I_y was regressed on body mass to remove size and age-related variation. Standardized residuals from this regression were compared by population using Mann–Whitney U and Kruskal–Wallis statistics due to smaller samples sizes when dividing populations.

In order to investigate developmental causality of variation in femoral shape (second research question), correlations between body mass corrected shape index residuals and femoral metrics, and femoral and tibial cross-sectional properties were explored using Pearson's correlation coefficients. All metrics and properties were corrected for size via regression on body mass in order to avoid spurious correlations due to the large amount of size variation in the ontogenetic data set.

11.2 Results

When analyzed without age subdivisions, populations do vary in shape ratio. Indian Knoll shows the highest (most elliptical) femoral ratios, followed by Kulubnarti, Mistihalj, Luis Lopes, California Amerindians, Dart, and Point Hope, respectively. Figure 11.1 shows the regression of femoral I_x/I_y on body mass ($r^2=0.393$, $p<0.001$) and boxplots of the residuals from this analysis across all age groups. Kruskal–Wallis tests of population differences across all ages are significant ($p<0.001$), and Mann–Whitney pair-wise comparisons indicated that shape indices are significantly higher in the Indian Knoll population relative to Point Hope, Dart, and Californian Amerindians.

When subdivided by age groups, population differences exist in all three age groups (Fig. 11.2; $p=0.01$, $p=0.002$, $p<0.001$, respectively). Table 11.2 shows the sample sizes, raw ratio means, residual means, and standard deviations. In age group 1, pair-wise comparisons reflect low strength ratios at Point Hope. In age group 2, pair-wise comparisons begin to detect high ratios at Indian Knoll and low ratios among the California Amerindians. Finally, in age group 3, pair-wise comparisons reflect both high ratios at Indian Knoll and Kulubnarti, and low ratios at Point Hope, Dart, and California Amerindian (Fig. 11.3).

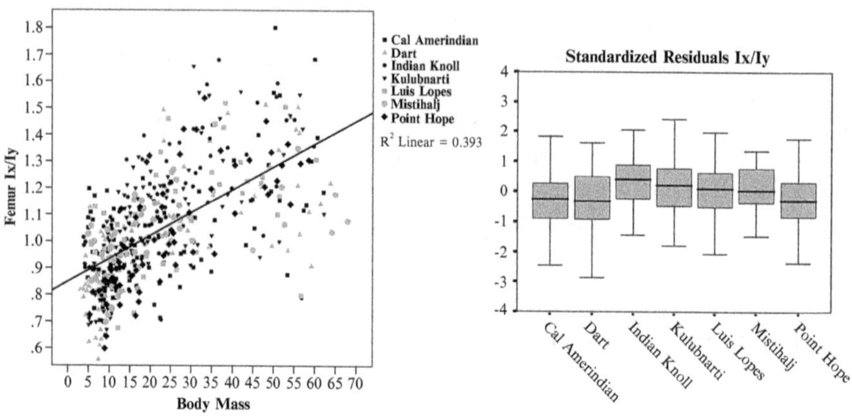

Fig. 11.1 A scatterplot of the regression of femoral I_x/I_y on body mass ($r^2=0.393$, $p<0.001$) and boxplots of the residuals from this analysis across all age groups

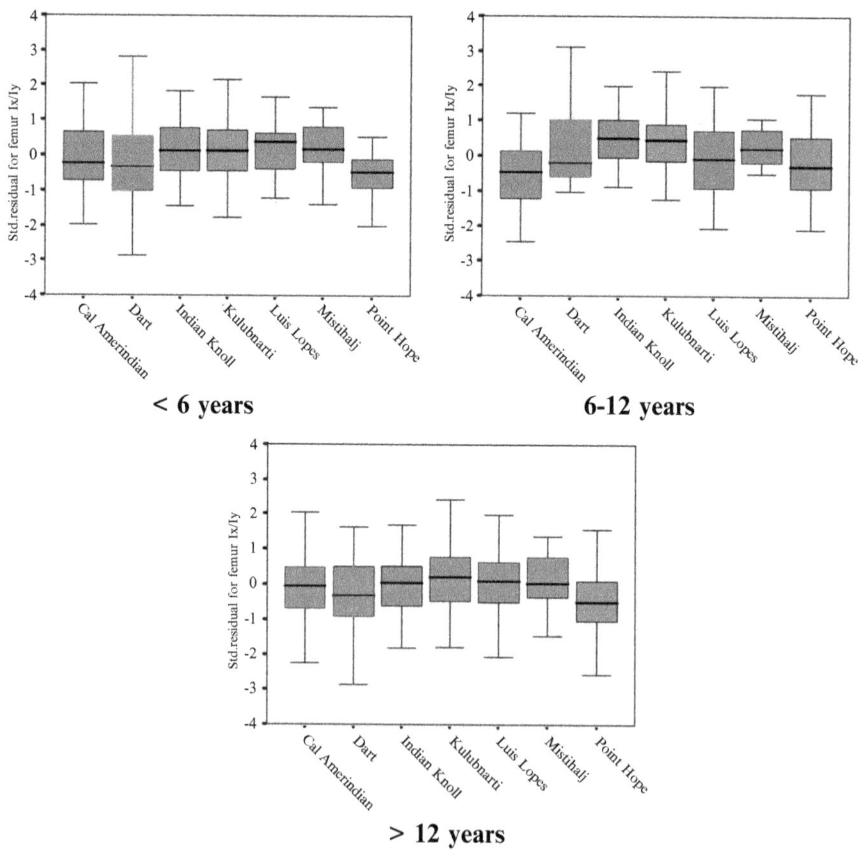

Fig. 11.2 Boxplots illustrating population-specific standardized residual medians for the three separate age categories

Table 11.2 Sample sizes, raw ratio values, standardized residual means, and standard deviations (standardized residual means are interpreted in the text)

Age category (years)	Sample	*N*	Raw ratio mean	Residual mean	Std. deviation
0.0–5.9	Cal. Amerindian	46	0.915	−0.122	0.134
	Dart	36	0.842	−0.446	0.146
	Indian Knoll	43	0.938	0.012	0.119
	Kulubnarti	51	0.943	−0.031	0.158
	Luis Lopes	19	0.946	−0.013	0.128
	Mistihalj	29	0.935	−0.028	0.121
	Point Hope	27	0.854	−0.537	0.116
6.0–11.9	Cal. Amerindian	25	0.995	−0.412	0.137
	Dart	10	1.100	0.274	0.189
	Indian Knoll	31	1.162	0.633	0.158
	Kulubnarti	29	1.141	0.564	0.185
	Luis Lopes	16	1.114	0.185	0.219
	Mistihalj	15	1.105	0.277	0.095
	Point Hope	23	1.082	0.122	0.195
12.0–17.9	Cal. Amerindian	20	1.263	−0.123	0.242
	Dart	25	1.218	−0.262	0.181
	Indian Knoll	19	1.350	0.736	0.215
	Kulubnarti	18	1.242	0.133	0.218
	Luis Lopes	12	1.214	−0.188	0.188
	Mistihalj	8	1.161	−0.928	0.162
	Point Hope	15	1.283	0.146	0.098
All ages	Cal. Amerindian	91	1.014	−0.202	0.213
	Dart	71	1.011	−0.279	0.240
	Indian Knoll	93	1.097	0.367	0.224
	Kulubnarti	98	1.056	0.178	0.216
	Luis Lopes	47	1.071	0.010	0.207
	Mistihalj	52	1.019	−0.079	0.154
	Point Hope	65	1.034	−0.146	0.222

Results of the analysis of correlations between femoral shape and other metrics of the femur and tibia are shown in Fig. 11.4. Out of the 12 analyses, nine comparisons reached statistical significance.

11.3 Discussion and Conclusion

Differences in strength ratios among populations do exist and manifest relatively early in growth (prior to the age of 6). However, the differences that appear the earliest are not high ratios of I_x/I_y, but low ratios at Point Hope, indicating that relatively round femoral shafts are an early development among this population. This pattern may be a product of cold-adapted body proportions. A wide body of research has

	Cal Amerindian	Dart	Indian Knoll	Kulubnarti	Luis Lopes	Mistihalj
			AGE CATEGORY 1			
Dart	0.209					
Indian Knoll	0.380	0.031				
Kulubnarti	0.481	0.060	0.725			
Luis Lopes	0.394	0.095	0.897	0.719		
Mistihalj	0.367	0.053	0.877	0.568	0.958	
Point Hope	0.041	0.442	0.001	0.004	0.006	0.001
			AGE CATEGORY 2			
Dart	0.126					
Indian Knoll	0.002	0.266				
Kulubnarti	0.001	0.496	0.593			
Luis Lopes	0.087	0.732	0.137	0.345		
Mistihalj	0.004	0.452	0.254	0.561	0.363	
Point Hope	0.219	0.367	0.034	0.075	0.648	0.132
			AGE CATEGORY 3			
Dart	0.727					
Indian Knoll	0.005	0.004				
Kulubnarti	0.024	0.016	0.116			
Luis Lopes	0.133	0.118	0.095	0.877		
Mistihalj	0.089	0.041	0.074	0.687	0.911	
Point Hope	0.842	0.985	0.002	0.012	0.077	0.051

Fig. 11.3 Results of Mann–Whitney pair-wise comparisons within age groups (p-values)

successfully explored the application of Bergmann's rule (1847) and Allen's rule (1877) to ecogeographic variation in human body proportions (Hiernaux and Froment 1976; Ruff 1991, 1994, 2002; Holliday and Falsetti 1995; Holliday 1997a, b, 1999; Katzmarzyk and Leonard 1998; Pearson 2000; Auerbach 2007; Temple and Matsumura 2011). In general, populations from high latitudes frequently display relatively wide bodies, high body masses for stature, short limbs relative to trunk length and foreshortened distal extremities, whereas populations from low latitudes have relatively narrow bodies, low body masses in relation to stature, long limbs, and long distal limb extremities. Cold-adapted body proportions are present in adults from Point Hope, including wide bi-iliac breadths and low crural indices (Holliday and Hilton 2010). Furthermore, recent research indicated that ecogeographic body proportion patterning remains constant during growth and that individuals with cold-adapted body proportions as adults are likely to show similar

		Femoral Ix/Iy
Tibial Ix/Iy	r	0.191
	p-value	<0.001
	N	452
Crural index	r	0.105
	p-value	0.026
	N	450
Femoral Curviture subtense	r	0.230
	p-value	<0.001
	N	484
Subtense Position ratio	r	0.134
	p-value	0.004
	N	449
Neckshaft Angle	r	0.117
	p-value	0.009
	N	493
Bicondylar Angle	r	0.098
	p-value	0.045
	N	421
Femoral TA	r	0.117
	p-value	0.008
	N	516
Femoral CA	r	0.147
	p-value	0.001
	N	516
Femoral J	r	0.021
	p-value	0.629
	N	513
Tibial TA	r	0.201
	p-value	<0.001
	N	452
Tibial CA	r	0.206
	p-value	<0.001
	N	452
Tibial J	r	0.075
	p-value	0.109
	14	452

Fig. 11.4 Correlation coefficients, *p*-values, and sample sizes for correlations between I_x/I_y and other metrics and strength properties

patterns during ontogeny (Cowgill et al. 2012). This is further supported by the low but significant correlation between femoral shaft ratios and crural indices detected in this analysis.

Several analyses have linked differences in bi-iliac breadth to variation in femoral shape indices (Ruff 1995; Weaver 2003, Cowgill et al. 2010; Shaw and Stock 2011). While most studies have focused on femoral shape at the subtrochanteric level, there is evidence that pelvic breadth may affect midshaft shape as well (Weaver 2003; Cowgill et al. 2010; Shaw and Stock 2011). Neandertals have long been noted to have relatively round femoral midshafts, particularly in light of their reconstructed activity levels, which, based on a variety of different indicators, were likely to be high. In an analysis using geometric morphometics of the hip, Weaver (2003) concluded that Neandertal midshaft femoral shape was primarily a consequence of wide bodies due to climate-related adaptations. In addition, Pearson and colleagues (2014) found a weak correlation between bi-iliac breadth and midshaft femoral shapes in samples that did not contain Inuits or other high latitude populations. It is likely that this correlation would be considerably higher if Inuits were included, as recent research suggests that the extreme adaptations of Arctic foragers may drive higher correlations in studies of human body proportions, and a sampling range of 50° of latitude is necessary to detect human ecogeographical patterning (Foster and Collard 2013). Furthermore, in an analysis of immature femora that combined information from cross-sectional geometry with data from force plate analyses, Cowgill et al. (2010) found that the relatively wide pelvic breadth of immature individuals in general was part of a suite of morphological characteristics that resulted in ML-reinforced femoral midshaft cross sections.

It would be an error, however, to reduce this to an issue of either body shape or mobility as the primary determinant of midshaft femoral shape, as the actual reality is likely to be more complex. Midshaft cross sections of adults from Point Hope are not significantly different in shape when compared to active populations of differing latitudes (Shackelford 2014). In fact, in this analysis, most of the statistically significant results highlighting relatively round femoral midshafts at Point Hope are in the earliest age category. Given that many of the youngest individuals in this age category were unlikely to be highly mobile, even with populational mobility taken into account, it is possible that body shape provides the basic shape "plan" for the femoral midshaft, which can be altered through increased mechanical loading during adolescence and adulthood.

While mobility is likely to be one of several factors that shape the femur, other research has indicated that adult shape indices based on external shaft dimensions do not correlate well with mobility inferred from subsistence strategy (Wescott 2006). This study of the development of shape indices over growth supports the conclusion that the ultimate causes of femoral shape variation are likely to be more complex than previously assumed. Among the three groups with the highest ratios, one is a hunter-gatherer population, one an agriculturalist group, and one a pastoralist society. Among the three lowest standardized ratios, one is an urban/industrial group and the other two are hunter-gatherer populations. These are only broad subsistence categories that many not reflect the more subtle variation in activity patterns and mobility. However, it does highlight the fact that the ultimate etiology of midshaft femoral shape is complex and that a simple correlation between mobility, subsistence strategy, and femoral shape is unlikely to exist.

By investigating correlations over the course of growth, this research hoped to identify biological factors other than mobility that might explain population variation in shaft shape. Significant correlations exist between femoral shaft shape and several other variables: tibial shaft shape (I_x/I_y), crural index, femoral curvature subtense, position of maximum curvature, femoral and tibial total and cortical area. A biomechanical explanation for variation in shaft shape is indicated by a number of these relationships, including the data on femoral curvature, which suggests femora are anteroposteriorly reinforced as femoral diaphyses become more curved, and in individuals where the position of maximum curvature is located more distally along the femoral shaft. In fact, femoral curvature has been linked to mobility levels in the past (Shackelford and Trinkaus 2002). However, it is important to note that while these relationships are statistically significant, the amount of variation explained by the factors included in this analysis is very, very low (ranging from approximately 1–5 %).

Given this, the development of variation in femoral shaft shape is likely to be complex and multifactorial. While patterns do occur early in human growth, the correspondence between these patterns and subsistence, and possibly mobility, are not simple. They are likely complicated by additional factors including body proportions, specifically bi-iliac breadth. In addition, other characteristics of the lower limb, such as femoral curvature, femoral neck angulation, and overall robusticity, are implicated in the ultimate creation of variation in midshaft femoral shape.

References

Alexeeva TI, Bogatenkov DV, Lebedinskaya GV (2003) Anthropology of medieval Vlakhs in comparative study: on date of Mistikhaly burial site. Scientific World, Moskva (in Russian, summery in English)

Allen JA (1877) The influence of physical conditions on the genesis of species. Rad Rev 1:108–140

Auerbach BM (2007) Human skeletal variation in the New World during the Holocene: effects of climate and subsistence across geography and time. Ph.D. Dissertation, Johns Hopkins University, Baltimore

Bergmann C (1847) Ueber die Verhältnisse der wärmeökonomie der Thiere zu ihrer Grösse. Göttinger Studien 3:595–708

Bird DW, Bliege Bird R (2000) The ethnoarchaeology of juvenile foragers: shellfishing strategies among Meriam children. J Anthropol Archaeol 19:461–476

Bird DW, Bliege Bird R (2002) Children on the reef: slow learning or strategic foraging? Hum Nat 13:269–297

Bird DW, Bliege Bird R (2005) Mardu children's hunting strategies in the western desert, Australia. In: Hewlett BS, Lamb ME (eds) Hunter-gatherer childhoods: evolutionary, developmental and cultural perspectives. Aldine, New Brunswick

Bird-David N (2005) Studying children in "hunter-gatherer" societies: reflections from a Nayaka perspective. In: Hewlett BS, Lamb ME (eds) Hunter-gatherer childhoods: evolutionary, developmental and cultural perspectives. Aldine, New Brunswick

Blurton Jones NG, Hawkes K, O'Connell JF (1989) Modeling and measuring costs of children in two foraging societies. In: Standen V, Foley RA (eds) Comparative socioecology: the behavioural ecology of humans and other mammals. Basil Blackwell, London

Blurton Jones NG, Hawkes K, Draper P (1994) Differences between Hadza and !Kung children's work: original affluence or practical reason. In: Burch ES, Ellanna LJ (eds) Key issues in hunter-gatherer research. Providence, Berg, pp 189–215

Bradley C (1984) Sexual division of labor and the value of children. Behav Sci Res 19:159–185

Bradley C (1993) Women's power, children's labor. Cross Cult Res 27:70–96

Cain M (1977) The economic activities of children in a village in Bangladesh. Popul Dev Rev 3:201–227

Cardoso HFV (2005) Patterns of growth and development of the human skeleton and dentition in relation to environmental quality: a biocultural analysis of a sample of 20th century Portuguese subadult documented skeletons. PhD thesis, McMasters University, Hamilton

Cowgill LW (2010) The ontogeny of Holocene and Late Pleistocene human postcranial strength. Am J Phys Anthropol 141:16–37

Cowgill LW (2014) Postcranial growth and development of immature skeletons from Point Hope, Alaska. In: Auerbach BM, Cowgill LW, Hilton CE (eds) The foragers of Point Hope: The biology and archaeology of humans on the edge of the Alaskan Arctic. Cambridge University Press, Cambridge

Cowgill LW, Eleazer CD, Auerbach BM, Temple DH, and Okazaki K (2012) Developmental variation in ecogeographic body proportions. Am J Phys Anthropol 148:557–570

Cowgill LW, Warrener A, Pontzer H, Ocobock C (2010) Waddling and toddling: the biomechanical effects of an immature gait. Am J Phys Anthropol 143:52–61

Eschman PN (1992) SLCOMM version 1.6. Eschman Archaeological Services, Albuquerque, NM

Foster F, Collard M (2013) A reassessment of bergmann's rule in modern humans. PLoS One 8(8): e72269

Hawkes K, O'Connell JF, Blurton Jones NG (1995) Hadza children's foraging: juvenile dependency, social arrangements, and mobility among hunter-gatherers. Curr Anthropol 36:688–700

Hiernaux J, Froment A (1976) The correlations between anthropobiological and climatic variables in sub-Saharan Africa: revised estimates. Hum Biol 48:757–767

Holliday TW (1997a) Postcranial evidence of cold adaptation in European Neandertals. Am J Phys Anthropol 104:245–258

Holliday TW (1997b) Body proportions in Late Pleistocene Europe and modern human origins. J Hum Evol 32:423–447

Holliday TW (1999) Brachial and crural indices of European Late Upper Paleolithic and Mesolithic humans. J Hum Evol 36:549–566

Holliday TW, Falsetti AB (1995) Lower limb length of European early modern humans in relation to mobility and climate. J Hum Evol 29:141–153

Holliday TW, Hilton CE (2010) Body proportions of circumpolar peoples as evidenced from skeletal data: Ipiutak and Tigara (Point Hope) versus Kodiak Island Inuit. Am J Phys Anthropol 142:287–302

Holt BM (2003) Mobility in Upper Paleolithic and Mesolithic Europe: evidence from the lower limb. Am J Phys Anthropol 122:200–215

Katzmarzyk PT, Leonard WR (1998) Climatic influences on human body size and proportions: ecological adaptations and secular trends. Am J Phys Anthropol 106:483–503

Korkiakangas P (1990) Work and play in the lives of country children. Ethnol Scand 20:85–93

Larsen H, Rainey F (1948) Ipiutak and the Arctic whale hunting culture. Anthropological papers of the American Museum of Natural History 42. American Museum of Natural History, New York

Larsen CS, Ruff CB, Kelly RL (1995) Structural analysis of the Stillwater postcranial human remains: behavioral implications of articular joint pathology. In: Larsen CS, Kelly RL (eds) Bioarchaeology of the stillwater marsh: prehistoric human adaptation in the western Great Basin. American Museum of Natural History, New York

Laughlin WS (1968) The demography of hunters: an Eskimo example. In: Lee RB, DeVore I (eds) Man the hunter. Aldine, Chicago

Liversidge HM, Molleson T (2004) Variation in crown and root formation and eruption of human deciduous teeth. Am J Phys Anthropol 123:172–180

Minturn L, Hitchcock JT (1963) The Rajuts of Khalopur, India. In: Whiting GG (ed) Six cultures: studies of child rearing. Wiley, New York

Moberg M (1985) Household production and the value of children: a microeconomic approach to Third World population trends. Anthropol UCLA 14:41–52

Morrison JB (1968) Bioengineering analysis of force actions transmitted by the knee joint. Biomed Eng 3:164–170

Nag M, White B, Peet RC (1978) An anthropological approach to the study of the economic value of children in Java and Nepal. Curr Anthropol 19:293–306

Nagurka ML, Hayes WC (1980) An interactive graphics package for calculating cross sectional properties of complex shapes. J Biomech 13:59–64

Nydegger WF, Nydegger C (1963) Tarong: an Illocas barrio in the Philippines. In: Whiting GG (ed) Six cultures: studies of child rearing. Wiley, New York

O'Neill MC, Ruff CB (2004) Estimating human long bone cross-sectional geometric properties: a comparison of noninvasive methods. J Hum Evol 47:221–235

Pearson OM (2000) Activity, climate, and postcranial robusticity: implications for modern human origins and scenarios of adaptive change. Curr Anthropol 47:63–99

Pearson OM, Petersen TR, Sparacello VS, Daneshvari SR, Grine FE (2014) Activity, "body shape", and cross-sectional geometry of the femur and tibia. In: Carlson KJ, Marchi D (eds) Reconstructing mobility: environmental, behavioral, and morphological determinants. Springer, New York

Porter KA (1996) The agency of children, work, and social change in the South Pare Mountains, Tanzania. Anthropol Work Rev 17:8–19

Rainey F (1971) The Ipiutak culture: excavations at Point Hope, Alaska. Addison-Wesley modular Publications in anthropology 8, Addison-Wesley, Reading, pp. 1–32

Ratha SN, Mahakud NK (1988) Kisan children at work: data from a west Orissa village. Man India 68:369–379

Romney K, Romney R (1963) The mixtecans of Juxtlhuaca, Mexico. In: Whiting GG (ed) Six cultures: studies of child rearing. Wiley, New York

Ruff CB (1987) Sexual dimorphism in human lower limb bone structure: relationship to subsistence strategy and sexual division of labor. J Hum Evol 16:391–416

Ruff CB (1991) Climate and body shape in hominid evolution. J Hum Evol 21:81–105

Ruff CB (1994) Morphological adaptation to climate in modern and fossil hominids. Yearb Phys Anthropol 37:65–107

Ruff CB (1995) The biomechanics of the hip and birth in early Homo. Am J Phys Anthropol 98:527–574

Ruff CB (2000) Body size, body shape and long bone strength in modern humans. J Hum Evol 38:269–290

Ruff CB (2002) Variation in human body size and shape. Annu Rev Anthropol 31:211–232

Ruff CB (2003) Ontogenetic adaptation to bipedalism: age changes in femoral to humeral length and strength proportions in humans, with a comparison to baboons. J Hum Evol 45:317–349

Ruff CB (2007) Body size prediction from juvenile skeletal remains. Am J Phys Anthropol 133:698–716

Sakaue K (1998) Bilateral asymmetry of the humerus in Jomon people and modern Japanese. Anthropol Sci 105:231–246

Saunders SR, DeVito C (1991) Subadult skeletons in the Raymond Dart anatomical collection: research potential. Hum Evol 6:421–434

Scheuer L, Black S (2000) Developmental juvenile osteology. Academic, San Diego

Shackelford LL (2007) Regional variation in the postcranial robusticity of Late Upper Paleolithic humans. Am J Phys Anthropol 133:655–668

Shackelford LL (2014) Bone strength and subsistence activities at Point Hope. In: Auerbach BM, Cowgill LW, Hilton CE (eds) The foragers of Point Hope: the biology and archaeology of humans on the edge of the Alaskan Arctic. Cambridge Press, Cambridge

Shackelford LL, Trinkaus E (2002) Late Pleistocene human femoral diaphyseal curvature. Am J Phys Anthropol 118:359–370

Shaw CN, Stock JT (2011) The influence of body proportions on femoral and tibial midshaft shape in hunter-gatherers. Am J Phys Anthropol 144:22–29

Smith BH (1991) Standards of human tooth formation and dental age assessment. In: Kelley MA, Larsen CS (eds) Advances in dental anthropology. Wiley-Liss, New York, pp 143–168

Temple DH, Matsumura H (2011) Do body proportions among prehistoric foragers from Hokkaido conform to ecogeographic expectations? Evolutionary implications of size and shape variation among northerly hunter-gatherers. Int J Osteoarchaol 21:268–282

Trinkaus E, Ruff CB, Esteves F, Coelho JMS, Silva M, Mendova M (2002a) The lower limb remains. In: Zilhão J, Trinkaus E (eds) Portrait of the artist as a child. Intituto Português de Arqueologia, Lisbon, pp 435–465

Trinkaus E, Ruff CB, Esteves F, Coelho JMS, Silva M, Mendova M (2002b) The upper limb remains. In: Zilhão J, Trinkaus E (eds) Portrait of the Artist as a Child. Intituto Português de Arqueologia, Lisbon, pp 466–488

Tucker B, Young AG (2005) Growing up Mikea: children's time allocation and tuber foraging in southwestern Madagascar. In: Hewlett BS, Lamb ME (eds) Hunter-gatherer childhoods: evolutionary, developmental and cultural perspectives. Aldine, New Brunswick

Van Gerven DP, Beck R, Hummert JR (1990) Patterns of enamel hypoplasia in two medieval populations from Nubia's Batn el Hajar. Am J Phys Anthropol 82:413–420

Van Gerven DP, Sheridan SG, Adams WY (1995) The health and nutrition of a medieval Nubian population: the impact of political and economic change. Am Anthropol 97:468–480

Weaver TD (2003) The shape of the Neandertal femur is primarily the consequence of a hyperpolar body form. Proc Natl Acad Sci USA 100:6926–6929

Webb WS (1946) Indian Knoll, Site Oh 2, Ohio County, Kentucky: The University of Kentucky Reports in Anthropology and Archaeology. Department of Anthropology, Lexington, pp 111–365

Wescott DJ (2006) Effect of mobility on femur midshaft external shape and robusticity. Am J Phys Anthropol 130:201–213

White B (1975) The economic importance of children in a Javanese village. In: Nag M (ed) Population and social organization. Mouton, The Hague, pp 127–146

Winters HD (1974) Introduction to the new edition. In: Webb WS (ed) Indian knoll. University the Tennessee Press, Knoxville

Wyer J (1986) Child labor in Brazilian agriculture. Crit Anthropol 6:63–80

Chapter 12
Territory Size in *Canis lupus*: Implications for Neandertal Mobility

Christopher S. Walker and Steven E. Churchill

Abstract Inferences about Neandertal home range sizes have historically been reconstructed with reference to lithic raw material transport distances. Here we use data on northern latitude social carnivores to predict home range sizes for Neandertal groups. Given that Neandertals must have relied heavily on animal protein and fat in the plant food-poor environments of Pleistocene Europe, their home range sizes and levels of logistical mobility were likely largely determined by prey abundance and distribution. We use the gray wolf (*Canis lupus*) to develop a model that relates climatic variables and predator group aggregate mass to home range size. Pack size data were combined with average wolf mass values to produce mass-specific territory sizes (in $km^2\ kg^{-1}$), which in turn allowed for the prediction of home range areas for Neandertal groups of varying sizes. Results indicate that even at fairly small social group sizes (less than 33 individuals) Neandertals likely required and maintained large territories (≈1,400–5,400 km^2), which is consistent with results of studies of lithic raw material procurement patterns. The concordance between these two types of estimates lends support to the idea that lithic raw material procurement was embedded in subsistence mobility in the European Mousterian.

Keywords Wolf • Home range • Mousterian • Late Pleistocene • Paleoanthropology

C.S. Walker (✉)
Department of Evolutionary Anthropology, Duke University, Biological Sciences Building, Box 90383, Durham, NC 27708, USA
e-mail: christopher.walker@duke.edu

S.E. Churchill
Department of Evolutionary Anthropology, Duke University, Biological Sciences Building, Box 90383, Durham, NC 27708, USA

Evolutionary Studies Institute, University of the Witwatersrand, Private Bag 3, Wits 2050, South Africa
e-mail: churchy@duke.edu

K.J. Carlson and D. Marchi (eds.), *Reconstructing Mobility: Environmental, Behavioral, and Morphological Determinants*, DOI 10.1007/978-1-4899-7460-0_12,

12.1 Introduction

Mobility is arguably the key variable in the adaptive strategies of foraging people, as it is through the management of camp and individual movement across the landscape that hunter-gatherers deal with problems of biodepletion and temporal and geographic patchiness of resources, and through which they exploit the resource structure of their environment (Binford 1980, 2001; Kelly 1983). Two key sets of questions have emerged in studies of Neandertal mobility: (1) what combination of camp moves (residential mobility) and individual movement (logistical mobility) did Neandertals commonly employ to exploit resources in their foraging territories (i.e., their "home ranges"), and (2) how large were Neandertal home ranges, how variable was territory size across time and space, and what factors conditioned this variability?

Despite focused research attention, little consensus has emerged about Neandertal settlement systems. Site size, faunal indicators of season of site occupation, inferences about site function, lithic indicators of hunting intensity, and lithic raw material procurement distances have all been brought to bear on the question. These indicators suggest to some that Neandertals employed a settlement system based on high residential and low logistical mobility [a circulating (Wallace and Shea 2006) or "forager" (Binford 1980) system], at least at some times and in some places (Mellars 1996; Çep and Waiblinger 2001; Zilhão 2001; Martínez-Moreno et al. 2004). However, these same indicators suggest that, at certain times and places, Neandertals might have been employing a radiating mobility system (Wallace and Shea 2006) also known as a "logistical collector" system (Binford 1980) involving few residential moves and high logistical mobility (Lieberman 1993; Lieberman and Shea 1994; Patou-Mathis 2000; Marks and Chabai 2001; Wallace and Shea 2006). Still others detect evidence of mixed strategies, seasonally alternating between circulating and radiating strategies (Richter 2001), or systems that were complex, temporally variable, or impossible to discern based on current evidence (see Lieberman 1998; Hovers 2001; Adler and Tushabramishvili 2004; Rivals et al. 2009).

For modern human groups, resource distribution and hunting dependence are central to understanding ranging behavior, with sparse resources and high reliance on hunting both increasing potential home range sizes (Roebroeks 2003). Since these ecological parameters are unknown for Mousterian foragers (although reasonable inferences can be made: see below), indirect proxies such as raw material transport distances have been used to infer territory size. Raw material transport distances are hypothesized to reflect territory use when raw material procurement is embedded in subsistence and settlement systems (see Binford 1979; Milliken 1988), as is believed to be the case for Neandertals (Kuhn 1995). The maximum transport distance of a material from source to camp has thus been used as an indicator of potential maximum territory size (Fernandes et al. 2008), and the prevalence of resources from a given distance has been taken as a reflection of home range size and mobility. Mousterian assemblages, particularly in western Europe, tend to be predominantly made on very local (within about 5 km of the site: Mellars 1996) sources, although most sites contain a small but significant portion of artifacts produced from raw materials sourced at intermediate distances (5–30 km: Mellars 1996; Kuhn 2011). However, materials procured from more distant locations (100–200 km at some Neandertal sites: Conard and Adler 1997; Féblot-Augustins 1997; Slimak and Giraud 2007; Spinapolice 2012), although rare, have been taken by some to suggest the maintenance of expansive territories among Mousterian foragers, at least in northern and central Europe (taking into account zooarcheological data as well: Féblot-Augustins 1993; Patou-Mathis 2000). Still, the preponderance of local and near-local flint in Mousterian assemblages suggests moderately-sized territories and low mobility overall (Féblot-Augustins 1993; Mellars 1996; Fernandes et al. 2008; Fernández-Laso et al. 2011). High energy budgets, combined with a high energetic cost of transport and slow walking speeds (Steudel-Numbers and Tilkens 2004; Wall-Scheffler 2014), may have also constrained the ability of Neandertals to maintain large territories (see MacDonald et al. 2009). In Marine Isotope Stage (MIS) 5a unit 1 of Baume-Vallée (south-central France), flint was found from as far away as 46 km, suggesting a potential territory of 6,644 km^2 (Fernandes et al. 2008). Similarly, at the nearby MIS 3 site of Le Rond de Saint-Arcons, raw materials were imported from up to 40 km away, indicating a potential maximum territory area of 5,024 km^2 (Fernandes et al. 2008). A larger territory—13,000 km^2—has been suggested for Neandertals of the Aquitaine Basin (south-western France) based on raw materials in MIS 7-6 levels at Vaufrey and other sites (Geneste 1988; Féblot-Augustins 1993). At the (likely MIS 3) eastern European site of Karabi Tamchin in Crimea, the majority of flint was harvested from 25 km away, implying an approximate territory of only 1,962 km^2 (Burke 2006). However, Burke (2006) proposes that Crimean Neandertals needed the entire 27,000 km^2 Crimean peninsula to harvest their mobile primary prey (*Equus hydruntinus*). This territory estimate, however, is based on population sizes between 175 (the lowest viable breeding population: Wobst 1974) and 300 Neandertals (an estimate derived from Steele's (1996) equation predicting range size from group size based on carnivore and primate models), and thus are not reflective of the home range size of a smaller social group within that population. In central and north-central Europe, where raw material transport distances are typically larger, flint movement generally suggests territory sizes on the order of 10,000 km^2 [for example, at Raj Cave (Poland) and Külna (Czech Republic): Féblot-Augustins 1993]. Even the largest of these estimates are moderate in the context of more recent, comparable hunter-gatherer groups. Nunamiut Eskimo and Crow Indian groups, both of which relied primarily on hunting of terrestrial game (87 and 80 % of total diet, respectively), exploited territories upwards of 60,000 km^2 (Kelly 1983). Other northern latitude groups that relied more heavily on marine resources, such as the Baffinland Inuit (25,000 km^2) and Netsilingmiut (6,000 km^2), utilized somewhat smaller, but still sizeable, territories. Given a diet with moderately high to high dependency on meat and fat (see Kuhn and Stiner 2006; Speth 2012), one might expect home range sizes closer to those observed in modern northern latitude foragers who were consuming mostly terrestrial game (although the modern groups may have existed at larger sizes: see Kelly 1995).

Morphologically, Neandertals exhibit high levels of lower limb long bone diaphyseal robusticity (but not outside the norm for highly-mobile modern human foragers), which has been interpreted by some to reflect high mobility (Trinkaus 1983a, b). Might this suggest that they were exploiting large territories?

Carnivores have also been used to model Neandertal territory size, which results in what seem to be unrealistically small estimates (ranging from 834 km^2 for a group of 25–7,356 km^2 for a population of 300: Steele 1996). Data on actual mobility in carnivores (such as day range sizes), however, is scarce and, despite efforts to empirically delineate the determinants of carnivore mobility (Gittleman and Harvey 1982; Carbone et al. 2005), the resulting models are problematic because (1) they require input variables which are highly variable across space and which, for Pleistocene Eurasia, would themselves be highly estimated; (2) they generally involve proportional equations, but lack proportionality constants, such that direct estimates are not obtainable; and (3) the results are generally applicable only at a broad interspecific scale (since the empirical relationships are determined on carnivores that range in size over orders of magnitude). This latter concern is perhaps the most troubling, since the large intraspecific variation in mobility observed in some of the larger-bodied species (Gittleman and Harvey 1982; Höner et al. 2005; Jędrzejewski et al. 2007) causes the observed interspecific relationships to breakdown at the family taxonomic level and below. As with humans, mobility is a variable that can be adjusted by carnivores to adapt to local ecological factors, and as with humans, this variability makes it difficult to characterize the behavior of any given species, even within a given biome. Consequently, any carnivore models of Neandertal mobility are, at best, tools of estimation.

Nevertheless, there are some noteworthy observations from carnivore studies with potential ramifications for understanding Neandertal mobility. First, across the order Carnivora, home range size varies positively with metabolic need (and thus body size) and with the proportion of meat in the diet (Gittleman and Harvey 1982). Accordingly, we can expect relatively large mammals like the Neandertals, with a moderately- to heavily- terrestrial animal-based diet (see Kuhn and Stiner 2006), to have had to range widely to sustain themselves. Second, in both lions and wolves, it has been shown that territory sizes and core areas vary inversely with the encounter rates of ungulates (Gittleman and Harvey 1982; Jędrzejewski et al. 2007). Although primary productivity decreased during colder climatic cycles in the Pleistocene, attendant changes in plant biomass structure (specifically, a reduction in allocation of primary productivity to inedible supportive structures), as woodland gave way to open steppe, would have increased animal production. Thus, Neandertal territory size and mobility (as well as those of the other members of the large-bodied carnivore guild) would likely have decreased during colder, more open conditions. Finally, following the observation by Carbone and Gittleman (2002) that it takes a standing crop of 10,000 kg of prey to support 90 kg of carnivore, regardless of predator body size, there must exist a relationship between the aggregate mass of a carnivore group, secondary biomass in prey species, and territory size. A similar relationship would be expected among Neandertals, although the extent to which they were feeding at a lower trophic level would weaken this relationship. Following this logic, the aim of this paper is to construct a new model for Neandertal territory size based upon *Canis lupus*, one of the primary large-bodied carnivores of northern latitudes, that incorporates environmental variables and that, consequently, can be used as a predictive tool across the entirety of the Neandertal range and climatic events. Note that this model is based on the empirical relationship between prey biomass and predator biomass, across all carnivores, and is, thus, robust to any potential differences between wolves and Neandertals in hunting tactics or prey choice.

12.2 Gray Wolf Ecology and Behavior

Though nonhuman primates are the closest living relatives of the genus *Homo*, ecological differences between extant nonhuman primates and extinct hominins have led some to suggest that the best models for hominid biogeography are social carnivores, a group that typically exhibits high habitat tolerance (Schaller and Lowther 1969). Behavioral similarities between wolves and Pleistocene hominins, including a varied diet, social flexibility, and a capacity for endurance locomotion, make *Canis* a particularly compelling comparative genus (Arcadi 2006). Of the extant members of *Canis*, the gray wolf (*C. lupus*) most closely fills the niche once occupied by Neandertals and is used here as the model species. Like Neandertals, wolves occupy northern latitudes, can survive in both extreme cold and milder climates, prey predominately on large ungulates, and exhibit social flexibility (see Arcadi 2006; Kuhn and Stiner 2006).

Gray wolves are endurance hunters, known to travel as far as 200 km in a day in search of prey (though average 20–30 km), and up to 20 km actively pursuing prey (Mech 1970; Mech and Korb 1978; Arcadi 2006). They are capable of sustained trotting and average 8.7 km h^{-1} during travel (Mech 1994). Moderate-to large-sized game such as deer, moose, and bison form the bulk of the gray wolf diet; however, gray wolves have also been known to prey upon smaller game such as mice, rabbit, and fish, and will eat plant matter as well (Mech and Peterson 2003; Peterson and Ciucci 2003). When prey is abundant, gray wolves prefer juvenile or elderly large game and typically only resort to prime-aged adult large game when prey choice is limited (Fritts and Mech 1981). Gray wolves can take a diversity of vertebrate prey due to social flexibility in hunting group size and their relatively small body size (Arcadi 2006). Thus, an individual gray wolf is small enough to subsist upon small prey, but multiple wolves can join together to hunt larger animals. Bears are the primary competition for gray wolves with specific species varying by location. Smaller bears are typically dominated by a pack of gray wolves; however, gray wolves are often subordinate to more massive bears such as larger individuals of *Ursus arctos* (Ballard et al. 2003). Gray wolves also overlap with felines such as mountain lions and lynx in many areas, but, as smaller-bodied solitary animals, they pose less of a threat than bears and, further, as both opportunistic scavengers and hunters, gray wolves also often compete with scavengers ranging from birds to foxes to coyotes (Ballard et al. 2003).

Like extant primates (and, most likely, extinct hominins), wolves and other canids exhibit flexibility with regard to social dynamics, including group composition and size (Arcadi 2006). Typical gray wolf packs range between 5 and 10 individuals, but can balloon to upwards of 20 individuals (Fuller et al. 2003). Numerous group types have been observed, including all male groups, single male/single female groups, large fission/fusion groups, and polygynous groups (Mech and Boitani 2003). As in humans, both sexes disperse in gray wolves (Lehman et al. 1992; Hill et al. 2011).

Gray wolf territory size is highly variable, ranging from just a few square kilometers to tens of thousands of square kilometers (US Fish and Wildlife Service 2008). For the majority of carnivores, territory size increases with body mass and the amount of meat in the diet (Gittleman and Harvey 1982). Studies specific to gray wolf home ranges have generally found correlations between territory size and variables relating to prey distribution (Van Ballenberghe et al. 1975; Fritts and Mech 1981; Messier 1985; Fuller et al. 1992; Ballard et al. 1997; Jędrzejewski et al. 2007). Fuller and colleagues (1992) found that wolf territory size is, as might be expected, negatively correlated with increasing ungulate density. Greater ungulate densities result in less travel required to find prey and, consequently, smaller ranges. Others have pointed to moose density, specifically, as the primary determinant of pack size, and, in turn, territory size (Messier 1985). Similarly, Van Ballenberghe and colleagues (1975) note a correlation between high prey density, large packs, and small territories. Some research, however, suggests that pack size and territory size are not correlated and that latitude (a proxy inversely related to terrestrial ecosystem productivity) is the single best predictor of territory size in gray wolves (Jędrzejewski et al. 2007). Alternative findings include a negative correlation between prey density and dispersal events (Fritts and Mech 1981) and a positive correlation between territory size and prey migration (Ballard et al. 1997). Not only is territory size smaller when prey are abundant, but new, typically small, territories are formed during times of plenty (Fritts and Mech 1981). Further, while gray wolves do not commonly follow the migrations of prey such as caribou, they will pursue migrating herds during periods of low prey density (Ballard et al. 1997). This creates a temporary, yet dramatic, spike in range and can lead to new territory formation (Ballard et al. 1997). Lastly, regardless of territory size, within a given normal (nonmigratory) territory, often a more commonly used core territory exists (Jędrzejewski et al. 2007).

As a starting point, we might predict that mass-specific territory size (that is, the area required to support one kilogram of wolf) is inversely proportional to prey biomass density, and that prey biomass density will account for most of the variance in wolf territory size once pack size (and thus pack aggregate mass) is controlled for. We might also expect that climatic variables (mean winter temperature and precipitation) and geography (latitude), as a proxy measure of environmental variables, will be highly intercorrelated with prey biomass and will fall out of the model. Assuming that a significant relationship between prey biomass density, wolf pack aggregate mass, and territory size can be established, we can then use this relationship to predict the territory sizes of Neandertal social groups of varying sizes (varying aggregate mass) living in varied prey biomass densities.

12.3 Materials and Methods

Data on 104 wolf packs from ten locations across North America and Europe were collected from the literature (Table 12.1). Only wolf packs for which reliable environmental and prey data could be obtained were included in the study. Six variables were analyzed in the construction of the model: wolf territory size (km^2), group aggregate mass [kg; calculated by multiplying pack size by the regional average mass of mature wolves, derived from Kolenosky (1972), Fritts and Mech (1981), Messier (1985), Ballard et al. (1997), and Jędrzejewski et al. (2007), and ranging between 27.8 kg for wolves in east-central Ontario and 44.4 kg for wolves in Alaska, with a global mean of 34.5 kg], prey biomass (kg km^{-2}), latitude (°N), average winter temperature (K), and average winter precipitation (mm). Some evidence suggests that productivity of the prey population is a primary determinant of wolf territory size (Carbone and Gittleman 2002; Fuller et al. 2003), however, prey biomass has been found to be negatively correlated with territory size in wolves (Jędrzejewski et al. 2007), and is used here due to a lack of published secondary productivity (the addition of new animal cells per unit time per unit area) data for modern wolves, and because Pleistocene estimates of secondary productivity would be less reliable than those of prey biomass. Data on wolf territory size, pack size, and prey biomass were collected from Ballard et al. (1987), (1997), Fritts and Mech (1981), Fuller (1989), Hayes and Harestad (2000), Jędrzejewski et al. (2007), Kolenosky (1972), Mech

Table 12.1 Wolf pack summary data

Country	Region	Mean territory (km^2)	Latitude	Packs	Source
Canada	East-central Ontario	224	50 °N	2	Kolenosky (1972)
Canada	Southwestern Quebec	351	47 °N	14	Messier (1985)
Canada	Yukon	849	62 °N	28	Hayes and Harestad (2000)
Poland	Bialowieza Forest	201	52.5 °N	4	Jędrzejewski et al. (2007)
Romania	Brasov	164	45.5 °N	2	Promberger et al. (1998)
USA	North-central MN	116	47.75 °N	4	Fuller (1989)
USA	Northeastern MN	243	48 °N	11	Mech (1973)
USA	Northwestern MN	344	48.5 °N	8	Fritts and Mech (1981)
USA	Northwestern Alaska	1,868	66 °N	6	Ballard et al. (1997)
USA	South-central Alaska	1,193	63 °N	25	Ballard et al. (1987)

(1973), Messier (1985), and Promberger et al. (1998). Temperature and precipitation data were obtained from online databases of the Western Regional Climate Center (http://www.wrcc.dri.edu/spi/divplot1map.html), Canada's National Climate Data and Information Archive (http://www.climate.weatheroffice.ec.gc.ca/index.html), the Alaska Climate Research Center (http://climate.gi.alaska.edu/), and the European Climate Assessment and Dataset (http://eca.knmi.nl/). Mean winter temperature and precipitation data are specific to the year in which wolf data were collected and come from weather stations nearest the wolf study areas. Climate estimates for Late Pleistocene Europe are from Stage 3 Project published datasets (ftp://ftp.essc.psu.edu/pub/emsei/pollard/Stage3/).

Data were analyzed using a pairwise correlation test and backward elimination regressions with Durbin-Watson tests for autocorrelation in JMP Pro 10 (JMP®, Version 10 Pro. SAS Institute Inc., Cary, NC). A Pairwise correlation test was used to examine the relationship between all model variables and to identify potential sources of cross-correlation. Backward elimination regressions were then utilized to identify an equation, with the fewest possible variables, that best explains change in wolf territory size. A Durbin-Watson test was applied to each regression model to identify the presence or absence of autocorrelation. Lastly, late Pleistocene European climate estimates and Neanderthal aggregate mass estimates were introduced to the model to produce a probable range of potential Neandertal territory sizes.

Many researchers believe that Neandertals were organized into small groups, perhaps on the order of 8–12 individuals (e.g., Smith 2007; Vallverdú et al. 2010; Lalueza-Foxa et al. 2011). Among modern, non-sedentary foragers, modal group sizes tend to be around 25–30 individuals (Kelly 1995). We thus sought to model Neandertal home range sizes for social groups between about 10–30 individuals. According to Ruff and colleagues (1997), the body size of an average Neandertal was 77.9 ± 4.7 kg for males and 66.4 ± 4.8 kg for females. For the purpose of group composition projections, 36 kg (one-half the mean male/female body size) was chosen to represent the size of a juvenile Neandertal. Neandertal aggregate mass values of 500, 1,000, and 1,500 kg were used as sample estimates for input into the model. The number of Neandertals represented by these aggregate masses would vary depending on the age and sex composition of the group, but if one makes some simplifying assumptions [dependency ratios varied between 1.25–3.00 (Binford 2001), all dependents were juveniles (see Caspari and Lee 2004), and sex ratios among producers were equal] these values should roughly correspond with groups of 10–11, 19–23, and 29–33 individuals, respectively. Average winter temperature during MIS 3 Warm Interval (59,000–44,000 years before present) and the last glacial maximum (LGM) (25,000–16,000 years before present) estimated by the Stage 3 project (Figs. 12.1 and 12.2, respectively; ftp://ftp.essc.psu.edu/pub/emsei/pollard/Stage3/) were incorporated into the model for two regions: southern France and the Crimean Peninsula, where territory estimates based upon archeological evidence have been published (Burke 2006; Fernandes et al. 2008). While Neandertals appear to have been fully extinct by the LGM (MIS 2), we used the LGM average winter temperature estimates from the Stage 3 project as broadly representative of climatic conditions endured by Neandertals during the preceding glacial maxima (MIS 4 and 6).

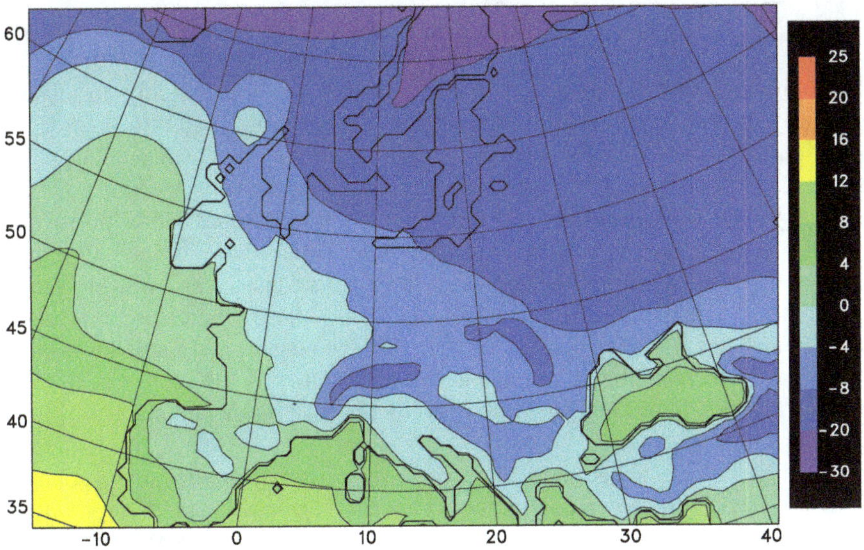

Fig. 12.1 Marine Isotope Stage 3 warm interval (59,000–44,000 B.P.) European Winter Climate Map; Credit: The Stage 3 Project (http://www.esc.cam.ac.uk/research/research-groups/oistage3/stage-three-project-simulations)

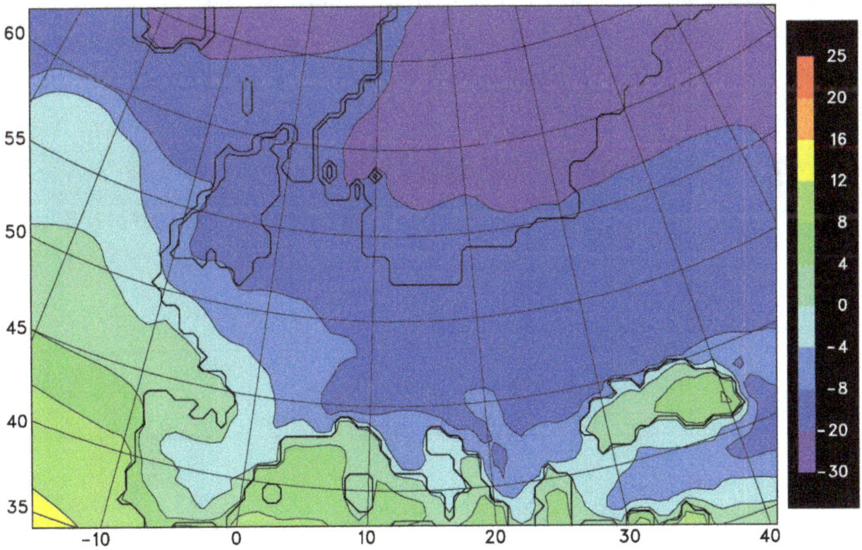

Fig. 12.2 Last Glacial Maximum (25,000–16,000 B.P.) European Winter Climate Map; Credit: The Stage 3 Project (http://www.esc.cam.ac.uk/research/research-groups/oistage3/stage-three-project-simulations)

12.4 Results

Many of the six variables analyzed in the present study are highly correlated (Table 12.2). Wolf pack territory size and latitude ($r=0.92$; $p<0.001$), aggregate mass and latitude ($r=0.93$; $p<0.001$), and aggregate mass and territory size ($r=0.92$; $p<0.001$) are the most highly correlated variables in the study. Given that aggregate mass, latitude, and territory size are all significantly positively correlated, wolf packs at higher latitudes control larger territories and attain greater aggregate masses. Prey biomass and aggregate mass are significantly, but moderately, negatively correlated ($r=-0.63$; $p=0.049$), indicating less wolf mass in areas of high prey density. Further, average winter precipitation was only significantly correlated with average winter temperature ($r=0.72$; $p=0.02$), suggesting that precipitation plays little to no role in wolf aggregate mass and territory dynamics.

The first step of the backward elimination regression incorporated all variables (Primary Model). Together, latitude, prey biomass, aggregate mass, average winter temperature, and average winter precipitation explain about 90 % of the variation ($r^2=0.90$) in wolf pack territory size. Since average winter precipitation was most weakly correlated with territory size, it was removed from the regression, yielding an r^2 value of 0.90. Ultimately, the strongest two variable equation in explaining territory size incorporated aggregate mass and average winter temperature (Fig. 12.3; $r^2=0.90$ Durbin-Watson Statistic = 2.53; Root-Mean-Square Error = 204.49). In a simple linear regression, aggregate mass alone accounts for 84 % of the variation ($r^2=0.84$) in territory size, but average winter temperature was the only other single variable in the study that increased r^2 beyond 0.87.

Primary model predictive equation:

$$\text{Territory Size}\left(\text{km}^2\right)=5{,}258.7+\left(\text{Aggregate Mass}\left(\text{kg}\right)\times 3.65\right)$$
$$-\left(\text{Average Winter Temperature}\left(\text{K}\right)\times 20.91\right)\pm 205\ \text{km}^2$$

Table 12.2 Correlation coefficients (r)

Variable	Territory size (km^2)	Latitude (°N)	Aggregate mass (kg)	Average winter precipitation (mm)	Average winter temperature (K)
Latitude (°N)	**0.92**	–	–	–	–
Aggregate mass (kg)	**0.92**	**0.93**	–	–	–
Average winter precipitation (mm)	−0.48	−0.58	−0.39	–	–
Average winter temperature (K)	**−0.81**	**−0.82**	**−0.68**	**0.71**	–
Prey biomass (kg/km^2)	−0.63	−0.47	**−0.63**	0.19	0.57

All variables included in models. Bold values are significant ($p<0.05$)

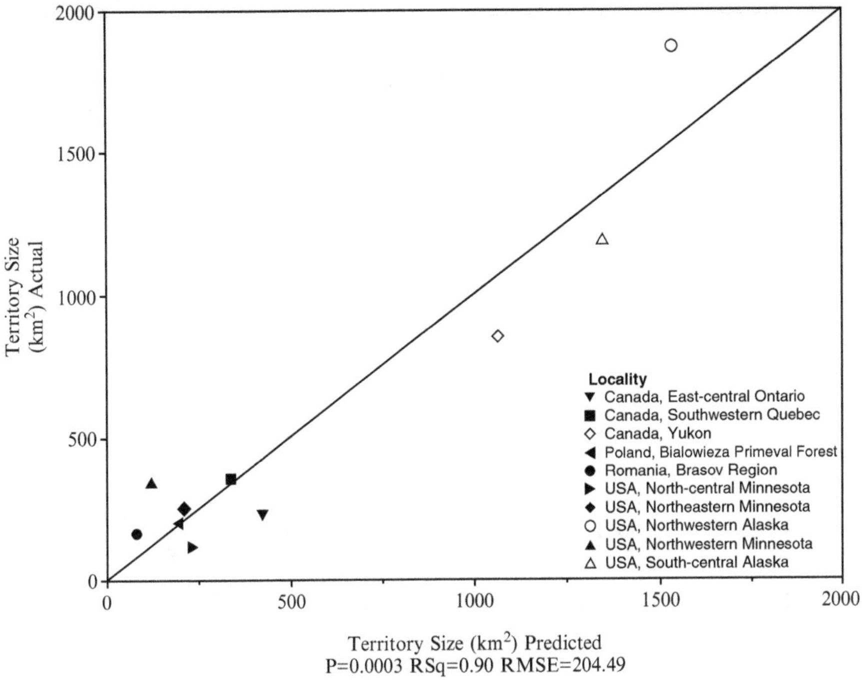

Fig. 12.3 Primary model—actual vs. predicted territory size. Independent variables: aggregate mass and average winter temperature. Solid symbols < 300 kg aggregate mass, open symbols > 300 kg aggregate mass. *RMSE* root-mean-square error

The second strongest regression model (Secondary Model) incorporated latitude and prey biomass (Fig. 12.4). Latitude was the second best single variable predictor of wolf territory size ($r^2 = 0.84$), whereas prey biomass was the fourth best ($r^2 = 0.39$), only topping average winter precipitation ($r^2 = 0.23$). Together latitude and prey biomass explained 89 % of the variation ($r^2 = 0.89$) in wolf pack territory size (Durbin-Watson Statistic = 1.88; Root-Mean-Square Error = 217.34). One potential complication with using latitude in a model for Neandertal territory size, however, is that factors tied to latitude that are likely contributing to the high correlation with territory size, such as temperature, biome type, and even decreased human population, have been reshaped due to climate change (that is to say, ecological conditions at a given latitude differed between the Pleistocene and today). Consequently, only the primary model is particularly useful in estimating Neandertal home ranges. Tables 12.3 and 12.4 summarize output of the primary model. At the smallest group size (aggregate mass = 500 kg; 3–4 adults and 6–8 children), the model predicts MIS 3 warm phase territory sizes of between 1,375 km^2 and 1,543 km^2 in southern France and 1,375 km^2 and 1,459 km^2 in the Crimean Peninsula. At a group size of approximately 8–13 adult and 16–25 children (1,500 kg), the model predicts territory sizes of 5,025–5,193 km^2 and 5,025–5,109 km^2 in southern France and the

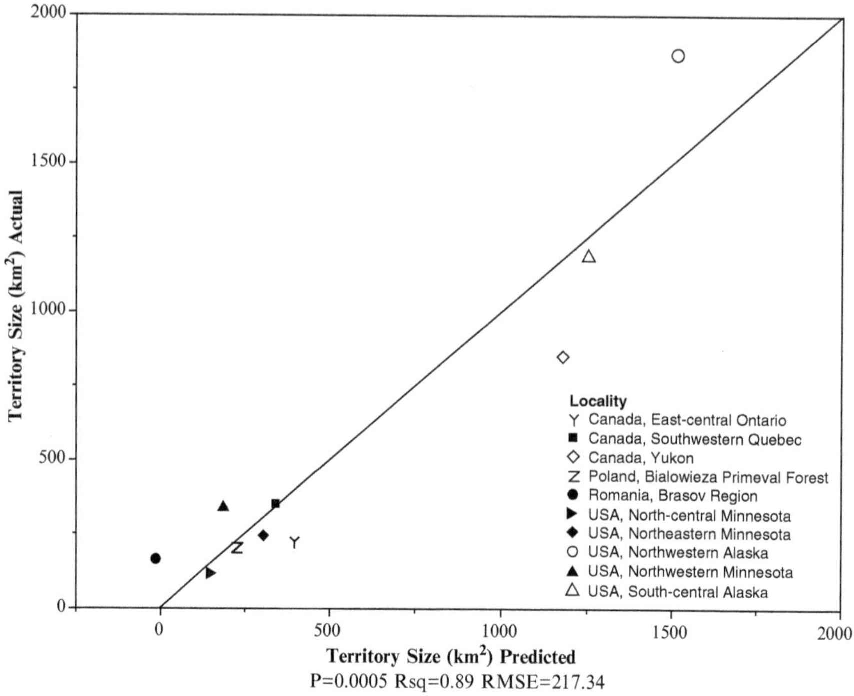

Fig. 12.4 Secondary model—actual vs. predicted territory size. Independent variables: latitude and prey biomass. Solid symbols < 50 °N Latitude, letters = 50–60 °N Latitude, open symbols > 60 °N Latitude. *RMSE* root-mean-square error

Table 12.3 Predicted Neandertal territory size during Marine Isotope Stage 3 Warm Interval (59,000–44,000 years B.P.)

	Southern France estimated territory size (km^2)		Crimea estimated territory size (km^2)	
Neandertal aggregate mass	0 °C/273 K	−8 °C/265 K	0 °C/273 K	−4 °C/269 K
500 kg (≈3–4 adults, 6–8 children)	1,375	1,543	1,375	1,459
1,000 kg (≈6–8 adults, 11–17 children)	3,200	3,368	3,200	3,284
1,500 kg (≈8–13 adults, 16–25 children)	5,025	5,193	5,025	5,109

Massif Central in southern France contained two temperature zones during the winter months (December, January, February) of MIS 3 Warm Interval: 0 °C to −4 °C and −4 °C to −8 °C. The homologous landmass to modern day Crimea was dominated by temperatures ranging between 0 °C and −4 °C during the same period

Crimean Peninsula, respectively. During the colder LGM, groups of 500, 1,000, and 1,500 kg are predicted to have maintained territories of 1,459–1,543 km^2, 3,284–3,368 km^2, and 5,109–5,193 km^2, respectively, in southern France; and 1,459–1,794, 3,284–3,619, and 5,109–5,444 km^2, respectively, in Crimea.

Table 12.4 Predicted Neandertal territory size during last glacial maximum (25,000–16,000 years B.P.)

	Southern France estimated territory size (km²)		Crimea estimated territory size (km²)	
Neandertal aggregate mass	–4 °C/269 K	–8 °C/265 K	–4 °C/269 K	–20 °C/253 K
500 kg (≈3–4 adults, 6–8 children)	1,459	1,543	1,459	1,794
1,000 kg (≈6–8 adults, 11–17 children)	3,284	3,368	3,284	3,619
1,500 kg (≈8–13 adults, 16–25 children)	5,109	5,193	5,109	5,444

During the winter months (December, January, and February) of the last glacial maximum (25,000–16,000 years B.P.), temperatures in Massif Central, southern France ranged between –4 °C to –8 °C. The area most closely approximating modern day Crimea experienced temperatures between –4 °C and –20 °C

12.5 Discussion and Conclusions

Any equation attempting to model life in the past must include variables that are both relevant and attainable for the period. Although latitude is highly correlated with territory size in wolves, its usefulness in a model for Neandertals is questionable. Latitude is used primarily as a proxy for ecological productivity in the modern world, but it is fixed, unlike the climates and ecological productivity of a given region. Modern day European high latitude grasslands are likely less productive than Pleistocene steppe because they experience decreased insolation and higher albedo, and thus shallower summer thaws and shorter growing seasons (Guthrie 1982). Consequently, a model that omits latitude (primary model) is favorable over the secondary model. Average winter temperature can explain many of the environmental and ecological factors explained by latitude, but it is not fixed. Group aggregate mass is an integral part of the equation due to its relationship to carrying capacity, per group member kill rate, and prey biomass. Pairwise correlations indicate that, in wolves, prey biomass is negatively correlated with both group aggregate mass and territory size and that group aggregate mass is positively correlated with territory size, suggesting that wolves cope with decreasing available prey by increasing territory size and pack size. Since large packs are able to capture more prey than small packs, but less prey per individual (McNay and Ver Hoef 2003), increasing pack size minimizes the risk of not catching prey at the expense of increased mobility and decreased per capita dividends.

Applied to Neandertals, the wolf model suggests that Neandertals required and maintained large territories (ca. 1,400–5,400 km²). These home ranges are large relative to those of most tropical foragers using predominantly circulating settlement systems (Kelly 1983; Binford 2001), and are large relative even to those of most mid-latitude groups for whom data are available (Kelly 1983), although many of these groups relied to some degree on marine resources (and, thus, their home range sizes were determined less by terrestrial productivity and were smaller than would be expected if they had relied entirely on terrestrial resources). While there appear to exist no simple relationships between home range size and settlement system, a few implications of these results merit discussion. First, the home range sizes estimated by our model imply territories with radii on the order of 20–40 km [we note that these values are consistent with strontium isotope evidence that a single Neandertal individual was interred about 20 km from the area in which he/she spent some childhood years (Richards et al. 2008). However, 20 km is not far for mobile foragers, and the results of Richards and colleagues would be consistent with a territory of almost any size greater than about 300 km^2]. Thus, return-trip movement across a territory (either from a residential camp positioned near the center, or from one positioned near the margin) was unlikely to be within day-trip distance. Average day-trip distances among modern foragers (on foot) vary by sex and group, ranging between about 6–17 km for females and 12–25 km for males, and generally occur at walking speeds between 2.2–2.6 km h^{-1} (Binford 2001). Consequently, logistical forays away from camp, if taken, may have required travel times of up to a week if moving across the home range diameter and back. At the smaller end of our home range size estimates, full exploitation of the territory could likely be accomplished via day-trips or single overnight excursions away from a residential camp, and thus a settlement system based on limited residential moves (a radiating mobility system) may have sufficed to ameliorate problems of biodepletion and spatial patchiness of resources. At larger territory sizes, however, residential moves likely became increasingly necessary to position foragers closer to temporally limited resource patches or to limit logistical mobility while uniformly exploiting the home range (and thus avoiding problems of biodepletion). A similar relationship between territory size and residential moves would not be expected to hold for modern foragers with transportation technology (such as dog sleds, snowmobiles, or canoes). In the absence of transportation technology, Neandertals undoubtedly faced energetic and burden-carrying constraints on logistical foraging radii (Steudel-Numbers and Tilkens 2004; Wall-Scheffler 2014), which may have forced a heavy reliance on camp moves as a personnel positioning strategy (see MacDonald et al. 2009).

Second, we find interesting the high degree of concordance between our estimates, based on the ecology of gray wolves and some generous assumptions about Neandertal ecology, and estimates which have been independently derived from lithic raw material movement distances. This supports the suggestion that raw material procurement was embedded in the subsistence rounds of Neandertal groups (see Binford 1979). The preponderance of local flint at Mousterian sites is likely to be a reflection of the tendency of Neandertal groups to collect flint during hunting or gathering trips within the "foraging radius" of the sites (Féblot-Augustins 1999; Mellars 1996). The less frequent occurrence of flint from more distant locations likely reflects the movement of some curated tools, made at distant locations from raw material that was locally-collected there, during residential moves or logistical excursions. While it appears that at some times and in some places Neandertals scheduled raw material procurement separately from subsistence rounds (Milliken 2007; Spinapolice 2012), which might suggest a more complex, radiating mobility strategy, they were generally not doing so. We wish to avoid making too much of these results, since human foraging systems are highly dynamic across space and time, and thus it is inherently foolish to make gross generalizations about Neandertal mobility and settlement systems. Nonetheless, territory sizes estimated by the wolf model and by lithic studies, and the concordance between these two independent estimates, seem to us consistent with settlement systems that may have relied more heavily on residential than logistical movement (circulating mobility), in territories that, by virtue of their large size, demanded fairly high mobility levels of their occupants.

Acknowledgements Thanks to Damiano Marchi and Kristian J. Carlson for the opportunity to participate in the symposium and to contribute to this volume, and to David Mech for providing useful guidance during the data collection process. We are also grateful to two anonymous reviewers whose insights and comments served to improve this chapter.

References

Adler DS, Tushabramishvili N (2004) Middle Palaeolithic patterns of settlement and subsistence in the southern Caucasus. In: Conard N (ed) Settlement dynamics of the Middle Paleolithic and Middle Stone Age, vol II. Kerns, Tübingen, pp 91–132

Arcadi AC (2006) Species resilience in Pleistocene hominids that traveled far and ate widely: an analogy to the wolf-like canids. J Hum Evol 51:383–394

Ballard WB, Whitman JS, Gardner CL (1987) Ecology of an exploited wolf population in south-central Alaska. Wildlife Monogr 98:1–54

Ballard WB, Ayres LA, Krausman PR, Reed DJ, Fancy SG (1997) Ecology of wolves in relation to a migratory caribou herd in northwest Alaska. Wildlife Monogr 135:3–47

Ballard WB, Carbyn LN, Smith DW (2003) Wolf interactions with non-prey. In: Mech LD, Boitani L (eds) Wolves: behavior, ecology, and conservation. University of Chicago Press, Chicago, pp 259–271

Binford LR (1979) Organization and formation processes: looking at curated technologies. J Anthropol Res 35:255–273

Binford LR (1980) Willow smoke and dog's tails: hunter-gatherer settlement systems and archaeological site formation. Am Antiquity 45:4–20

Binford LR (2001) Constructing frames of reference: an analytical method for archaeological theory building using hunter-gatherer and environmental data sets. University of California Press, Berkeley

Burke A (2006) Neanderthal settlement patterns in Crimea: a landscape approach. J Anthropol Arch 25:510–523

Carbone C, Gittleman JL (2002) A common rule for the scaling of carnivore density. Science 295:2273–2276

Carbone C, Cowlishaw G, Isaac NJB, Rowcliffe JM (2005) How far do animals go? Determinants of day range in mammals. Am Nat 165:290–297

Caspari R, Lee SH (2004) Older age becomes common late in human evolution. Proc Natl Acad Sci U S A 101:10895–10900

Çep B, Waiblinger J (2001) The use of cave and open-air sites in southern Germany. In: Conard N (ed) Settlement dynamics of the Middle Paleolithic and Middle Stone Age, vol II. Kerns, Tübingen, pp 251–263

Conard NJ, Adler DS (1997) Lithic reduction and hominid behavior in the Middle Paleolithic of the Rhineland. J Anthropol Res 53:147–175
Féblot-Augustins J (1993) Mobility strategies in the late Middle Palaeolithic of central Europe and western Europe: elements of stability and variability. J Anthropol Arch 12:211–265
Féblot-Augustins J (1997) La Circulation des Matières Premières au Paléolithique. ERAUL, Liège
Féblot-Augustins J (1999) Raw material transport patterns and settlement systems in the European Lower and Middle Palaeolithic: continuity, change and variability. In: Roebroeks W, Gamble C (eds) The Middle Palaeolithic occupation of Europe. Leiden University, Leiden, pp 193–214
Fernandes P, Raynal JP, Moncel MH (2008) Middle Paleolithic raw material gathering territories and human mobility in the southern Massif Central, France: first results from a petro-archeological study on flint. J Archaeol Sci 35:2357–2370
Fernández-Laso MC, Chacón Navarro MG, García-Antón MD, Rivals F (2011) Territorial mobility of Neanderthal groups: a case study from Level M of Abric Romaní (Capellades, Barcelona, Spain). In: Conard NJ, Richter J (eds) Neanderthal lifeways, subsistence and technology: one hundred and fifty years of Neanderthal study. Springer, Dordrecht, pp 187–202
Fritts SH, Mech LD (1981) Dynamics, movements, and feeding ecology of a newly protected wolf population in northwestern Minnesota. Wildlife Monogr 80:3–79
Fuller TK (1989) Population dynamics of wolves in north-central Minnesota. Wildlife Monogr 105:3–41
Fuller TK, Berg WE, Radde GL, Lenarz MS, Joselyn GB (1992) A history and current estimate of wolf distribution and numbers in Minnesota. Wildlife Soc B 20:42–55
Fuller TK, Mech LD, Cochrane JF (2003) Wolf population dynamics. In: Mech LD, Boitani L (eds) Wolves: behavior, ecology, and conservation. University of Chicago Press, Chicago, pp 161–191
Geneste JM (1988) Les industries de la grotte Vaufrey: technologie du débitage, économie et circulation de la matière première. In: Rigaud JP (ed) La Grotte Vaufrey: Paléoenvironnement, chronologie, activités humaines. Société Préhistorique Française, Paris, pp 441–517
Gittleman JL, Harvey PH (1982) Carnivore home-range size, metabolic needs and ecology. Behav Ecol Sociobiol 10:57–63
Guthrie RD (1982) Mammals of the mammoth steppe as palaeoenvironmental indicators. In: Hopkins DM, Matthews JV, Schweger CE, Young SB (eds) Palaeoecology of Beringia. Academic, New York, pp 307–326
Hayes RD, Harestad AS (2000) Demography of a recovering wolf population in the Yukon. Can J Zoolog 78:36–48
Hill KR, Walker RS, Božičević M, Eder J, Headland T, Hewlett B et al (2011) Co-residence patterns in hunter-gatherer societies show unique human social structure. Science 331:1286–1289
Höner OP, Wachter B, East ML, Runyoro VA, Hofer H (2005) The effect of prey abundance and foraging tactics on the population dynamics of a social, territorial carnivore, the spotted hyena. Oikos 108:544–554
Hovers E (2001) Territorial behavior in the Middle Paleolithic of the Southern Levant. In: Conard NJ (ed) Settlement dynamics of the Middle Paleolithic and Middle Stone Age. Kerns, Tübingen, pp 123–152
Jędrzejewski W, Schmidt K, Theuerkauf J, Jędrzejewska B, Kowalczyk R (2007) Territory size of wolves Canis lupus: linking local (Białowieża Primeval Forest, Poland) and Holarctic-scale patterns. Ecography 30:66–76
Kelly R (1983) Hunter-gatherer mobility strategies. J Anthropol Res 37:277–306
Kelly R (1995) The foraging spectrum. Smithsonian Institution, Washington, DC
Kolenosky GB (1972) Wolf predation on wintering deer in east-central Ontario. J Wildlife Manage 36:357–369
Kuhn SL (1995) Mousterian lithic technology: an ecological perspective. Princeton University Press, Princeton

Kuhn SL (2011) Neanderthal technoeconomics: an assessment and suggestions for future developments. In: Conard NJ, Richter J (eds) Neanderthal lifeways, subsistence and technology: one hundred fifty years of Neanderthal study. Springer, Dordrecht, pp 99–110

Kuhn SL, Stiner MC (2006) What's a mother to do? Curr Anthropol 47:953–981

Lalueza-Foxa C, Rosas A, Estalrrich A, Gigli I, Campos PF, García-Tabernero A et al (2011) Genetic evidence for patrilocal mating behavior among Neandertal groups. Proc Natl Acad Sci U S A 108:250–253

Lehman N, Clarkson P, Mech LD, Meier TJ, Wayne RK (1992) A study of the genetic relationships within and among wolf packs using DNA fingerprinting and mitochondrial DNA. Behav Ecol Sociobiol 30:83–94

Lieberman DE (1993) The rise and fall of seasonal mobility among hunter-gatherers: the case of the southern Levant. Curr Anthropol 34:599–614

Lieberman DE (1998) Neandertal and early modern human mobility patterns. In: Akazawa T, Aoki K, Bar-Yosef O (eds) Neandertals and modern humans in western Asia. Plenum, New York, pp 263–275

Lieberman DE, Shea JJ (1994) Behavioral differences between archaic and modern humans in the Levantine Mousterian. Am Anthropol 96:300–332

MacDonald K, Roebroeks W, Verpoorte A (2009) An energetics perspective on the Neanderthal record. In: Hublin JJ, Richards MP (eds) The evolution of hominin diets: integrating approaches to the study of Palaeolithic subsistence. Springer Science + Business Media BV, New York, pp 211–220

Marks AE, Chabai VP (2001) Constructing Middle Paleolithic settlement systems in Crimea: potentials and limitations. In: Conard NJ (ed) Settlement dynamics of the Middle Paleolithic and Middle Stone Age. Kerns, Tübingen, pp 179–204

Martínez-Moreno J, Mora R, de la Torre I (2004) Methodological approach for understanding Middle Palaeolithic settlement dynamics at La Roca dels Bous (Noguera, Catalunya, northeast Spain). In: Conard N (ed) Settlement dynamics of the Middle Paleolithic and Middle Stone Age, vol II. Kerns, Tübingen, pp 393–414

McNay ME, Ver Hoef JM (2003) Predation of moose and caribou by a regulated wolf population. Alaska Department of Fish and Game Report

Mech LD (1970) The wolf: the ecology and behavior of an endangered species. Natural History Press, New York

Mech LD (1973) Wolf numbers in the Superior National Forest of Minnesota. USDA Forest Service research paper NC-97. St. Paul, Minnesota: North-Central Forest Experiment Station, pp. 1–10

Mech LD (1994) Travel speeds of arctic wolves. J Mammal 75:741–742

Mech LD, Boitani L (2003) Wolf social ecology. In: Mech LD, Boitani L (eds) Wolves: behavior, ecology, and conservation. University of Chicago Press, Chicago, pp 1–34

Mech LD, Korb M (1978) An unusually long pursuit of a deer by a wolf. J Mammal 59:860–861

Mech LD, Peterson RO (2003) Wolf-prey relations. In: Mech LD, Boitani L (eds) Wolves: behavior, ecology, and conservation. University of Chicago Press, Chicago, pp 131–160

Mellars P (1996) The Neanderthal legacy: an archaeological perspective from Western Europe. Princeton University Press, Princeton

Messier F (1985) Social organization, spatial distribution, and population density of wolves in relation to moose density. Can J Zool 63:1068–1077

Milliken S (1988) Hunter-gatherer land use in the Late Glacial south-east Italy. Oxford J Archaeol 17:269–286

Milliken S (2007) Neanderthals, anatomically modern humans, and 'modern human behaviour' in Italy. Oxford J Archaeol 26:331–358

Patou-Mathis M (2000) Neanderthal subsistence behaviors in Europe. Int J Osteoarch 10: 379–395

Peterson RO, Ciucci P (2003) The wolf as a carnivore. In: Mech LD, Boitani L (eds) Wolves: behavior, ecology, and conservation. University of Chicago Press, Chicago, pp 104–130

Promberger C, Ionescu O, Mertens A, Minca M, Predoiu G, Promberger-Furpab B et al (1998) Carpathian large carnivore project—annual report 1997/1998. Munich Wildlife Society, Ettal

Richards M, Harvati K, Grimes V, Smith C, Smith T, Hublin JJ, Panagopoulou E (2008) Strontium isotope evidence of Neanderthal mobility at the site of Lakonis, Greece using laser-ablation PIMMS. J Archaeol Sci 35(5):1251–1256

Richter J (2001) For lack of a wise old man? Late Neanderthal land-use patterns in the Altmühl River valley, Bavaria. In: Conard NJ (ed) Settlement dynamics of the Middle Paleolithic and Middle Stone Age. Kerns, Tübingen, pp 205–219

Rivals F, Moncel MH, Patou-Mathis M (2009) Seasonality and intra-site variation of Neanderthal occupations in the Middle Palaeolithic locality of Payre (Ardèche, France) using dental wear analyses. J Arch Sci 36:1070–1078

Roebroeks W (2003) Landscape learning and the earliest peopling of Europe. In: Rockman M, Steele J (eds) Colonization of unfamiliar landscapes: the archaeology of adaptation. Routledge, London, pp 99–115

Ruff CB, Trinkaus E, Holliday TW (1997) Body mass and encephalization in Pleistocene Homo. Nature 387:173–176

Schaller GB, Lowther GR (1969) The relevance of carnivore behavior to the study of early hominids. Southwest J Anthropol 25:307–341

Slimak L, Giraud Y (2007) Circulations sur plusieurs centaines de kilomètres durant le Paléolithique moyen. Contribution à la connaissance des sociétés néandertaliennes. C R Palevol 6:359–368

Smith RF (2007) An individual-based comparative advantage model: did economic specialization mediate the fluctuating climate of the late Pleistocene during the transition from Neanderthals to modern humans? PhD Dissertation. Rutgers University, New Brunswick, NJ

Speth JD (2012) Middle Palaeolithic subsistence in the Near East: zooarchaeological perspectives—past, present and future. Before Farming 2012:1

Spinapolice EE (2012) Raw material economy in Salento (Apulia, Italy): new perspectives on Neanderthal mobility patterns. J Arch Sci 39:680–689

Steele J (1996) On predicting hominid group sizes. In: Steele J, Shennan S (eds) The archaeology of human ancestry. Routledge, New York, pp 230–252

Steudel-Numbers K, Tilkens M (2004) The effect of lower limb length on the energetic cost of locomotion: implications for fossil hominins. J Hum Evol 47:95–109

Trinkaus E (1983a) Neandertal postcrania and the adaptive shift to modern humans. In: Trinkaus E (ed) The Mousterian legacy: human biocultural changes in the Upper Pleistocene. Oxford: British Archaeological Reports International Series 164, pp. 165–200

Trinkaus E (1983b) The Shanidar Neandertals. Academic, New York

US Fish & Wildlife Service (2008) Grey wolf (*Canis lupus*). Department of the Interior, Washington, DC

Vallverdú J, Vaquero M, Cáceres I, Allué E, Rosell J, Saladié P et al (2010) Sleeping activity area within the site structure of archaic human groups. Curr Anthropol 51:137–145

Van Ballenberghe V, Erickson AW, Byman D (1975) Ecology of the timber wolf in Northeastern Minnesota. Wildlife Monogr 43:1–43

Wallace IJ, Shea JJ (2006) Mobility patterns and core technologies in the Middle Paleolithic of the Levant. J Arch Sci 33:1293–1309

Wall-Scheffler CM (2014) The balance between burden carrying, variable terrain and thermoregulatory pressures in assessing morphological variation. In: Carlson KJ, Marchi D (eds) Reconstructing mobility: environmental, behavioral, and morphological determinants. Springer, Heidelberg

Wobst HM (1974) Boundary conditions for Paleolithic social systems: a simulation approach. Am Antiquity 39:147–178

Zilhão J (2001) Middle Paleolithic settlement patterns in Portugal. In: Conard N (ed) Settlement dynamics of the Middle Paleolithic and Middle Stone Age. Kerns, Tübingen, pp 597–608

Chapter 13
The Effects of Terrain on Long Bone Robusticity and Cross-Sectional Shape in Lower Limb Bones of Bovids, Neandertals, and Upper Paleolithic Modern Humans

Ryan W. Higgins

Abstract Theoretically, flexed joint postures on sloped surfaces should increase anteroposterior (AP) bending stresses in both human tibiae and bovid metacarpals; however, irregular terrain should also introduce increases in mediolateral (ML) bending stresses in distal limb bones via increased variation in the orientation of ground reaction forces on the feet and nonlinear locomotion. To investigate the effects of terrain type on lower leg bone cross-sectional geometry, this study uses Neandertal and Upper Paleolithic samples plus a large comparative sample of flat terrain (85 species), mountainous (19 species), and mixed terrain (5 species) bovids to see to what extent terrain affects relative AP to ML strength in lower leg bones. Section moduli at the midshaft were compared between groups occupying contrasting terrains. Results suggest that although locomotion on mountainous terrain routinely introduces elevated AP and ML bending stresses to distal limb segments, perhaps with greater relative increases in the ML direction, the signal for locomotion in non-flat terrain in the cross-sectional properties of the human tibia is an increase in overall bending strength and a slight increase in AP relative to ML bending strength due to lateral buttressing by the fibula.

Keywords Terrain • Robusticity • Tibia • Neandertal • Metacarpal • Bovid

13.1 Introduction

Experimental research suggests that bone responds to mechanical loading in vivo in ways that allow bone to better resist deformation or breaking from similar mechanical loading in the future; this is referred to as "Wolf's Law," or in its more modern

R.W. Higgins (✉)
Center for Functional Anatomy and Evolution, Johns Hopkins School of Medicine, 1830 East Monument Street, Room 301, Baltimore, MD 21205, USA
e-mail: rhiggin9@jhmi.edu

K.J. Carlson and D. Marchi (eds.), *Reconstructing Mobility: Environmental, Behavioral, and Morphological Determinants*, DOI 10.1007/978-1-4899-7460-0_13,

interpretation, "Bone functional adaptation" (BFA) (see Ruff et al. 2006a for a review). Long bones (e.g., humeri, femora, tibiae, etc.) accomplish BFA by modifying cortical thickness along their diaphyses during bone remodeling. When functional anatomists model a long bone diaphysis like an engineer's beam, cross-sectional dimensions can be used to calculate correlates that represent a bone's ability to resist deformation (i.e., rigidity) and failure (i.e., strength) from bending stresses. Correlates I_x, I_y, (i.e., second moments of area) and J (i.e., polar moment of area) estimate anteroposterior (AP), mediolateral (ML), and twice average (polar) rigidity in bending of a beam, respectively. Correlates Z_x, Z_y, and Z_p (i.e., section moduli) estimate corresponding strength properties. J and Z_p are also good correlates of resistance to deformation and failure due to torsional stress when a beam is fairly uniform in cross section. The application of beam theory to paleontological, bioarchaeological, and osteological remains provides a useful means for estimating the mechanical competence of bone diaphyses under particular habitual loading regimes and has become a means for inferring activity (i.e., the mechanical environment of the bone) in past human populations.

This chapter is interested in how mobility patterns may affect the cross-sectional dimensions of lower leg bones at the midshaft. Mobility is intended here as the habitual loading regimes related to walking in different terrains. Lower leg bones appear to be a more likely candidate to express an effect of terrain due to the more overriding effect of body shape more proximally (Scott 1985; Ruff et al. 2006b). The midshaft is chosen because it is theoretically the location experiencing the greatest bending stresses in long bone diaphyses (Martin et al. 1998) and thus, changes due to BFA can be expected to be most visible at this location.

While animals (including humans) from rugged-mountainous areas tend to consistently exhibit relatively robust lower leg bones (Scott 1985; Ruff 1999, 2008; Ruff et al. 2006b; Marchi et al. 2011), it is currently not clear if there is an effect of terrain on the cross-sectional shape of these leg bones (Z_x/Z_y). Elevated habitual ML bending stresses may be introduced on uneven terrain by an increase in more transversely-oriented steps (i.e., nonlinear locomotion) and/or more variation in orientation of ground reaction forces on the feet during locomotion (Carlson and Judex 2007; Marchi and Shaw 2011; Marchi et al. 2011; Carlson 2014). Compatibly, Scott (1985) found increases in ML bending strength in the lower leg bones of mountainous bovids; however, the possibility that this is an artifact of greater ML bending moments associated with relatively broader body shapes cannot currently be ruled out.

It has been suggested that ML hypertrophy in the Neandertal lower limb bones may be related to greater lateral movement of the body on irregular surfaces (Trinkaus 1986, 1992; Trinkaus et al. 1999). However, increased ML robusticity in the lower limb bones of Neandertals appears to be a primitive condition, shared with other early hominins, which may be related to relatively broad body shapes compared to modern humans (Trinkaus et al. 1999; Trinkaus and Ruff 2012). Moreover, Mousterian artifacts, which are occasionally associated with Neandertal remains, have been found in a variety of terrains including flat, hilly, and mountainous (van Andel et al. 2003). The tibia of the Neolithic Tyrolean "iceman," whose mummified

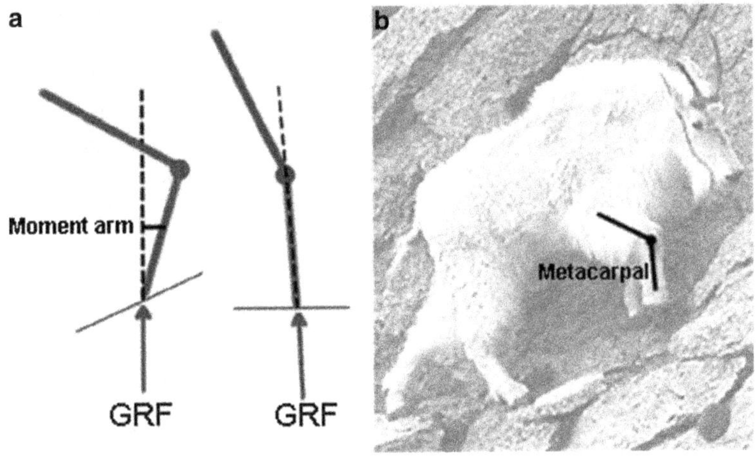

Fig. 13.1 (**a**) General schematic showing that more flexed joint postures needed to negotiate sloped surfaces should increase anteroposterior bending stresses on lower limb bones that flex posteriorly due to an increased moment arm for the ground reaction force. (**b**) Picture showing the location of the bovid lower leg bone used in this study. GRF: ground reaction force

corpse was discovered in a mountainous area in the Tyrolean Alps, displays exceptional AP bending strength, which is attributed to great mobility over mountainous terrain (Ruff et al. 2006b). Correspondingly, flexed joint postures on sloped terrains should theoretically increase AP bending stresses in distal limb segments in limbs with hinge joints that flex posteriorly (e.g., human distal lower limb or the bovid distal forelimb) (Fig. 13.1a,b). Two groups occupying mountainous regions [i.e., represented by a Neolithic Liguria (Italy) sample and an Iron Age Abruzzo (Italy) sample] displayed robust and elliptically-shaped tibial midshafts, which is consistent with increases in AP loading; however, their fibulae were robust, which is consistent with increases in ML bending moments (Marchi et al. 2011). Thus, it is not clear if locomotion in mountainous terrain results in increases in ML relative to AP bending strength in the lower limb segment, or the opposite; and, it is not clear if this relationship is affected by body shape, and/or differences in anatomy [e.g., tibia plus fibula versus a single metacarpal bone in the distal limb; and/or mechanical efficiency at the joints (see Shackelford 2014)].

Understanding the effects of terrain on cross-sectional properties of lower leg bones may be helpful for inferring mobility patterns in past populations. To further investigate whether or not there is an effect of terrain on the cross-sectional shape of lower leg bones, this chapter differs from previous studies by having the following objectives. First, bovids will be reanalyzed using a variable that represents body shape to determine if the lower leg bones of mountainous bovids have greater ML bending strength than is expected for their body shape (i.e., broad versus narrow). The family Bovidae provides an excellent opportunity to study how locomotion in different terrains affects loading regimes experienced by the bones of distal limb segments since (a) it is a large family with many extant bovid taxa that can easily be categorized as flat terrain or rocky slope-dwelling species and multiple species comparisons are important for making inferences about adaptation (Garland and Adolph 1994), and (b) their lower limb segments contain only a single weight-bearing bone, which is ideal for beam theory analyses. Second, since BFA reflects activity patterns of individuals in life and Neandertal remains have been found in different terrains, Neandertals will not be treated as one group, rather Neandertal remains will be assigned to terrain groups (i.e., flat, mixed, mountainous) reflecting the regional landscapes surrounding their individual discovery sites. Upper Paleolithic modern human remains will be similarly grouped by terrain. Subsequently, differences in cross-sectional properties of tibiae grouped by terrain and hominin type will be assessed. The two variable grouping strategies represent contrasting assumptions regarding clan territories. Grouping by hominin taxa assumes that clans within a taxon have similar landscape usage, while grouping by terrain assumes that different clans within a taxon may adopt territories that differ in landscape. In addition, terrain categories here assume that the death site represents, on average, the territorial landscape of the clan (of which the individual belonged); and although this may be problematic, it is the best clue available for the osteological remains assessed here. In addition, since differences in male and female activity patterns are associated with differences in long bone cross-sectional properties (Ruff 1987), sexual dimorphism will also be considered in the human sample.

Here it is generally predicted that animals who occupy mountainous terrains will exhibit increases in overall bending strength (Z_p) and greater increases in ML relative to AP strength (as represented by low Z_x/Z_y ratios) at the midshafts of their distal limb bones after accounting for body shape due to increases in elevated habitual ML bending stresses and nonlinear locomotion on uneven surfaces.

13.2 Materials and Methods

A total of 109 bovid taxa are included in the analyses. They are listed in Tables 13.1 and 13.2 along with sample sizes per taxon, terrain classifications, and body mass data. Species and subspecies averages for metacarpal, radial, and humeral lengths and AP and ML breadths at the metacarpal midshaft come from Scott (1985). Length measurements represent the greatest distance between articular surfaces taken to the nearest mm. Breadth measurements are accurate to the nearest tenth of a mm. Weighted body mass (BM) means come from the following literature sources: Scott (1979), the journal *Mammalian Species* (1969–2010), Nowak (1999), and Kingdon and Pagel (1997). Body mass estimates are sex-specific for a taxon when (1) the metacarpal measurements represent only one sex and (2) a sex-specific BM average for that taxon is available in the literature. When metacarpal measurements for a taxon include both sexes, a weighted body mass average is used. Terrain classifications are from Scott (1979) who took into account usual habitat, feeding behavior, and predator avoidance. Comparisons in this study focus mainly on two terrain groups: "flat" (primarily level forest, plains, and high plateau dwelling) and

Table 13.1 Sample size and body mass data for the flat terrain bovids

Species	Male						Female						Undetermined/Combined						
	N[a]	W[b]	S	H	MS	K	N	W	S	H	MS	K	N	W	S	H	MS	K	BM[c]
Small and medium bovids																			
Cephalophus callipygus	1						1						0	20.1	18.0	20.1		19.5	19.4
Cephalophus dorsalis	6						4			11.6			1		20.0	20.6		19.5	20.0
Cephalophus leucogaster	3						4						1		20.0	12.7		17.5	16.7
Cephalophus natalensis	4			14.0			1			10.0			1		16.0	14.0		13.0	14.3
Cepahlophus niger	0			22.7			0			18.1			1		11			20	20.4
Cephalophus nigrifrons	4						3						0		18	13.9		16.0	16.0
Cephalophus sylviculator	3						2						0		61	68	62.5	62.5	63.5
Cephalophus monticola	6						4						1		5.5			6.25	5.9
Cephalophus maxwelli	1						0						0		5.5		9	8	7.5
Cephalophus zebra	1						0						0		11			17.5	14.3
Madoqua saltiana	0						1						1		3.5				3.5
Madoqua s. phillipsi	2						1						0		3.5				3.5
Madoqua guetheri	1				3.72		2				4.55		1		3.5			4.6	4.2
Madoqua kirki	6				4.59		2				5.13		0		5.0	5.3		5.5	4.8
Neotragus pygmaeus	1						0						0		3.5	4.0		3.8	3.8
Neotragus batesi	3					3.75	2					4.35	0		6.4	4.0			4.0

(continued)

Table 13.1 (continued)

Species	Male						Female						Undetermined/Combined						
	N[a]	W[b]	S	H	MS	K	N	W	S	H	MS	K	N	W	S	H	MS	K	BM[c]
Neotragus moschatus	2			5.0			3			5.4			1		4.5			5.0	5.2
Tragelaphus scriptus	5	53.5	64.0	41.0		55.0	3	33.0	52.0	26.5		42.0	0						47.8
Tragelaphus euryceros	5		227	405		322.5	6		182	253		231.5	1				185		266.2
Sylvicapra grimmea	8					16.3	6					18.8	0		13.0				17.3
Ourebia ourebi	10			13.2		17.0	1			14.3		17.2	1		18.0	12.8			15.2
Redunca redunca	4		45.0			54.0	4		38.5			40.0	0			40.0			44.4
Redunca arundinum	3		68.0	50.7		77.5	2		57.0	48.9		67.5	0			42.9			62.3
Raphiceros campestris	7			10.5			2			11.4			0		13.5	11.8		11.5	10.7
Tetraceros quadricornis	1						1						0		17.0	22.7	19.5		19.7
Redunca fulvorufula	3		32.0	30.2		30.0	4		29.5	30.1		27.0	0			28.0			29.7
Pelea capreolus	0		23.0	23.0			1		20.5				0			25.0		24.0	20.5
Tragelaphus imberbis	3		91.0			100.0	3		64.0			63.0	0						79.5
Tragelaphus strepsiceros	6	252.5	260	228		252.5	3	167.5	170	157		167.5	0						219.7
Tragelaphus spekei	6		91.0	113		105	2		57			62.5	2						90.0
Tragelaphus buxtoni	2		216			240	4		150			175	0						180.3
Tragelaphus angasi	3		114	109.5		120	1		68	62		76	0						103.0
Hippotragus equinus	2		280			271	0		260	259		251.5	2			219.3			270.8
Hippotragus niger	7		235	228		235	4		218			210	2			181			225.5
Aepyceros melampus	7	62.5	61.0	55.9			7	42.5	45.5	43.3			0			46.4			51.8

Alcelaphus lichensteini	0	180.5		177		171.5	0	103		163.3		150.5	1	136	141		157.6
Boselaphus tragocamelus	2	241	250	270			2	169	170	119			2				203.2
Kobus leche	1		100	111.8		107.5	0		73	86.6		77.5	1				99.6
Kobus megaceros	2		91			105	3		73			75	0				83.6
Kobus ellipsiprymnus	3		227			250	0		182			180	0		160.5		238.5
Kobus defassa	1		239				2		182				1				203.4
Kobus vardoni	2		68	73.8		79	0		50	64		63	0		59		73.6
Litocranius walleri	7		45			41.5	3		41			36.5	0		25		41.9
Ammodorcas clarkei	4		31	29.5			1			22.7			0			25.5	28.7
Kobus kob kob	1		70			103	0		45.5			68.5			64.2		86.5
Kobus kob leucotis	5		82				0		54.5				0		64.2		82.0
Kobus kob thomasi	2		91				0		64				0		64.2		91.0
Damaliscus korrigum	7		123				3		114				1				120.1
Damaliscus dorcas	9		73	65.5		72.5	2		66	61		62.5	3		64		68.5
Damaliscus hunteri	2		91				2		86				1		159		88.5
Damaliscus lunatus	1		155.0	136.7		140.0	2.0		145.0	124.7		112.5	0.0		122.9		132.9
Connochaetes taurinus	9		239	210		227.5	5		193	170.5		200	7		182.2		210.3
Connochaetes gnou	0			180	148.5	160	1				116	135	3	136			139.5
Gazella soemmeringi	2		45.5				0						2			42.0	42.0
Gazella granti	16		75	73.2		70.8	4		50			52.5	1		47.4		68.3
Gazella thomsoni	10		23.3	24.0			2		18.0	17.6			0		18.8		22.7
Gazella tilonura	0		27				0						1				27.0
Antidorcas marsupialis	5		34.0	41.0	36.3	44.5	5		28.0	37.2	32.8	31.5	1				35.6

(continued)

Table 13.1 (continued)

Species	Male						Female						Undetermined/Combined						
	N[a]	W[b]	S	H	MS	K	N	W	S	H	MS	K	N	W	S	H	MS	K	BM[c]
Antilope cervicapra	4		45.5	40.9			3		29.5	35.3			1			22.7			38.5
Gazella dama	1		75				0						0			73		57.5	75
Gazella leptoceros	1		27				0						0					16	27
Gazella dorcas	3		23		16.5		0				12.6		1			23		17.5	18.9
Gazella pelzelni	3		11.6				0						0						11.6
Gazella subgutturosa	4		23	23	27.4		0			14	23.2		0						24.5
Addax nasomasculatus	1		118		112.5	117.5	0		104.0		75	75	4			70			103.5
Oryx gazelle	10		177	205.75		210	5		164.0	195.0		202.5	0			176.0			194.1
Oryx tao	0		205				0		193.0				2						199.0
Oryx leucoryx	1		77	92.7			1		73.0	87.7			0						82.6
Saiga tatarica	2		45.5				1						0	47.5					45.5
Gazella spekei	2		13.3				0						0						13.3
Gazella gazella	2		27		24.9		6		22		18.1		0			16.9			21.5
Panthalops hodgsoni	0	45.5	50		38.8		0	27.5			25.9		2						35.7
Procapra guttorosa	0		20.5		31.5		2		16		24		0						20.0
Large Bovids																			
Bos gaurus	5		1,000				1		510				3	825		900			825.0
Bos banteng	2						1						0	650					650.0
Bos sauveli	1		714				1		365				2						539.5
Bubalis bubalis	2		725				2		375				0						550.0
Bubalis mindorensis	1						0				220		1	260					260.0

Taurotragus derbianus	3		704			678.5	7		493			400	0	519.9
Taurotragus oryx	4		590		550	671	1		432		392.5	450	1	559.0
Syncerus caffer caffer	6		318	497.5			4			534			1	512.4
Bison bison	4		865	750.5	725.5		2			408	431.5		0	660.1
Ovibos moschatus	3		364				2			196			3	290.5
Bos grunniens	1	1,000	525	360			1	333					2	480.7

[a]Sample sizes for the long bone measurement

[b]Sample provenience: W: *Walker's Mammals of the World*, 6th Edition; Nowak (1999); S: *Adaptation and Allometry in Bovid Postcranial Proportions*, Scott (1979); H: *CRC Handbook of Mammalian Body Mass*; Silva & Downing (1995); MS: *Mammalian Species* (1969–2009); K: *Kingdon Field Guide to African Mammals*; Kingdon (1997)

[c]Species or subspecies body mass weighted average in kilograms

Table 13.2 Sample sizes and body mass data for the mixed and mountainous terrain bovids

	Male						Female						Undetermined/Combined						
Species	N[a]	W[b]	S	H	MS	K	N	W	S	H	MS	K	N	W	S	H	MS	K	BM[c]
Mixed																			
Ovis musimon	1	32	45.5				1						0						32.0
Ovis orientalis	6		59				4						0						59.0
Ovis ammon hodgsoni	0						0						2						–
Ovis ammon mongolica	1				134		0				90		0						134.0
Ovis ammon polii	11		136		111.9		1		91		55.3		0						108.7
Mountainous																			
Gazella cuvieri	2		27			27.5	0					17.5	0						27.3
Oreotragus oreotragus	7			11.3			3			13			2		13.5	12.1		13	11.9
Dorcatragus megalotis	0						1						0		9			10.5	9.8
Naemorhedus goral	7			32	25		5				29		1		27	27	28.5		28.7
Capricornis sumatraensis	5						3						0		102	87.5			94.8
Capricornis crispus	0				35.9		2			30	38.4		1		29.5		37.5		34.5
Oreamnos americana	2			75	91.1		0			70	36.4		0		113.5	90.9			83.1
Rupicapra rupicapra	6	40					0						1		45.5	15.3			42.8
Capra ibex	7		87	87.4	69.5	92.5	3			53	31	75	0			100			74.8

Capra aegagrus hicrus	0			60.2			1			33.4			0		20	33.4
Capra caucasica	1	82.5	87				0	55					0			84.8
Capra pyrenaica	2		**77**				0						0	57.5		67.3
Capra falconeri	0	95	100	88.7			1	36	54.5				0			45.3
Hemitragus jehmlahicus	2		91				3		77				0		14.7	82.6
Pseudois nayaur	2		59		67.5		0				45		1			60.2
Ammotragus lervia	2	122.5	113.5	110		120	1	47.5	59			47.5	1			92.1
Budorcas taxicolor	2		250				1						2	275		275.0
Ovis canadensis	3	108		100	95		2	72		57.9	71		2		136	86.4
Ovis canadensis nelsoni	4	72			82		1	43			48		0			74.7
Ovis dalli	7				77.4		3				48.4		0		55.7	68.7

[a]Sample sizes for the long bone measurement

[b]Sample provenience: W: *Walker's Mammals of the World*, 6th Edition; Nowak (1999); S: *Adaptation and Allometry in Bovid Postcranial Proportions*, Scott (1979); H: *CRC Handbook of Mammalian Body Mass*; Silva & Downing (1995); MS: *Mammalian Species* (1969–2009); K: *Kingdon Field Guide to African Mammals*; Kingdon (1997)

[c]Species or subspecies body mass weighted average in kilograms

"mountainous" (primarily hilly, rocky, or steep terrain dwelling). The terrain classification for the Atlas Mountain gazelle is updated from Scott (1979) in this study to reflect this gazelle's mountainous geographic range and observations that it climbs slopes over 45° (Cuzin 2003). Mixed terrain species are also included in this investigation for comparison, however, it should be noted that the degree of flatness or ruggedness of their average habitat is variable, seasonal, and not well-known. "Large bovid" flat terrain species also studied by Scott (1979) are treated as a separate group since they share unique differences in their appendicular dimensions compared to other flat terrain bovids (Scott 1979). Sample sizes for the bovid groups are as follows: 74 flat, 19 mountainous, 5 mixed terrain taxa, and 11 large bovids.

Lower limb bone lengths (in millimeters) and body mass estimates (in kilograms) for the Neandertals and (Early and Middle) Upper Paleolithic modern humans were taken from Trinkaus and Ruff (2012), and classified by terrain. Terrain classifications were determined as follows (Table 13.3): the label "flat" was given to osteological remains found in geographic regions that were predominately flat; the label "mixed" was given to remains found in elevated sites in relatively flat areas or in sites in non-flat landscapes characterized by numerous low ridges and valleys; and lastly, the label "Mts" was given to remains found in elevated sites (e.g., rockshelters in cliffs or mountain-sides), in rugged, sloped landscapes.

Long bones have marrow cavities and consequently the most precise way to estimate their cross-sectional properties is to use "hollow beam theory" equations, as was done with the hominin data (Trinkaus and Ruff 2012). Nevertheless, because periosteal contours have a greater impact on bone rigidity and strength in bending than endosteal contours (Stock and Shaw 2007; Sparacello and Pearson 2010), when only midshaft AP and ML diameters are available, as is the case with the bovid data, solid beam theory should be capable of identifying meaningful mechanical trends among the metacarpals of the different genera and species of bovids. The solid beam equations for Z_x, Z_y, and Z_p used for the bovid data are described in Table 13.4.

Linear regression analyses are used to reveal trends in the bovid samples. To investigate whether mountainous bovid metacarpals are more resistant to failure due to bending and torsion, section moduli (Z_x, Z_y, and Z_p) are regressed over body mass (BM) × bone length (L), which accounts for (1) a standard bending force applied to weight-bearing bones (i.e., BM), (2) the beam length, which affects bending moments about the beam, and (3) appropriate scaling with section moduli (in mm^3) (Ruff 2000; Ruff et al. 2006b). To investigate shape differences in the cross sections of bovid metacarpals, Z_x is regressed over Z_y. To control for ML bending stresses related to broader body shapes, the ratio Z_x/Z_y is regressed over the body shape variable BM/Total Limb Length^3 (i.e., an isometric Body Mass Index); thus, bovids with large values will be relatively broad for their height. Total Limb Length (TL) represents (in mm): metacarpal length + radius length + humerus length. Data are natural log transformed to reduce heteroskedasticity (Zar 2010). ANCOVA is the preferred test for statistical differences in group line elevations; however, since assumptions (see Zar 2010) of the parametric ANCOVA tests were violated (i.e., slopes were dissimilar between groups being compared, results not shown here) in the total bovid

Table 13.3 Sample sizes and terrain descriptions for the hominin sample

Neandertal (n=6 m, 4f)	Sex	Terrain[a]	Description of the discovery site and the terrain of the region
Saint-Césaire 1[b]	M	Flat	Ground level rock shelter (La Roche-à-Pierrot) in a relatively flat area
Spy 2	M	Mix	50°28′56.6400″N, 004°40′08.7600″. Elevated terrace and cave site (18 m above Omeau River) in a relatively flat region
Chapelle-aux-Saints 1[a]	M	Mix	44°59′17.16″N, 1°43′33.96″E. Low hillside cave in area characterized by valleys and low ridges
Ferrassie 1	M	Mix	44°57′7″N, 0°56′17″E. Low rockshelter in an area characterized by valleys and low ridges
Ferrassie 2	F		
Amud 1[b]	M	Mts	32°52′0″N 35°30′0″E. Elevated cave (cliff face) in the mountainous terrain of Upper Galilee
Shanidar 2[b], 6	M	Mts	36°50′N, 44°13′E. Elevated cave (42 m) in the Zagros Mountain range
Shanidar 6	F		
Palomas 96	F	Mts	37°47′59″N, 0°53′45″W. Elevated cave in a rugged hillside, surrounded by a flat coastal plain landscape
Tabun C1	F	Mts	32°40′13.8″ N, 34°57′55.8″E. Elevated cave in Mount Carmel mountain range near flat coastal plain
Early Upper Paleolithic/ Late Upper Paleolithic (n=12 m, 5f)			
Tianyuan 1	M	Mts	39°39′28″N, 115°52′17″ E. Elevated cave site in Taihang Mountain range a short distance from the North China Plain
Minatogawa 1	M	Flat	Found in a limestone quarry in a relatively flat area near Gushikami village in southern Okinawa, Japan
Minatogawa 3	F		
Minatogawa 4	F		
Sunghir 1	M	Flat	56°10′33″N, 40°30′33″E. Open air site in flat plain
Veneri 1	M	Flat	40°04′11″N, 18°05′45″E. Ground level cave in flat plain
Veneri 2	F		
Dolni Vestonice 3	F	Flat	48°53′16″N, 016°38′57″E. Open air site on a relatively flat surface at the base of a hill; the surrounding area is flat
Dolni Vestonice 13	M		
Dolni Vestonice 14	M		
Dolni Vestonice 15	F		
Dolni Vestonice 16	M		
Ohalo 2	M	Mix	32°43′19.54″N, 35°34′19.72″E. Open air site (currently submerged in the Sea of Galilee) in an area characterized by both mountainous and flat terrain

(continued)

Table 13.3 (continued)

Early Upper Paleolithic/ Late Upper Paleolithic (n=12 m, 5f)	Sex	Terrain[a]	Description of the discovery site and the terrain of the region
Cro-Magnon 1 (4332)	M	Mts	Rock shelter in a limestone cliff in an area characterized by ridges and valleys
Arene Candide IP	M	Mts	Cave in the slope of Mount Caprazoppa in an area characterized by ridges and valleys
Grotte-des-Enfants 4[b]	M	Mts	43°46′59″N, 7°37′00″E. Cave is on a seaside cliff in an area characterized by ridges and valleys
Paviland 1	M	Mts	51° 33′ 0.31″N, 4° 15′ 18.67″W. Cave is on a rugged, seaside cliff

[a]Flat=remains found in flat terrain; Mix=remains found in elevated sites in relatively flat areas or in sites in non-flat landscapes characterized by numerous low ridges and valleys; Mts: remains found in mountainous or rugged sites

[b]Estimates of body mass or tibia length are not available for these individuals, thus they are only used in Z_x/Z_y comparisons and are not used in relative section moduli (Z_p, Z_x, Z_y/BM×L) comparisons

Table 13.4 Solid beam theory equations

y=AP midshaft diameter

x=ML midshaft diameter

Second moment of areas

(1) $\boldsymbol{I}_x = \frac{y^3 \times x \times \pi}{64}$: estimates AP bending rigidity

(2) $\boldsymbol{I}_y = \frac{x^3 \times y \times \pi}{64}$: estimates ML bending rigidity

(3) $\boldsymbol{J}=I_y+I_x$: estimates overall rigidity in bending and torsion

Section moduli

(4) $\boldsymbol{Z}_x = \frac{I_x}{0.5y}$: estimates strength in AP bending

(5) $\boldsymbol{Z}_y = \frac{\boldsymbol{I}_y}{0.5\text{x}}$: estimates strength in ML bending

(6) $\boldsymbol{Z}_p = \frac{I_x + I_y}{0.25(x+y)}$: estimates overall strength in bending and torsion

sample, nonparametric "Quick Tests" were used as an alternative. A Quick Test first counts the number of data points for each group that fall above and below a common regression line and then evaluates group elevation differences, using a 2 × 2 contingency table and a Fisher's Exact Test (Tsutakawa and Hewett 1977).

For the hominin sample, regression slopes were often not significantly different from zero for the groups. Consequently, regression-based methods (i.e., ANCOVA and Quick Tests) were not appropriate in all cases for comparing relative bending strength trends between the hominin groups. For consistency, the author prefers to use the same statistical test for all hominin comparisons. Thus, Student's t-tests are used to statistically evaluate group differences in relative strength ratios [i.e., Z_p, Z_x, or Z_y/(BM×L)] and the cross-sectional shape ratio (Z_x/Z_y) in all hominin comparisons.

Of particular interest in the bovid sample is a case study using a closely-related group of bovids, i.e., gazelles, who are similar in body size, body shape, and anatomy. Here the lower leg bone of the one mountainous member (*Gazella cuvieri*) is predicted to exhibit increases in overall strength (Z_p) and increases in ML relative to AP strength compared to 13 flat terrain gazelles. Since there is only one mountainous taxon in the gazelle analyses, statistical significance will be tested using 95 % prediction intervals. All statistical tests performed in this paper are two tailed and α is set at 0.05.

13.3 Results

13.3.1 The Total Bovid Sample

Results for the bovid regression comparisons are listed in Table 13.5 and corresponding plots can be found in Figs. 13.2a–c. *P*-values listed below represent terrain group regression elevation differences evaluated for statistical significance using Quick Tests. Results from comparisons using the total bovid sample suggest that the metacarpals of mountainous bovids generally exhibit a significant ($p<0.0001$) increase in overall bending strength (Z_p over BM×L) compared to those of flat terrain bovids (Table 13.5; Fig. 13.2a). This is due to significant ($p<0.0001$) increases in both relative AP and ML bending strength (Table 13.5). In regard to the mixed terrain and the large bovids, they appear to be intermediate in relative Z_p falling equally above and below the common regression lines (Table 13.5; Fig. 13.2a); this is also the case for Z_x, and Z_y (Table 13.5). In the cross-sectional shape analysis (i.e., Z_x regressed over Z_y), the metacarpals of mountainous bovids exhibit significant ($p<0.0001$) increases in ML relative to AP bending strength compared to those of flat terrain bovids (Table 13.5; Fig. 13.2b). Mixed terrain individuals resemble the flat terrain individuals in the Z_x over Z_y analysis and fall mostly (4:1) above the regression line while large bovids more closely resemble the mountainous terrain individuals generally falling below (2:9) the common regression line (Table 13.5; Fig. 13.2b).

In regard to body shape, most large bovids appear to be relatively broad in body shape exhibiting high values for the ratio BM/TL^3, while mountainous bovids tend be moderate and flat terrain bovids tend to be moderate to narrow in body shape (Fig. 13.2c). When the ratio Z_x/Z_y is regressed over BM/TL^3, flat terrain and mixed terrain bovids fall equally above and below the common regression line, and all but 3 out of 19 mountainous bovids fall below the line (Table 13.5, Fig. 13.2c). This suggests that the metacarpals of mountainous bovids, which are significantly different from those of flat terrain bovids ($p<0.001$), do indeed have more ML bending strength than is expected for their body shapes. In contrast, the large flat terrain bovids exhibit less ML bending strength than is expected for their broad body shapes (i.e., 10 falling above and 1 falling below the line) (Table 13.5, Fig. 13.2c).

Table 13.5 Quick Test comparisons for the total bovid sample

Regressions[a] (equation; R^2; p-value)	Terrain[b]	Above line	Below line	Comparisons and significance				
				Flat-Mix	Flat-Mts	Mix Mts	Flat-Large	Mts-Large
Z_p over BM × L (y=0.93x–1.88; 0.93; ***)	Flat	18	56	NS	***	NS	NS	*
	Mix	2	3					
	Mts	16	3					
	Large[c]	5	6					
Z_x over BM × L (y=0.91x–2.53; 0.95; ***)	Flat	22	52	NS	***	NS	NS	*
	Mix	2	3					
	Mts	16	3					
	Large	4	7					
Z_y over BM × L (y=0.93x–2.55; 0.92; ***)	Flat	18	56	NS	***	*	NS	*
	Mix	2	3					
	Mts	17	2					
	Large	5	6					
Z_x over Z_y (y=0.97x–0.05; 0.99; ***)	Flat	52	22	NS	***	***	**	NS
	Mix	4	1					
	Mts	0	19					
	Large	2	9					
Z_x/Z_y over BM/TL^3 (y=–0.30x–4.67: 0.31; ***)	Flat	44	30	NS	***	NS	*	***
	Mix	3	2					
	Mts	3	16					
	Large	10	1					

[a]Variables are natural logged transformed in all regressions
[b]See Table 13.3 for variable explanation
[c]Large bovids
NS nonsignificant, *$p<0.05$, **$p<0.01$, ***$p<0.001$

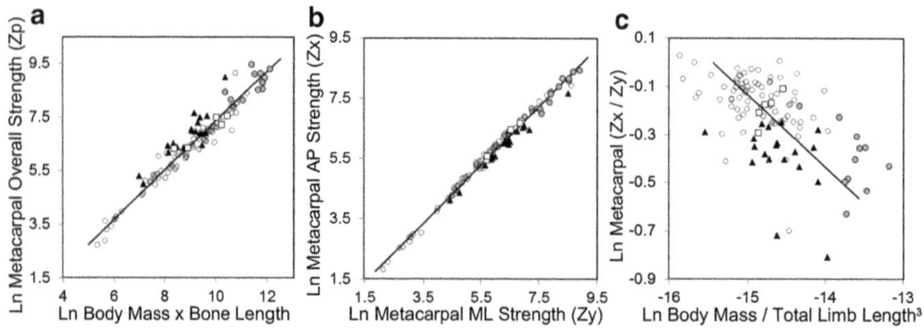

Fig. 13.2 (**a**) Relationships in relative overall bending strength [polar section modulus/(body mass × metacarpal length)], (**b**) between AP (Z_x) and ML (Z_y) bending strength, and (**c**) between AP/ML bending strength and body shape [i.e., body mass/(humerus + radius + metacarpal length)^3] in flat terrain bovids (*open circles*), mountainous bovids (*black triangles*), mixed terrain bovids (*open squares*), and large bovids (*large gray circles*)

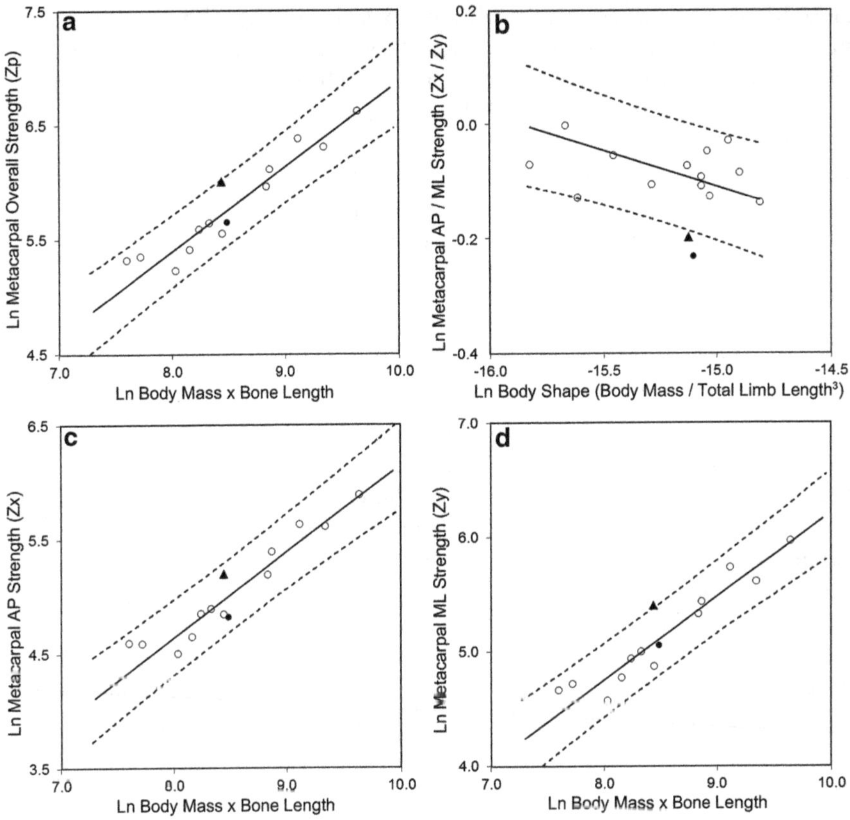

Fig. 13.3 Regressions and 95 % prediction intervals for flat terrain gazelles (*open circles*) with mountainous gazelle (*black triangle*) added for comparison. (**a**) Plot for relative overall bending strength [i.e., polar section modulus/(body mass × metacarpal length)]. (**b**) Plot for AP/ML bending strength versus body shape [i.e., body mass/(humerus + radius + metacarpal length)^3]. Plots for relative (**c**) AP and (**d**) ML bending strength [section moduli/(body mass × metacarpal length)]. The *black circle* represents a flat terrain gazelle (i.e., *Eudorcas tilonura*) with an exceptionally low Z_x/Z_y ratio

The Gazelle Sample

To compare the relative bending strength properties (i.e., section moduli over BM × L) and cross-sectional shape properties (Z_x/Z_y over BM/TL^3) of gazelle metacarpal midshafts, regression lines with 95 % prediction intervals were plotted for 13 flat terrain gazelle taxa, and data points representing the mountainous gazelle taxon were subsequently added to these plots for comparison. Similar to the trends observed using the total bovid sample, the metacarpal midshaft of the mountainous gazelle exhibits (a) high (although not statistically significant) overall bending strength (Z_p) (Fig. 13.3a) and (b) statistically significant increases in ML compared to AP bending. The latter is true whether or not taking into account the gazelle's body shape (see Fig. 13.3b for the comparison that controls for body shape). Unexpectedly, one flat terrain gazelle (i.e., *Eudorcas tilonura*) exhibited exceptional increases in ML relative AP bending strength (Fig. 13.3b, black circle); however, comparisons of relative AP and ML bending strength reveal that this is due to relatively decreased AP strength and not relatively increased ML strength (Fig. 13.3c,d). Corresponding to study expectations, the metacarpal of the mountainous gazelle does indeed exhibit increases in relative ML bending strength at the midshaft (Fig. 13.3d).

13.3.2 The Hominin Sample

Student's *t*-tests were performed to test group differences in relative strength ratio means [section moduli/(BM × L)] and ratio means for AP strength relative to ML strength (Z_x/Z_y). Similar to the bovid results, mountainous terrain individuals exhibited increases in overall AP and ML relative bending strength, and means for mix terrain individuals were intermediate between means for the mountainous and the flat terrain groups. The relative strength ratio means for the terrain groups are as follows: 115.0 mts, 103.6 mix, 86.3 flat for Z_p, 78.7 mts, 71.7 mix, 59.2 flat for Z_x, and 50.7, 47.3, 40.9 for Z_y. A post hoc decision to combine the mountainous and mixed terrain groups for the statistical comparisons was made by the author for the following reasons: both groups have small sample sizes ($n=7$ and $n=4$, respectively), both are similar in their relative strength properties, and the resulting non-flat and flat terrain groups are characterized by identical sample sizes ($n=11$) and similar proportions of males and females. Results for the hominin Student's *t*-test comparisons are listed in Table 13.6 and corresponding plots can be found in Figs. 13.4, 13.5, 13.6, and 13.7.

When the non-flat and flat terrain groups were compared using the total hominin sample (Neandertals, Upper Paleolithics, males and females), the tibial midshafts of individuals whose remains were found in non-flat terrain (i.e., mountainous and mixed terrain) had significantly greater overall strength ratios ($p<0.005$), greater AP strength ratios ($p<0.05$), and greater ML strength ratios ($p<0.001$). Comparisons between Neandertals and modern humans and between males and females showed no significant differences in group means. Since body shape differences have been documented for Neandertals versus modern humans, and males versus females, this suggests that body shape does not have a large effect on the cross-sectional properties of the human tibia. Thus, terrain appears to affect relative strength properties in the tibial midshaft while hominin type and sex, and by extension body shape, does not appear to have a significant effect in this sample.

In comparisons of AP relative to ML bending strength at the tibial midshaft using the total hominin sample, males exhibit significant increases in AP relative to ML bending strength compared to females ($p=0.05$; Fig. 13.7a). Tibiae of the non-flat terrain group show increased AP relative to ML bending strength compared to tibiae of the flat terrain group (Fig. 13.7c); however, in this comparison the group means

Table 13.6 *T*-test comparisons for the hominin sample

	Mean (*n*)	Std. error	Significance
Z_p/BM × L			
Non-flat (Mts + Mix)[a]	110.9 (11)	4.75	** (***)[b]
Flat	86.3 (11)	6.06	
Neandertal	108.1 (6)	3.44	NS (NS)
Mod. Hum.	95.0 (16)	6.04	
Males	102.4 (13)	6.10	NS (NS)
Females	93 (9)	7.06	
Z_x/BM × L			
Non-flat (Mts + Mix)	76.2 (11)	4.17	* (**)
Flat	59.2 (11)	4.86	
Neandertal	72.9 (6)	5.02	NS (NS)
Mod. Hum.	65.7 (16)	4.62	
Males	71.1 (13)	4.76	NS (NS)
Females	62.8 (9)	5.52	
Z_y/BM × L			
Non-flat	49.4 (11)	2.21	** (**)
Flat	40.9 (11)	1.99	
Neandertal	50.0 (6)	2.07	NS (NS)
Mod. Hum.	43.4 (16)	2.10	
Males	45.5 (13)	2.33	NS (NS)
Females	44.7 (9)	2.70	
Z_x/Z_y			
Non-flat	1.6 (15)	0.07	NS (*)
Flat	1.4 (12)	0.07	
Neandertal	1.6 (10)	0.09	NS (NS)
Mod. Hum.	1.5 (17)	0.06	
Males	1.6 (18)	0.06	* (0.08)
Females	1.4 (9)	0.07	

NS nonsignificant, *$p<0.05$, **$p<0.01$, ***$p<0.001$

[a]See Table 13.3 for variable explanation

[b]Significance level in parentheses represents comparisons with the exclusion of outlier, Veneri 1, from the non-flat group; see text for details

are not significantly different due to the inclusion of an "extreme outlier" (defined as a data point with a value that is more than three times the interquartile range for the group; Zar 2010) in the flat terrain group (i.e., Veneri 1). If the outlier is excluded, there is a visible and significant trend ($p<0.05$) of increased AP relative to ML bending strength in the non-flat group compared to the flat terrain group (Table 13.6). The non-flat group and the flat terrain group have similar proportions of male and females, so the difference in group means does not appear to be biased by sex differences in locomotor behavior. Comparisons between terrain group means using samples that are specific in hominin type and sex (e.g., Neandertal males) would be best for controlling for the influences of terrain. However, when the hominin sample is broken up into these smaller samples the sample sizes for the non-flat or flat terrain groups (which are often zero or one for one of the terrain groups) are too small

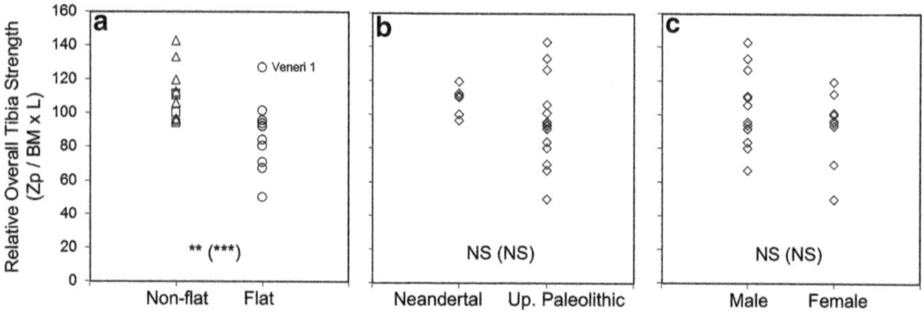

Fig. 13.4 Plots displaying relative overall bending strength [Z_p/(body mass × tibia length)] for the hominin tibia sample grouped by (**a**) terrain, (**b**) hominin type, and (**c**) sex. Mountainous, mixed, and flat terrain individuals are represented by triangles, squares, and circles, respectively. *NS* nonsignificant, $^{**}p<0.01$, $^{***}p<0.001$. Significance level in parentheses represents comparisons with the exclusion of outlier, Veneri 1, from the non-flat group

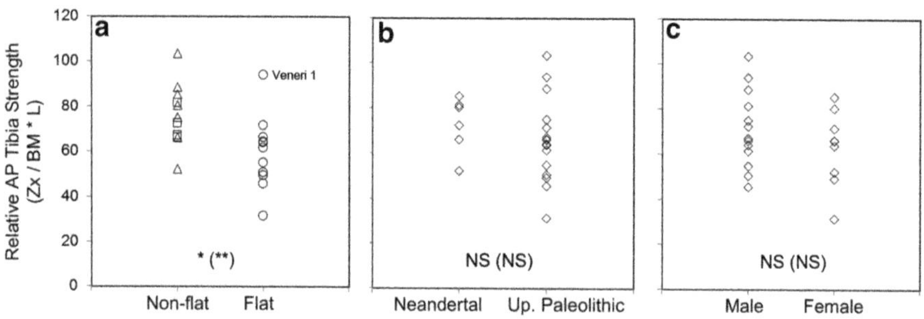

Fig. 13.5 Plots displaying relative AP bending strength [Z_x/(body mass × tibia length)] for the hominin tibia sample grouped by (**a**) terrain, (**b**) hominin type, and (**c**) sex. See Fig. 13.4 for explanation of variables. *NS* nonsignificant, $^{*}p<0.05$, $^{**}p<0.01$. Significance level in parentheses represents comparisons with the exclusion of outlier, Veneri 1, from the non-flat group

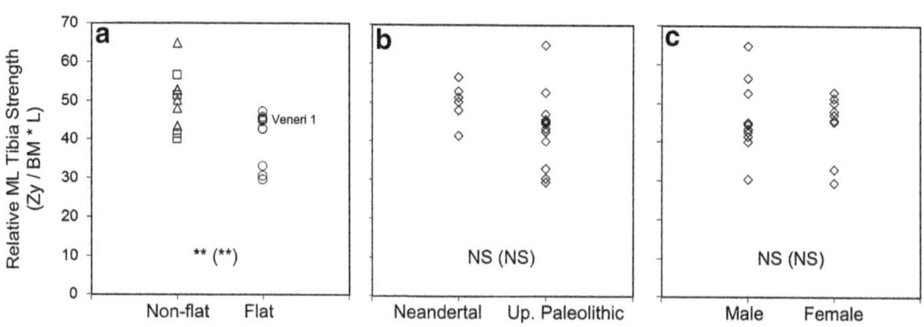

Fig. 13.6 Plots displaying relative ML bending strength [Z_y/(body mass × tibia length)] for the hominin tibia sample grouped by (**a**) terrain, (**b**) hominin type, and (**c**) sex. See Fig. 13.4 for explanation of variables. *NS* nonsignificant, $^{**}p<0.01$. Significance level in parentheses represents comparisons with the exclusion of outlier, Veneri 1, from the non-flat group

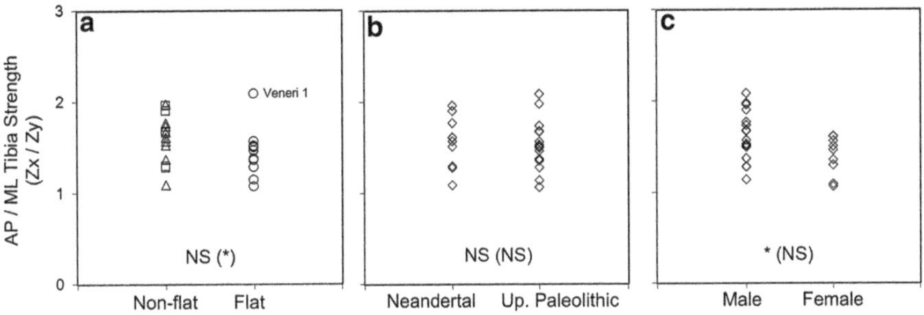

Fig. 13.7 Plots displaying ratios for AP relative to ML bending strength (Z_x/Z_y) for the hominin tibia sample grouped by (**a**) terrain, (**b**) hominin type, and (**c**) sex. See Fig. 13.4 for explanation of variables. *NS*: nonsignificant; $*p<0.05$. Significance level in parentheses represents comparisons with the exclusion of outlier, Veneri 1, from the non-flat group

for informative comparisons. The exception is the Upper Paleolithic male sample, which consists of five non-flat terrain individuals and six flat terrain individuals. Similar to the results using the total hominin sample, when comparisons are made using only the Upper Paleolithic male sample, the non-flat terrain group exhibits increases in relative Z_p, Z_x, and Z_y and greater AP relative to ML bending strength, but differences in means are only near significant in this smaller sample for relative Z_p ($p=0.09$) and Z_x ($p=0.10$) (Fig. 13.8a,b). However when Veneri 1, which is 2.9 times the interquartile range for the Upper Paleolithic male flat group, is excluded, the differences in means are significant in comparisons for Z_p ($p<0.05$), Z_x ($p=0.01$), and Z_x/Z_y ($p<0.05$), but not for Z_y ($p=0.18$) (Fig. 13.8a–c).

13.4 Discussion

Distal hinge joints that flex posteriorly are a feature of both the bovid forelimb and the hominin lower limb. Theoretically, flexed joint postures at these hinge joints on sloped surfaces should similarly increase AP bending stresses in both human tibiae and bovid metacarpals. However, mountainous terrain should also introduce increases in ML bending stresses in both human tibia and bovid metacarpal via variation in the orientation of ground reaction forces on the leg in nonlinear locomotion (Carlson and Judex 2007; Marchi and Shaw 2011; Marchi et al. 2011). The results for the relative strength comparisons (Z_p, Z_x, and Z_y) for both the bovid and hominin samples support both expectations; at midshaft of their lower leg bones, mountainous groups exhibited increases in overall bending strength, due to increases in both AP (Z_x) and ML (Z_y) bending strength.

In regard to increases in ML relative to AP bending strength, results for the bovid and hominin samples greatly differ. Bovid leg bone comparisons suggest that locomotion over rugged-mountainous terrain results in greater increases in ML relative to AP bending strength at the midshaft (Figs. 13.2b, 13.7a, 13.8d), while trends in

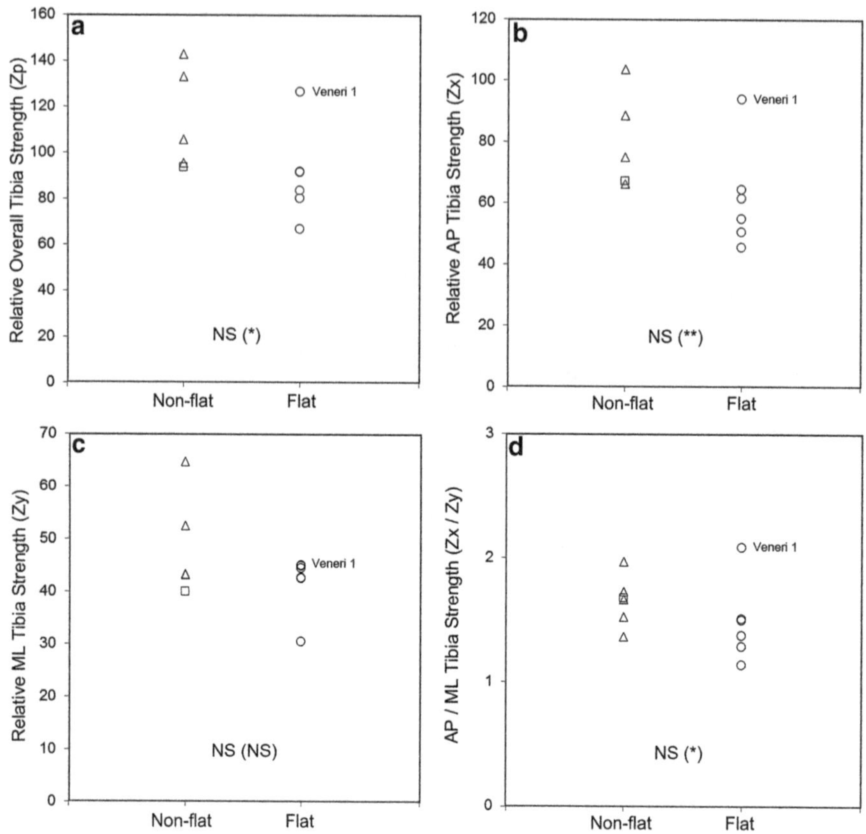

Fig. 13.8 Plots for relative (**a**) overall, (**b**) AP, and (**c**) ML bending strength and (**d**) AP relative to ML bending strength for Upper Paleolithic males. See Fig. 13.4 for explanation of variables. *NS* nonsignificant, $^{*}p<0.05$, $^{**}p<0.01$. Significance level in parentheses represents comparisons with the exclusion of outlier, Veneri 1, from the non-flat group

hominin tibia (although not significant with the inclusion Veneri 1 in the flat terrain group) better support a modest increase in AP relative to ML bending strength in non-flat terrain. This difference may be due to the fact that, although the tibia is the major weight-supporting bone in the hominin leg, it is accompanied laterally by the fibula, which adds ML bending strength to the human leg, which is supported by previous studies that have found increases in fibular robusticity in (1) humans occupying mountainous areas and (2) field hockey players who perform repetitive and dynamic changes in direction probably increasing ML loading of the leg (Marchi et al. 2011; Marchi and Shaw 2011; Sparacello et al. 2014; but see Shackelford 2014). There is no laterally paired bone to the bovid metacarpal; and thus, the bovid metacarpal, as a single beam is likely an excellent indicator of how terrain differences affect routine bending stresses acting on the distal limb segments of these animals.

This study suggests that a potential indicator for human locomotion in non-flat terrain is an increase in overall bending strength and an increase in AP relative to ML bending strength; however, the latter effect may be seen as a modest or significant trend in this sample depending on the inclusion or exclusion of Veneri 1 in the non-flat group. This study also supports previous findings of sex differences in tibial shape (Z_x/Z_y) with males displaying greater increases in AP relative to ML bending strength, which is likely related to sexual division of labor (Ruff 1987).

Previous studies on bovid long bone cross-sectional properties were not clear on the effect of body shape on ML relative to AP bending strength in metacarpal diaphyses. The results of the present study provide evidence that increases in ML bending strength in the limb bones of mountainous bovids are not merely due to augmented ML bending moments introduced by broad body shapes. Thus, there is evidence that locomotion in mountainous terrain may introduce elevated ML stresses to the distal limb segments of mountainous bovids.

The bovid and hominin samples not only differ in anatomy, but also differ in that the bovid sample represents a large cross-species comparative sample where the locomotor behavior of the individual species and subspecies have been directly observed and documented (Jarmon 1974; Scott 1979; Cuzin 2003). In contrast, the smaller archaeological human sample represents only two taxa and the terrain classifications are limited to evaluations of the topography immediately surrounding the discovery sites of the remains rather than observed behavior. Thus, while it is reasonable to assume that bovid species samples generally fit the overall locomotor behavior classification of the taxa, it is impossible to know whether the site of a hominin individual at death reflects his or her locomotor behavior over a significant period of his or her lifetime. This may be problematic for the non-flat and flat terrain groupings in the hominin sample. Consequently, it is possible that outliers, which more closely resemble the opposite group in cross-sectional morphology, may be present. This may be the case for Veneri 1, an Early Upper Paleolithic male whose remains were found in a ground level cave in a flat geographic area. His very robust and elliptical tibia is quite unusual compared to those of the other flat terrain individuals used in this study, and it better resembles the morphology of tibiae from the non-flat group in this analysis and human individuals and populations from mountainous areas described in previous studies (Ruff et al. 2006b; Marchi et al. 2011) rather than the flat terrain group in the analysis. Unfortunately, current data cannot determine the locomotor behavior of Veneri 1 individual in the months before his death; however, the data do suggests that his tibia is quite unusual compared to the other flat terrain individuals. The Spy 2, Paviland 1, and Palomas 96 individuals were all found in elevated sites near remarkably flat terrain. How much of their individual locomotor activities were spent on flat or sloped surfaces is also unknown. Perhaps, individuals who used elevated shelters for home bases walked down from and up to these shelters several times a day regardless of what and where he/she hunted (i.e., mammoths or ibex), gathered (i.e., highland or lowland vegetation), or collected for raw materials.

Lastly, grouping specimens by taxa assumes that clans within a hominin taxon have similar landscape usage, while grouping by the regional terrain surrounding the death site is consistent with the possibility that clans within a species may adopt territories that differ in landscape. Here, nonsignificant differences between taxonomic groups and significant differences between hominin terrain groups in the cross-sectional properties of the tibia support the latter. Furthermore, the results from the comparisons using terrain groups support the assumption that the death site may be a good clue, on average, regarding the mean territorial landscape adopted by the clan to which the individual belonged.

13.5 Conclusion

Theoretically, flexed joint postures on sloped surfaces should increase AP bending stresses in both human tibiae and bovid metacarpals. In addition, mountainous terrain should introduce increases in ML bending stresses in both bovid and human lower leg bones via increased variation in the orientation of ground reaction forces on the feet and nonlinear locomotion. The results for both the bovid and hominin samples support both expectations; yet, the effects of increased ML- and AP-directed bending stresses are expressed differently between groups who occupy contrasting terrains in the two samples. The leg bone of mountainous bovids exhibit significantly greater increases in ML relative to AP bending strength compared to flat terrain bovids while the tibiae of hominins from non-flat show a modest increase in AP relative to ML bending strength. The difference is credited here to the lateral buttressing of the tibia by the fibula, which strengthens the hominin distal lower limb segment in bending in the ML direction. In contrast, the metacarpal, as a single unsupported beam, is likely a less confounded indicator of differences in habitual loading regimes experienced by the distal limb segment of bovids.

In comparisons using both Neandertal and Upper Paleolithic individuals, terrain groupings better explain variation in relative bending strength properties of the tibia than either hominin taxon or sex. This study supports previous studies in suggesting that a potential signal for locomotion in non-flat terrain in the cross-sectional properties of the human tibial midshaft is an increase in overall bending strength and a slight increase in AP relative to ML bending strength. However, it cannot be discounted that the distal limb bones as a unit/complex (tibia + fibula) may actually experience relatively greater increases in routine and elevated ML-directed stresses in non-flat terrain, as is suggested by the bovid metacarpals and is consistent with previous studies that found significant increases in fibular robusticity in humans occupying mountainous areas.

References

Bowyer RT, Leslie DM Jr (1992) *Ovis dalli*. Mamm Species 393:1–7

Carlson KJ (2014) Linearity in the real world—an experimental assessment of non-linearity in terrestrial locomotion. In: Carlson KJ, Marchi D (eds) Reconstructing mobility: environmental, behavioral, and morphological determinants. New York: Springer

Carlson KJ, Judex S (2007) Increased non-linear locomotion alters disphyseal bone shape. J Exp Biol 210:3117–3125

Cuzin F (2003) Les grands Mammifères du Maroc méridional (Haut Atlas, Anti Atlas, Sahara). Distribution, Ecologie et Conservation [Thesis]. Université Montpellier II, Montpellier

Fedosenko AK, Blank DA (2001) *Capra sibirica*. Mamm Species 675:1–13

Fedosenko AK, Blank DA (2005) *Ovis ammon*. Mamm Species 773:1–15

Garland T, Adolph SC (1994) Why not do two-species comparative studies: limitations on inferring adaptation. Phys Zool 67(4):797–828

Jarmon P (1974) The social organization of antelope in relation to their ecology. Behav 48:215–267

Kingdon J, Pagel M (1997) The Kingdon field guide to African mammals. Academic, San Diego

Kingswood SC, Blank DA (1996) *Gazella subgutturosa*. Mamm Species 518:1–10

Leslie DM Jr, Schaller GB (2008) *Pantholops Hodgsonii*. Mamm Species 817:1–13

Marchi D, Shaw CN (2011) Variation in fibular robusticity reflects variation in mobility patterns. J Hum Evol 61:609–616

Marchi D, Sparacello VS, Shaw CN (2011) Mobility and lower limb robusticity of a pastoralist Neolithic population from North-Western Italy. In: Pinhasi R, Stock J (eds) Human bioarchaeology of the transition to agriculture. Wiley-Liss, New York, pp 317–346

Martin RB, Burr DB, Sharkey NA (1998) Skeletal tissue mechanics. Springer, New York

Neas JF, Hoffmann RS (1987) *Budorcas taxicolor*. Mamm Species 277:1–7

Nowak RM (1999) Walker's mammals of the world, 6th edn. Johns Hopkins University Press, Baltimore

Rideout CB, Hoffmann RS (1975) *Oreamnos americanus*. Mamm Species 63:1–6

Ruff CB (1987) Sexual dimorphism in human lower limb bone structure: relationship to subsistence strategy and sexual division of labor. J Hum Evol 16:391–416

Ruff CB (1999) Skeletal structure and behavioral patterns of prehistoric Great Basin populations. In: Hemphill BE, Larsen CS (eds) Prehistoric lifeways in the Great Basin wetlands: bioarchaeological reconstruction and interpretation. University of Utah Press, Salt Lake City, pp 290–320, Chapter 14

Ruff CB (2000) Body size, body shape, and long bone strength in modern humans. J Hum Evol 38:269–290

Ruff CB (2008) Biomechanical analyses of archaeological human skeletal samples. In: Katzenberg M, Saunders A (eds) Biological anthropology of the human skeleton. Wiley, New York, pp 183–206

Ruff CB, Holt BM, Trinkaus E (2006a) Who's afraid of the big bad Wolff?: "Wolff's Law" and bone functional adaptation. Am J Phys Anthropol 129:484–498

Ruff CB, Holt BM, Sladek V, Berner M, Murphy WA, Nedden D et al (2006b) Body size, body proportions, and mobility in the Tyrolean "Iceman.". J Hum Evol 51:91–101

Scott KM (1979) Adaptation and allometry in bovid postcranial proportions [Thesis]. Yale University, New Haven

Scott KM (1985) Allometric trends and locomotor adaptations in the Bovidae. B Am Mus Nat Hist (USA) 179(2):199–288

Shackelford LL (2014) Variation in mobility and anatomical responses in the Late Pleistocene. In: Carlson KJ, Marchi D (eds) Reconstructing mobility: environmental, behavioral, and morphological determinants. Springer, New York

Sparacello VS, Marchi D, Shaw CN (2014) The importance of considering fibular robusticity when inferring the mobility patterns of past populations. In: Carlson KJ, Marchi D (eds) Reconstructing mobility: environmental, behavioral, and morphological determinants. New York: Springer

Sparacello VS, Pearson OM (2010) The importance of accounting for the area of the medullary cavity in cross-sectional geometry: a test based on the femoral midshaft. Am J Phys Anthropol 143(4):612–24

Stock JT, Shaw CN (2007) Which measures of diaphyseal robusticity are robust? A comparison of external methods of quantifying the strength of long bone diaphyses to cross-sectional geometric properties. Am J Phys Anthropol 134:412–423

Trinkaus E (1986) The Neandertals and modern human origins. Annu Rev Anthropol 15: 193–218

Trinkaus E (1992) Second day's discussion chaired by Vandermeersch B and Marks AE. In: Akazawa T, Aoki K, Kimura T (eds) The evolution and dispersal of modern humans in Asia. Hokusen-sha, Tokyo, pp 295–302, Chapter 16

Trinkaus E, Stringer CB, Ruff CB, Hennessy RJ, Roberts MB, Parfitt SA (1999) Diaphyseal cross-sectional geometry of the Boxgrove 1 Middle Pleistocene human tibia. J Hum Evol 37:1–25

Trinkaus E, Ruff CB (2012) Femoral and tibial diaphyseal cross-sectional geometry in Pleistocene *Homo*. Paleo Anth 13–62

Tsutakawa RK, Hewett JE (1977) Quick test for comparing two populations with bivariate data. Biometrics 33:215–219

van Andel H, Davies W, Weninger B, Joris O (2003) Archaeological dates as proxies for the spatial and temporal human presence in Europe: a discourse on the method. In: van Andel H, Davies W (eds) Neanderthals and modern humans in the European landscape during the last glaciation. McDonald Institute, Cambridge, pp 21–30

Weinberg PJ (2002) *Capra cylindricornis*. Mamm Species 695:1–9

Zar JH (2010) Biostatistical analysis, 5th edn. Prentice-Hall, Englewood Cliffs

Chapter 14
Linearity in the Real World: An Experimental Assessment of Nonlinearity in Terrestrial Locomotion

Kristian J. Carlson

Abstract Amongst early human ancestors, cross-sectional geometric properties of lower limb bones are particularly useful for reconstructing mobility patterns. Experimental studies of diaphyseal loads characterizing locomotor activities, however, demonstrate disconnect with theoretical loads predicted from bone morphology alone. This complicates population-level comparisons, and makes specific behavioral inferences tenuous. Lack of a consistent definition for mobility further complicates comparisons. To contribute towards a consensus definition of mobility, here I address one specific relevant factor— what are the effects of nonlinear locomotion or turning. Mice in custom-designed cages accentuating turning (condition 1) and linear movement (condition 2) were compared with mice (control) permitted to move freely in cages. Locomotor behavior of individuals was documented multiple times per day. At the end of the experiment, limb bones were harvested, scanned with high resolution CT, and subjected to structural analyses. Comparing growing BALB/cByJ female mice from a previous experiment and growing C57BL/6J female mice ($n = 30$, 10 per group) in the present experiment, permits comparisons of structural effects of movement regimes on femoral cortical areas, second moments of area, polar moment of area, and shape ratios, as well as activity profiles. C57BL/6J groups differed in activity level, while BALB/cByJ groups did not. Mice in turning groups tended to have more elliptical diaphyses, while linear and control mice differed comparatively less often. Distinctive diaphyseal shapes in turning mice support the idea that nonlinear movements (e.g., turning) have recognizable effects on long bone diaphyseal structure.

K.J. Carlson (✉)
Evolutionary Studies Institute, Palaeosciences Centre, University of the Witwatersrand, WITS 2050 Johannesburg, South Africa

School of Geosciences, University of the Witwatersrand, WITS 2050 Johannesburg, South Africa

Department of Anthropology, Indiana University, Bloomington, IN, USA
e-mail: kristian.carlson@wits.ac.za

K.J. Carlson and D. Marchi (eds.), *Reconstructing Mobility: Environmental, Behavioral, and Morphological Determinants*, DOI 10.1007/978-1-4899-7460-0_14,

This suggests limb loading, likely in side-to-side orientations, is relatively high during turning compared to linear movement.

Keywords Mobility • Femur • Bone functional adaptation • Mediolateral • Positional behavior • Turning • Mouse

14.1 Introduction

Reconstructing activity patterns of extinct (e.g., human ancestors) and extant organisms (e.g., modern human hunter-gatherers) typically does not benefit from direct observational data. This constrains understanding how organisms are adapted to their surrounding environment to inferences of function through form. In situations such as these, behavioral activity must be inferred from indirect evidence, usually from whatever musculoskeletal structure is preserved. Bone functional adaptations, as estimated through modelling long bones as beams and applying cross-sectional geometric analyses, are one means of inferring activity patterns in organisms that are no longer observable (Ruff et al. 1984, 1993; Ruff 1999, 2009; Trinkaus et al. 1999; Stock and Pfeiffer 2001, 2004; Carlson et al. 2007; Shaw and Stock 2009a, b).

Using cross-sectional geometric properties to infer activity profiles is grounded in experimental (e.g., Demes et al. 1998, 2001; Lieberman et al. 2004) and comparative studies (e.g., Ruff and Runestad 2002; Carlson 2005; Shaw and Stock, 2009a, b). The entire body of work demonstrates that loading patterns and the elicited anabolic responses in bone that are associated with movements are complexly related (see reviews in Pearson and Lieberman 2004; Ruff et al. 2006; Judex and Carlson 2009). Exacerbating the complexity of these relationships in characterizing specific movements is the narrowness with which behavioral variation is usually experimentally modelled. Bone surface strains measured in limb bones are usually restricted to those experienced during linear movements in laboratory settings (e.g., Demes et al. 1998, 2001), though a few studies have attempted to collect bone surface strain data from more diversified conditions, e.g., outdoor movements (Burr et al. 1996; Milgrom et al. 2000; Moreno et al. 2008).

Mobility is traditionally defined as movement across the landscape, typically measured in kilometers, and broadly compared across different time scales: daily, seasonal, annual, etc. (Kelly 1995). Such an assessment, however, ignores factors that appear to be biomechanically relevant to bone loads associated with movement on natural substrates. Vertically uneven terrain elicits gait alterations that serve to retain gait stability (Daley and Biewener 2006; Daley et al. 2006). These alterations are associated with greater work in lower limb joints (e.g., knee and hip), ultimately increasing metabolic costs (Voloshina et al. 2013). Principal bone surface strains observed during uphill or downhill movements generally exceed those observed during level movements at similar speeds (Burr et al. 1996). In addition to vertical complexity, natural terrestrial landscapes are characterized by horizontal complexity (e.g., navigating obstacles on the landscape such as trees). Burr and colleagues (1996) observed tibial surface strains during turning movements that exceeded those observed during linear movements, even when the linear movements incorporated vertical elevation change. Demes and colleagues (2006) noted greater external peak forces (mediolateral) during turns compared to those experienced during linear movements. Carlson and Judex (2007) observed modest differences in diaphyseal shapes of growing mice raised for 8 weeks in two enclosures modelling different amounts of horizontal complexity, one accentuating turning and one accentuating linear movement. The latter authors suggested that different distributions of bone mass in femoral diaphyses of the two groups were linked to different loading regimes associated with turning and linear movements. Collectively, these studies suggest that behavioral responses to horizontal complexity (e.g., turning to avoid obstacles) may be at least as impactful on bone loading conditions as vertical complexity (e.g., movement on inclined or declined surfaces associated with elevation change).

Studies of bone functional adaptations modelling vertical complexity in the landscape are comparatively more common (e.g., Ruff 1999; Marchi 2008; Higgins 2014; Sparacello et al. 2014; see review in Ruff and Larsen 2014) than studies emphasizing the effect of horizontal complexity on bone functional adaptations (e.g., Carlson and Judex 2007; Marchi and Shaw 2011; Marchi et al. 2011). Carlson and colleagues (2007) suggested that "mobility comparisons would benefit from including assessments of travel conditions not only in the x-dimension (e.g., distance travelled), but also in the y (e.g., direction changes due to obstacle density in the landscape) and z (e.g., elevation changes due to terrain) dimensions as well ... In this sense, inclusive consideration of travel conditions (Ruff 1999; Carlson and Judex 2007) ultimately may offer greater potential to compare bone functional adaptations between groups." Studies of cross-sectional geometric properties fully embracing these suggestions by addressing habitat x-, y-, and z-dimensions remain rare (e.g., Carlson et al. 2008b; Sparacello and Marchi 2008). Even though different scales of resolution must be considered when comparing population mobility (e.g., residential or logistical mobility: Kelly 1995; Binford 2001), accounting for full variability in three-dimensional landscape structure when modelling movement (i.e., mobility patterns) offers the best chance for building accurate inferences of activity patterns from bone functional adaptations in the musculoskeletal structure of limbs.

Mouse models offer several advantages over other animal models when experimentally modelling bone functional adaptations (Carlson and Byron 2008). Mice are small, amenable to a variety of experimental conditions, mature relatively quickly (i.e., they reach adulthood at 4 months; Flurkey et al. 2007), and have been extensively studied (Beamer et al. 2002; Wergedal et al. 2005). Different inbred strains of mice (e.g., BALB/cByJ, C57BL/6J, myostatin deficient) show varying levels of anabolic bone responses and distribution patterns (Judex et al. 2002; Schmitt et al. 2010; Green et al. 2011; Wallace et al. 2012, 2013).

The goal of the present study is to elucidate the anabolic bone response and distribution in the femoral diaphysis associated with increased turning frequency. In a previous experiment designed to test the effect of turning on femoral diaphyseal structure, femora of BALB/cByJ mice demonstrated a modest, but significant, difference in diaphyseal relative rigidity (I_y/I_x) ratios compared to femora of BALB/cByJ mice performing linear movement. It was not readily apparent, however, whether the differential response was due to either (or both) mediolateral (ML) or anteroposterior (AP) rigidity since group differences in each single structural characteristic were nonsignificant (Carlson and Judex 2007). The C57BL/6J mouse strain exhibits a more sensitive osteogenic response to given load conditions than the BALB/cByJ mouse strain (Judex et al. 2002). Thus, the modest anabolic bone response observed by Carlson and Judex (2007) should be magnified in C57BL/6J mice. Here, I test the hypotheses that C57BL/6J mice using tunnels that emphasize turning will have the largest relative rigidity (I_y/I_x) ratios in femoral diaphyses, while C57BL/6J mice using tunnels that emphasize linear movement will have the smallest relative rigidity (I_y/I_x) ratios. I also test the hypotheses that group differences in relative rigidity (I_y/I_x) ratios are driven by adjustments in both ML and AP rigidity, such that C57BL/6J mice using tunnels that emphasize turning will have the greatest ML rigidity (I_y, which is the numerator in the shape ratio, will be greatest) and C57BL/6J mice using tunnels that emphasize linear movement will have the greatest AP rigidity (I_x, which is the denominator in the shape ratio, will be greatest).

14.2 Materials and Methods

Thirty-five 4-week-old female C57BL/6J mice (The Jackson Laboratory, Bar Harbor, ME, USA) were assigned randomly to one of the three groups ($n=12$ in each experimental group; $n=11$ in the control group). Mice were single-housed in one of the two custom-designed experimental enclosures, or in a control enclosure that consisted of an open cage with a standard wireframe top (Fig. 14.1; see also Carlson et al. 2008a: Fig. 1). Experimental enclosures emphasized quadrupedal turning or linear movements, while simultaneously discouraging climbing; design details have been described elsewhere (see Carlson and Judex 2007). Control enclosures did not limit quadrupedal movement, climbing, or any other behaviors expressed by mice in standard cages. Ten individuals per group were single-housed for 85 days, until approximately 4 months of age, while five individuals ($n=2$ in each experimental group; $n=1$ in the control group) were euthanized following an earlier endpoint in the protocol (day 57), at approximately 3 months of age. The small number of individuals euthanized after 56 days facilitated comparison to BALB/cByJ individuals subjected to the earlier experimental protocol (Carlson and Judex 2007), but subsample sizes effectively ruled out statistical analyses of any observed trends. Thus, results of these comparisons are not reported. For the duration of the experiment, mice were subjected to 12 h:12 h (light:dark) cycles using automatic timers attached to overhead room lights. The experimental protocol was approved by the Institutional Animal Care and Use Committee (IACUC) of NYIT College of Osteopathic Medicine (Old Westbury, NY).

Fig. 14.1 Individual enclosures on racks (top left), with a close-up of the control setup (top right). Turning group individuals are depicted on the bottom left, while linear group individuals are depicted on the bottom right. All individuals are singly housed for the duration of the experimental protocol

Unless explicitly noted otherwise, behavioral and morphological data collection follow published protocols (Carlson and Judex 2007; Carlson et al. 2008a; Wallace et al. 2013). Briefly, room temperature was monitored daily; food and water were available ad libitum. Body weight and food consumption were measured weekly by weighing subjects and food pellets remaining in cages. Since food pellets were weighed at the beginning and end of a week, the difference was assumed to represent food consumption (grams). Fluctuations in body weight and food consumption were useful for monitoring health and stress level of subjects, particularly across groups. No group-wide differences in either were observed.

Activity was monitored for the duration of the experiment by collecting positional behavior data from individuals (Prost 1965) using an instantaneous focal sampling strategy (Altmann 1974) adapted for use with mice. Behavioral categories follow those published elsewhere (Carlson and Judex 2007; Carlson et al. 2008a; Wallace et al. 2013), and are broadly grouped as either locomotor (i.e., walk, run, climb, jump) or postural behaviors (i.e., lie, sit, stand), with the former aggregate

Table 14.1 Sample information

	n	Locomotor behavior (%)	Postural behavior (%)	Body weight (g)	Femoral mechanical length (mm)
Turning (C57BL/6J:16)	10	23.9 (5.6)	76.1 (5.6)	20.5 (0.6)	13.2 (0.2)
Turning (C57BL/6J:12)	2[a]	25.0 (2.5)	75.0 (2.5)	20.3 (2.4)	12.9 (0.5)
Turning (BALB/cByJ: 12)	10	14.0 (8.8)	86.0 (8.8)	22.7 (1.2)	12.6 (0.1)
Linear (C57BL/6J:16)	10	23.5 (9.9)	76.5 (9.9)	21.0 (1.2)	13.3 (0.3)
Linear (C57BL/6J:12)	2[a]	14.3 (5.1)	85.7 (5.1)	20.1 (0.3)	12.7 (0.1)
Linear (BALB/cByJ: 12)	10	14.2 (3.7)	85.8 (3.7)	22.0 (1.3)	12.7 (0.3)
Control (C57BL/6J:16)	10	44.5 (5.5)	55.5 (5.5)	21.3 (0.9)	13.3 (0.2)
Control (C57BL/6J:12)	1[a]	55.4 (–)	44.6 (–)	19.7 (–)	13.1 (–)
Control (BALB/cByJ: 12)	10	12.9 (4.2)	87.1 (4.2)	22.9 (1.6)	12.5 (0.4)

From left to right, cells contain mean and 1 SD (in parentheses)

[a]12-week old individuals sacrificed at day 57 (comparable to BALB/cByJ individuals) rather than day 86 in the protocol

category approximating activity level (Table 14.1). Daily observations were spread over the entire 24 h range of light and dark conditions for the duration of the protocol. A total of 99 behavioral data points were collected for each of the 30 individuals euthanized after 85 days, while 56 behavioral data points were collected for each of the five individuals euthanized after 56 days. Additional observations were conducted at 1-min intervals over the course of an hour to ensure that rare positional behaviors were not overrepresented in daily assessments ($n=35$ h per group). Individuals from each of the three groups were randomly selected for these 1-h assessments. Tunnel passes by individuals in both experimental groups also were recorded during the 1-h assessments. Only when an individual passed at least halfway through a tunnel was it counted as a pass, irrespective of whether the individual ultimately exited through the same or opposite opening that it used to enter the tunnel. A subsequent pass was not considered until the individual exited through either end of the tunnel.

Upon termination of the experimental period, subjects were euthanized, limbs were disarticulated, femora were dissected free of soft tissue, and individual bones were stored in 70 % ethanol. Multiple (4) left femora were scanned at the same time using a Scanco μCT 40 system (Scanco Medical AG, Bassersdorf, Switzerland). An isotropic voxel size of 12.0 μm (55 kVp, 145 μA, 300 ms integration time) was used, and a stack of serial DICOM files was produced from each scan. Renderings were generated from image stacks using the segmentation editor and label fields module in Avizo 7.0 software (Visualization Sciences Group, Mérignac Cedex, France). The same upper and lower thresholds were used for all specimens, and established using external dimensions of bones and noise minimization as selection criteria. In order to position and align renderings in silico, alignment and levelling criteria reported elsewhere (Carlson 2005) were followed using the measurement tool and trackball in Avizo 7.0 (Visualization Sciences Group, Mérignac Cedex, France). After aligning and levelling renderings, they were digitally sectioned orthogonal to their longitudinal axis at 35, 50, and 65 % femoral mechanical length (from distal to proximal) using the

Table 14.2 ANOVA results

	F	p	Fisher's LSD post-hoc analysis results[a]
Locomotor behavior (%)	27.034	<0.001	Control ≫ Turning (<0.001) Control ≫ Linear (<0.001)
Terminal weight	1.741	0.195	
Femoral mechanical length	1.558	0.229	

Degrees of freedom: Between groups = 2, Within groups = 27
Note that these results are for 16-week-old C57BL/6J mice only
[a]Carlson and Judex (2007) reported nonsignificant differences in 12-week-old BALB/cByJ mice for the same variables

surface cut module (Carlson and Judex 2007). Resulting TIFF images with scales were imported into Scion Image (release Beta 4.0.2) and subjected to structural analyses using custom-written macros modelled after the SLICE program (Nagurka and Hayes 1980). Standard cross-sectional properties were calculated for all femora: subperiosteal area (Ps.Ar), cortical area (Ct.Ar), second moments of area about AP (I_y) and ML (I_x) anatomical axes, principal moments of area (I_{max}, I_{min}), and the principal angle (Θ) (Parfitt et al. 1987). Percentage cortical area (% Ct.Ar) was calculated as Ct.Ar divided by Ps.Ar, then multiplied by 100. The polar moment of area (J) was calculated as the sum of I_{max} and I_{min}. Ratios were calculated from second (I_y/I_x) and principal (I_{max}/I_{min}) moments of area, with the latter shape ratio being the best reflection of true cross-sectional shape and the former ratio being a better estimate of femoral rigidity about AP and ML anatomical planes (Carlson 2005). Calculations were performed in SPSS 21.0 (SPSS, Inc., Chicago, IL, USA).

Group differences in activity level, body weight, and femoral mechanical length were assessed with an ANOVA since variable distributions were not significantly different from normal distributions (nonsignificant K-S tests). Only activity level (% locomotor behavior) differed significantly between groups (Table 14.2). Because of this, and because any of these variables may theoretically covary with one another or with other structural variables, a Type III general linear model (GLM) was used to assess the effect of group, activity level, body weight, and femoral mechanical length on the selected structural variables. Only analyses of %Ct.Ar, I_y, I_x, I_y/I_x, and I_{max}/I_{min} are reported here. GLMs used one fixed effect (group) and three covariates (body weight, femoral mechanical length, and activity level) to assess variation in the selected structural properties. Two-way interactions between all variables were initially included in each GLM. In order to determine the best GLM for selected structural properties or ratios, interaction terms were systematically eliminated one-by-one from the full model starting with the nonsignificant interaction exhibiting the highest p-value, then rerunning the GLM after which the nonsignificant interaction again exhibiting the highest p-value was eliminated, and so on. Whittling away interaction terms continued until only statistically significant ones remained. In a majority of the final GLMs, all interaction terms were excluded. Statistical significance was achieved when $p<0.05$; all tests were two tailed. Statistical analyses were performed with SPSS 21.0 (SPSS, Inc., Chicago, IL, USA).

14.3 Results

14.3.1 Comparison of Group Activity Levels, Body Weights, and Femoral Lengths

On average, control mice engaged in significantly more locomotor behavior than turning or linear mice; the latter two groups did not differ significantly (Tables 14.1 and 14.2). Interestingly, C57BL/6J groups consistently exhibited increased locomotor behavior compared to respective BALB/cByJ groups (Table 14.1). Amongst C57BL/6J subjects, turning mice passed through tunnels ($n=35$, 19.6 ± 15.9 times per hour) about half as often as linear mice ($n=35$, 42.1 ± 33.7 times per hour). However, since linear tunnels were shorter (approximately 34.5 cm) than curved tunnels (approximately 48.0 cm), the disparity in overall distance travelled per hour was mollified to some extent (i.e., turning group travelled 942.1 cm/h; linear group travelled 1,453.9 cm/h). Average travel distance of turning and linear mice did not differ significantly according to a nonparametric, rank order Brunner–Munzel Test for Stochastic Equality (1.3426, df$=54.881$, $p=0.185$). Group differences in body weight and femoral mechanical length were not statistically significant (Table 14.2).

14.3.2 Comparison of % Ct.Ar

Femora of the control group consistently exhibited about 1–3 % higher % Ct.Ar than femora of turning or linear groups, while femora of the linear group consistently exhibited the lowest % Ct.Ar at 35 % (Table 14.3), 50 % (Table 14.4), and 65 % regions of interest (ROIs) (Table 14.5). In assessing these trends for statistical significance, GLM results are noteworthy. The GLMs explained a statistically significant amount of variation in % Ct.Ar at each ROI (Table 14.6). At 50 and 65 % ROIs, group was the only statistically significant fixed effect or covariate. In GLMs for both these ROIs, however, there was a significant interaction between group and femoral mechanical length. While no significant group differences were observed in femoral mechanical length (Tables 14.1 and 14.2), the significant interaction between these variables calls into question the significant effect of group on % Ct.Ar, or at least means it cannot be straightforwardly interpreted. No individual fixed effects or covariates were statistically significant in the GLM for the 35 % ROI (Table 14.6).

14.3.3 Comparison of I_y and I_x

Turning and control groups differed in femoral I_y, on average, by less than 1 % at each ROI, while the linear group had approximately 5–11 % lower I_y values than either of the other groups (Tables 14.3, 14.4, and 14.5). In assessing these trends for

Table 14.3 Descriptive statistics for femoral structural properties at the 35 % region of interest

	n	Ps.Ar (mm^2)	Ct.Ar (mm^2)	% Ct.Ar	I_y (mm^4)	I_x (mm^4)	I_{max} (mm^4)	I_{min} (mm^4)	J (mm^4)	I_y/I_x	I_{max}/I_{min}
Turning (C57BL/6J)	10	1.763	0.923	52.4	0.291	0.129	0.296	0.123	0.420	2.268	2.411
		(0.064)	(0.029)	(1.8)	(0.026)	(0.009)	(0.027)	(0.008)	(0.030)	(0.213)	(0.229)
		−0.6 %[a]	−4.6 %	−2.2 %	−0.7 %	−7.9 %	−2.0 %	−6.1 %	−3.0 %	8.0 %	4.3 %
Turning (BALB/cByJ)[b]	10	1.566	0.891	56.9	0.240	0.104	0.241	0.104	0.345	2.302	2.309
		(0.068)	(0.064)	(2.5)	(0.032)	(0.008)	(0.032)	(0.008)	(0.039)	(0.230)	(0.228)
		−1.3 %	−1.9 %	−0.3 %	−1.2 %	−5.5 %	−1.2 %	−5.5 %	−2.3 %	4.3 %	4.4 %
Linear (C57BL/6J)	10	1.713	0.892	52.0	0.262	0.127	0.269	0.120	0.389	2.064	2.232
		(0.092)	(0.082)	(3.3)	(0.035)	(0.016)	(0.038)	(0.013)	(0.051)	(0.066)	(0.130)
		−3.4 %	−7.9 %	−2.6 %	−10.6 %	−9.3 %	−10.9 %	−8.4 %	−10.2 %	−1.7 %	−3.4 %
Linear (BALB/cByJ)	10	1.572	0.899	57.2	0.240	0.107	0.240	0.107	0.347	2.239	2.249
		(0.087)	(0.061)	(2.4)	(0.028)	(0.012)	(0.028)	(0.012)	(0.039)	(0.136)	(0.150)
		−0.9 %	−1.0 %	0 %	−1.2 %	−2.7 %	−1.6 %	−2.7 %	−1.7 %	1.4 %	1.7 %
Control (C57BL/6J)	10	1.773	0.968	54.6	0.293	0.140	0.302	0.131	0.433	2.100	2.311
		(0.073)	(0.052)	(0.9)	(0.032)	(0.013)	(0.035)	(0.011)	(0.042)	(0.160)	(0.210)
Control (BALB/cByJ)	10	1.586	0.908	57.2	0.243	0.110	0.244	0.110	0.353	2.207	2.212
		(0.118)	(0.089)	(3.5)	(0.039)	(0.015)	(0.039)	(0.015)	(0.053)	(0.107)	(0.108)

From top to bottom, cells contain mean, 1 SD, and percent difference from controls

[a]Percentages calculated as difference between experimental group within a strain and control group within the same strain relative to control group within the same strain: ((((Turning or Linear)–Control)/Control) × 100), except for % Ct.Ar which is reported as simple difference in percentages. Positive values indicate a greater value in the experimental group, while negative values indicate a greater value in the control group

[b]Data for BALB/cByJ mice reported in Carlson and Judex (2007)

Table 14.4 Descriptive statistics for femoral structural properties at the 50 % region of interest

	n	Ps.Ar (mm²)	Ct.Ar (mm²)	% Ct.Ar	I_y (mm⁴)	I_x (mm⁴)	I_{max} (mm⁴)	I_{min} (mm⁴)	J (mm⁴)	I_y/I_x	I_{max}/I_{min}
Turning (C57BL/6J)	10	1.701	0.893	52.5	0.242	0.137	0.255	0.124	0.380	1.764	2.051
		(0.066)	(0.023)	(1.3)	(0.019)	(0.009)	(0.020)	(0.008)	(0.026)	(0.099)	(0.106)
		−1.3%[a]	−3.0 %	−1.0 %	0.4 %	−8.7 %	−1.2 %	−6.8 %	−2.8 %	9.0 %	5.5 %
Turning (BALB/cByJ) [b]	10	1.540	0.854	55.5	0.216	0.106	0.217	0.106	0.322	2.029	2.053
		(0.075)	(0.052)	(2.3)	(0.026)	(0.010)	(0.026)	(0.010)	(0.034)	(0.162)	(0.166)
		−0.4 %	−2.8 %	−1.5 %	−0.5 %	−4.5 %	−0.5 %	−3.6 %	−2.1 %	3.7 %	3.6 %
Linear (C57BL/6J)	10	1.682	0.878	52.1	0.227	0.140	0.243	0.124	0.367	1.634	1.948
		(0.073)	(0.079)	(3.5)	(0.024)	(0.018)	(0.029)	(0.013)	(0.041)	(0.101)	(0.120)
		−2.4 %	−4.7 %	−1.4 %	−5.8 %	−6.7 %	−5.8 %	−6.8 %	−6.1 %	1.0 %	0.2 %
Linear (BALB/cByJ)	10	1.544	0.857	55.5	0.214	0.109	0.215	0.108	0.323	1.955	1.995
		(0.086)	(0.052)	(1.0)	(0.025)	(0.012)	(0.025)	(0.012)	(0.036)	(0.109)	(0.107)
		−0.1 %	−2.5 %	−1.5 %	−1.4 %	−1.8 %	−1.4 %	−1.8 %	−1.8 %	−0.1 %	0.7 %
Control (C57BL/6J)	10	1.723	0.921	53.5	0.241	0.150	0.258	0.133	0.391	1.618	1.944
		(0.081)	(0.044)	(1.0)	(0.021)	(0.018)	(0.024)	(0.014)	(0.036)	(0.133)	(0.127)
Control (BALB/cByJ)	10	1.546	0.879	57.0	0.217	0.111	0.218	0.110	0.329	1.956	1.981
		(0.123)	(0.052)	(2.2)	(0.029)	(0.016)	(0.029)	(0.016)	(0.044)	(0.079)	(0.073)

From top to bottom, cells contain mean, 1 SD, and percent difference from controls

[a]Percentages calculated as difference between experimental group within a strain and control group within the same strain relative to control group within the same strain: ((((Turning or Linear)−Control)/Control) × 100), except for % Ct.Ar which is reported as simple difference in percentages. Positive values indicate a greater value in the experimental group, while negative values indicate a greater value in the control group

[b]Data for BALB/cByJ mice reported in Carlson and Judex (2007)

Table 14.5 Descriptive statistics for femoral structural properties at the 65 % region of interest

	n	Ps.Ar (mm^2)	Ct.Ar (mm^2)	% Ct.Ar	I_y (mm^4)	I_x (mm^4)	I_{max} (mm^4)	I_{min} (mm^4)	J (mm^4)	I_y/I_x	I_{max}/I_{min}
Turning (C57BL/6J)	10	1.805	1.106	61.3	0.385	0.161	0.416	0.130	0.547	2.394	3.196
		(0.084)	(0.044)	(1.1)	(0.041)	(0.017)	(0.045)	(0.011)	(0.054)	(0.178)	(0.203)
		−2.9%[a]	−4.1 %	−0.7 %	−0.3 %	−13.4 %	−3.0 %	−9.1 %	−4.4 %	14.7 %	6.4 %
Turning (BALB/cByJ) [b]	10	1.609	1.035	64.4	0.311	0.114	0.315	0.110	0.425	2.738	2.862
		(0.059)	(0.050)	(2.2)	(0.029)	(0.008)	(0.028)	(0.008)	(0.034)	(0.220)	(0.205)
		−1.0 %	−2.3 %	−0.9 %	−1.6 %	−3.4 %	−1.9 %	−3.5 %	−2.1 %	1.7 %	0.7 %
Linear (C57BL/6J)	10	1.809	1.089	60.1	0.358	0.173	0.397	0.134	0.531	2.077	2.960
		(0.092)	(0.100)	(3.5)	(0.051)	(0.026)	(0.062)	(0.016)	(0.073)	(0.216)	(0.355)
		−2.6 %	−5.6 %	−1.9 %	−7.3 %	−7.0 %	−7.5 %	−6.3 %	−7.2 %	−0.5 %	−1.4 %
Linear (BALB/cByJ)	10	1.621	1.052	64.9	0.320	0.117	0.326	0.111	0.437	2.742	2.932
		(0.091)	(0.048)	(1.6)	(0.036)	(0.012)	(0.035)	(0.013)	(0.046)	(0.181)	(0.156)
		−0.2 %	−0.7 %	−0.4 %	1.3 %	−0.8 %	1.6 %	−2.6 %	0.7 %	1.9 %	3.2 %
Control (C57BL/6J)	10	1.858	1.153	62.0	0.386	0.186	0.429	0.143	0.572	2.088	3.003
		(0.087)	(0.056)	(1.1)	(0.042)	(0.024)	(0.048)	(0.014)	(0.059)	(0.238)	(0.225)
Control (BALB/cByJ)	10	1.625	1.059	65.3	0.316	0.118	0.321	0.114	0.434	2.691	2.841
		(0.119)	(0.058)	(2.5)	(0.034)	(0.018)	(0.036)	(0.016)	(0.051)	(0.157)	(0.150)

From top to bottom, cells contain mean, 1 SD, and percent difference from controls

[a]Percentages calculated as difference between experimental group within a strain and control group within the same strain relative to control group within the same strain: ((((Turning or Linear)–Control)/Control) × 100), except for % Ct.Ar which is reported as simple differences in percentages. Positive values indicate a greater value in the experimental group, while negative values indicate a greater value in the control group

[b]Data for BALB/cByJ mice reported in Carlson and Judex (2007)

Table 14.6 General linear model (GLM) results

	Model F	df[a]	FE Group p	C % Loc. Beh. p	C Weight p	C Femur length p	Model adjusted R^2
35 % Ct.Ar	3.709*	5, 29	0.586	0.276	0.472	0.110	0.318
35 I_y	2.429	5, 29	0.056	0.361	0.529	0.249	0.198
35 I_x	5.021**	5, 29	0.553	0.463	0.944	0.005**	0.409
35 I_y/I_x	2.344	5, 29	0.034*	0.569	0.320	0.130	0.188
35 I_{max}/I_{min}	1.420	5, 29	0.106	0.211	0.233	0.172	0.068
50 % Ct.Ar	5.661**	7, 29	<0.001**	0.357	0.688	0.110	0.529[b]
50 I_y	4.295**	5, 29	0.042*	0.706	0.647	0.009**	0.362
50 I_x	6.888**	6, 29	0.826	0.038*	0.061	<0.001**	0.549[c]
50 I_y/I_x	2.372	5, 29	0.073	1.000	0.610	0.195	0.191
50 I_{max}/I_{min}	1.896	5, 29	0.054	0.108	0.386	0.159	0.134
65 % Ct.Ar	2.498*	7, 29	0.020*	0.502	0.346	0.167	0.266[b]
65 I_y	2.658*	5, 29	0.153	0.672	0.771	0.017*	0.222
65 I_x	3.797*	5, 29	0.474	0.834	0.706	0.010**	0.325
65 I_y/I_x	2.754*	5, 29	0.011*	0.484	0.769	0.638	0.232
65 I_{max}/I_{min}	1.011	5, 29	0.219	0.513	0.881	0.655	0.002

*$p<0.05$, **$p<0.01$

FE fixed effect, C covariate

[a]Degrees of freedom (corrected model, corrected total) for each general linear model (GLM)

[b]Interaction effect between Group and Femoral mechanical length was statistically significant at the 0.05 level

[c]Interaction effect between % Locomotor behavior and Body weight was statistically significant at the 0.05 level

statistical significance, some GLM results are noteworthy. GLMs of I_y were statistically significant only at 50 % and 65 % ROIs (Table 14.6), while the GLM for the 35 % ROI was borderline nonsignificant ($p=0.064$). At the 50 % ROI, group and femoral mechanical length had significant effects on variation in I_y, while at the 65 % ROI only femoral mechanical length had a significant effect on variation in I_y (Table 14.6).

Turning and linear groups, on average, exhibited 6–13 % lower I_x values at ROIs than the control group (Tables 14.3, 14.4, and 14.5). Turning and linear groups differed comparatively less in I_x values across ROIs (i.e., by about 1–6 %), and did not differ consistently in one direction (i.e., the linear group was not always lower or always higher than the turning group) (Tables 14.3, 14.4, and 14.5). In assessing these trends for statistical significance, GLM results are noteworthy. At each ROI, GLMs explained a statistically significant amount of variation in I_x (Table 14.6). At 35 and 65 % ROIs, only one covariate, femoral mechanical length, had a statistically significant effect on I_x. At the 50 % ROI, a second variable, % locomotor behavior (a covariate in the model), also had a statistically significant effect. However, the statistically significant interaction between % locomotor behavior and body weight at the 50 % ROI calls into question the significant effect of either variable, or at least means they cannot be straightforwardly interpreted.

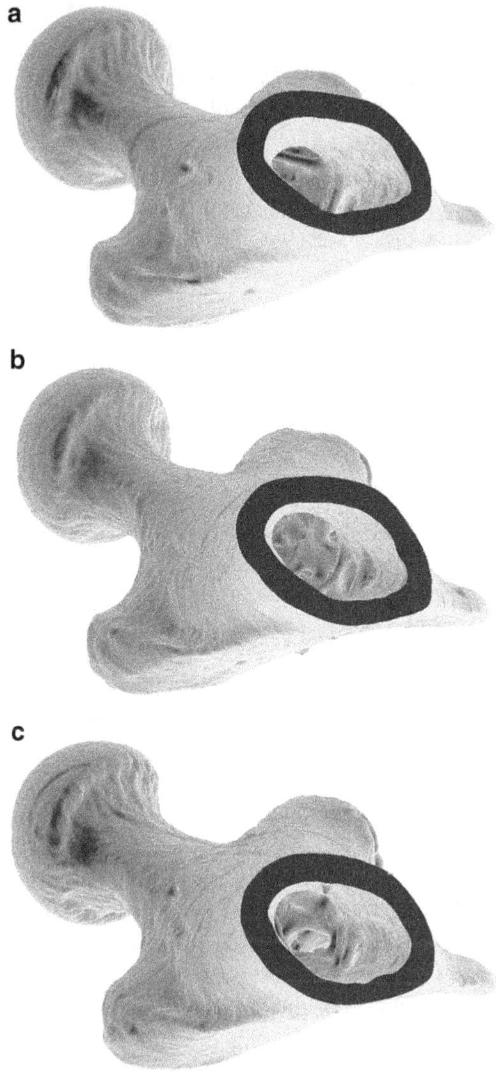

Fig. 14.2 Comparison of femoral midshaft diaphyses (50 % ROI) for turning (A), linear (B), and control (C) mice with relative rigidity (I_y/I_x) ratios nearest their respective group averages (see Table 14.4). The illustrated cross sections are from subject 10 (turning), subject 19 (linear), and subject 32 (control). Note the comparatively higher relative rigidity (I_y/I_x) ratio, and thus more elliptical cross section (uniformly dark gray area) of subject 10, corresponding to relatively more enhanced mediolateral rigidity

14.3.4 Comparison of Group I_y/I_x *and* I_{max}/I_{min} *Ratios*

The turning group, on average, exhibited about 8–15 % higher relative rigidity (I_y/I_x) ratios than control and linear groups (Tables 14.3, 14.4, and 14.5; Fig. 14.2). Control and linear groups differed by about 1–2 % in these ratios, and did not differ consistently in one direction (i.e., the linear group was not always lower or always higher than the control group) (Fig. 14.2). In assessing these trends for statistical significance, a few GLM results are noteworthy. Only at the 65 % ROI was the GLM for the I_y/I_x ratio statistically significant (Table 14.6). The group variable (fixed effect in the model) had a statistically significant effect on variation in I_y/I_x ratios at this ROI. It is worth noting that at the 35 % ROI, while the GLM was borderline nonsignificant ($p=0.072$), the effect of group was associated with a p-value of 0.034 (Table 14.6). Also, in the borderline nonsignificant ($p=0.070$) GLM for the 50 % ROI, the group variable itself had a p-value of 0.073 (Table 14.6). While GLMs at 35 and 50 % ROIs ultimately were not supported with statistical significance, and thus neither is the effect of group, perhaps in a larger sample of individuals these GLMs and the effect of group would become statistically significant. No other variables were statistically significant in GLMs of relative rigidity (I_y/I_x) ratios.

The turning group, on average, exhibited about 4–8 % higher shape (I_{max}/I_{min}) ratios than control and linear groups (Tables 14.3, 14.4, and 14.5; Fig. 14.2). Differences between the turning group and the control group in shape (I_{max}/I_{min}) ratios, therefore, were roughly half the magnitude of differences in relative rigidity (I_y/I_x) ratios. Control and linear groups differed less in shape (I_{max}/I_{min}) ratios, and not in a consistent fashion (i.e., the linear group was not always lower or always higher than the control group) (Tables 14.3, 14.4, and 14.5). Statistical support for these trends was not observed since each GLM was not statistically significant (Table 14.6).

14.4 Discussion

Compared to control individuals inhabiting open cages, C57BL/6J mice in both experimental groups (i.e., turning and linear mice) accrued less bone in the femoral diaphysis during growth. Less accrual is indicated both by lower absolute (Ct.Ar) and relative bone mass (%Ct.Ar) in turning and linear groups compared to the control group (Tables 14.3, 14.4, 14.5, and 14.6). Elevated activity level observed in the control group (Tables 14.1 and 14.2) is consistent with their greater accrual of bone mass in femoral diaphyses. Bearing in mind that group differences in body weight and femoral mechanical length were not statistically significant, it would seem that activity differences alone were behind group differences in femoral bone mass at the three ROIs. While group membership explained a statistically significant amount of variation in % Ct.Ar at two of the three femoral ROIs investigated in the study, a significant interaction between group and femoral mechanical length complicated this interpretation. Thus, the relationship between % Ct.Ar, group, and femoral mechanical length appears to be complicated, particularly since groups did not differ significantly in femoral mechanical length (Table 14.2). Further investigation of these relationships, perhaps in a larger sample where femoral mechanical length varied to a greater extent, may help clarify the nature of these complexities.

Despite the observed differences in bone accrual, growing C57BL/6J mice subjected to stereotyped movement regimes (turning and linear groups) still exhibited altered bone distributions within femoral diaphyses, often in the hypothesized direction (Tables 14.3, 14.4, 14.5, and 14.6). The turning group consistently exhibited the highest relative rigidity (I_y/I_x) ratios of any group, by at least 8 % at each ROI (Fig. 14.2). Differences between linear and control groups, on the other hand, were consistently less dramatically (Fig. 14.2). While statistical support for the effect of group was restricted to the 65 % ROI, larger samples very well could provide statistical support for the differences observed at 35 and 50 % ROIs as well. It is worth noting in this regard that Carlson and Judex (2007) observed in the BALB/cByJ turning group the same significant difference in relative rigidity (I_y/I_x) ratios at femoral midshafts (50 % ROI), but at a lower magnitude than in the present study (i.e., 5.6 % difference versus 8 % or more in the present study).

Turning and linear groups did not exhibit the highest ML (I_y) and AP (I_x) rigidities, respectively. The hypothesized reinforcement of femoral diaphyses in these directions, however, appears to have been impeded by overall lower accrual of bone mass in turning and linear groups compared to bone accrual in the control group. Clearly, however, increased frequency of turning behavior is driving some of the observed distributional difference in bone mass that is visible in relative rigidity (I_y/I_x) ratios. The same does not appear to be the case for increased frequency of linear movement though. Side-to-side external peak forces applied to the limbs are elevated during turning behaviors in primates, often exceeding fore-aft horizontal peak forces (Demes et al. 2006). The turning group consistently exhibited ML rigidities equivalent to those of the control group (i.e., average I_y values within 1 % of one another), despite the former exhibiting overall less bone mass accrued in cross sections. Both turning and linear groups, on the other hand, exhibited similarly-reduced AP rigidities (I_x) relative to those of the control group. This suggests preferential distribution of bone mass enhanced ML rigidity in the turning group, but did not enhance AP rigidity to the same extent in the linear group. Carlson and Judex (2007) reported a similar trend in BALB/cByJ mice as was observed in the present study, namely turning and control groups differed by less than 1 % in average ML rigidity. The BALB/cByJ linear group (Carlson and Judex 2007) exhibited a relatively more subtle reduction in ML rigidity compared to what was observed in the present study. Linear and turning groups in the present study, and that of Carlson and Judex (2007), exhibited reduced AP rigidity compared to that of the control group, though these differences were not significant. Results reported in the present study along with those reported by Carlson and Judex (2007) collectively refute the curious suggestion of Barak et al. (2011) who postulated that animals as small as mice may not experience strains during locomotion that are large enough to stimulate much of an osteogenic response to loading. Rather, the present study, along with Carlson and Judex (2007), observed the predicted form–function relationships in mouse femoral diaphyses, and on several occasions supported these trends with statistical significance, despite challenges associated with numerically small samples.

Interestingly, the C57BL/6J turning and linear groups differed significantly in trabecular and cortical structure of the proximal humerus, but not in trabecular and cortical structure of the distal femoral metaphysis (Wallace et al. 2013). The turning group exhibited proximal humeri with 12 % higher trabecular bone volume fraction,

11 % higher cortical area, and 12 % thicker cortices than the linear group, while no significant differences in their distal femoral diaphyses were reported despite similar, but more subtle, trends in trabecular bone volume fraction and cortical area (Wallace et al. 2013). While coordinated cortical and cancellous anabolic responses of C57BL/6 mice have been noted in certain mechanical loading conditions in laboratory settings (Sugiyama et al. 2010), an explanation for the apparent absence of coordinated anabolic responses in femoral cortical and cancellous bone of the experimental groups in the present study is unclear. Less dramatic anabolic responses of BALB/cByJ mice differentiating groups, in part, could be influenced by their apparent overall lower activity levels (Carlson and Judex 2007; Carlson et al. 2008a) compared to C57BL/6J mice (Table 14.1), or perhaps by greater osteogenic sensitivity of C57BL/6J mice compared to BALB/cByJ mice (Judex et al. 2002). Interestingly, despite exhibiting lower activity levels (Table 14.1), BALB/cByJ mice consistently exhibited higher % Ct.Ar than C57BL/6J mice in respective groups (Tables 14.3, 14.4, and 14.5). Greater overall bone mass (Ct.Ar) observed in growing C57BL/6J mice compared to growing BALB/cByJ mice (Carlson and Judex 2007), despite greater % Ct.Ar in BALB/cByJ mice, may be related to an additional four weeks of exposure to the experimental setup. Evaluation of a small number of C57BL/6J mice in each group subjected to the same endpoint (day 57) as BALB/cByJ mice (Carlson and Judex 2007) did not offer definitive insight into any of these scenarios (results not reported here).

Higher bone surface strains during turning compared to linear movement have been reported in humans (Burr et al. 1996) and goats (Moreno et al. 2008). Demes and colleagues (2006) observed higher ML-oriented external forces than AP-oriented external forces in two nonhuman primates during turning. Based on results of the present experiment, which along with the study of Carlson and Judex (2007) is the first to demonstrate bone functional adaptations to turning, turning behavior, or change in movement direction, has a clear impact on the structure of femoral diaphyses (i.e., enhances ML rigidity). Current standards for categorizing behaviors of free-ranging primates (Hunt et al. 1996) do not recognize turning behavior as a distinct category. In fact, turning is arguably an intrinsic component of all forms of locomotion performed by free-ranging animals, but frequencies of it remain unknown. When attempting to compare mobility within a taxon, or even across taxa, by using cross-sectional geometric properties to infer activity patterns, variation in the horizontal dimension of landscapes should be included. Comparisons of terrain "ruggedness" and "uneveness" represent a good start (Ruff 1999; Marchi 2008; Higgins 2014), but the present study and others (e.g., Marchi and Shaw 2011) demonstrate that not only variation in the vertical dimension may be functionally relevant. Indeed, by incorporating all three x-, y-, and z-dimensions into characterizations of terrain "ruggedness" or "uneveness," a richer understanding of the causes of variation in cross-sectional geometric properties will be achieved. Ultimately, this will improve resolution in reconstructed activity patterns of extinct (e.g., human ancestors) and extant taxa (e.g., modern human hunter-gatherers) based on musculoskeletal structure.

Acknowledgments I am grateful to the NYIT College of Osteopathic Medicine (Old Westbury, NY) Office of Research for funding this research. I also wish to thank the Institute for Human Evolution, University of the Witwatersrand for providing the funding that permitted travel to the 2011 AAPA conference. I wish to thank Kurt Amsler, Abe Furia, Liza Nicholson, Kathleen O'Rourke, Sunny Hwang, Brian Beatty, Ryan Palevsky, and several other NYCOM students and staff for their assistance with animal care at various points over the duration of this project. I would like to thank Stefan Judex for sharing his exceptional insights into bone functional adaptations throughout this project. I am extremely grateful to Mark Dowdeswell for advice on selecting and performing statistical analyses, and for other comments on the project. I wish to thank Damiano Marchi and one anonymous reviewer for constructive comments provided on earlier versions of this chapter.

References

Altmann J (1974) Observational study of behavior: sampling methods. Behaviour 49:227–267

Barak MM, Lieberman DE, Hublin JJ (2011) A Wolff in sheep's clothing: trabecular bone adaptation in response to changes in joint loading orientation. Bone 49:1141–1151

Beamer WG, Donahue LR, Rosen CJ (2002) Genetics and bone: using the mouse to understand man. J Musculoskel Neuron Interact 2:225–231

Binford LR (2001) Constructing frames of reference—an analytical method for archaeological theory building using hunter-gatherer and environmental data sets. University of California Press, Berkeley, CA

Burr DB, Milgrom C, Fyhrie D, Forwood M, Nyska M, Finestone A, Hoshaw S, Saiag E, Simkin A (1996) *In vivo* measurement of human tibial strains during vigorous activity. Bone 18:405–410

Carlson KJ (2005) Investigating the form-function interface in African apes, relationships between principal moments of area and positional behaviors in femoral and humeral diaphyses. Am J Phys Anthropol 127:312–334

Carlson KJ, Byron CD (2008) Building a better organismal model: the role of the mouse—introduction to the symposium. Integr Comp Biol 48:321–323

Carlson KJ, Judex S (2007) Increased non-linear locomotion alters diaphyseal bone shape. J Exp Biol 210:3117–3125

Carlson KJ, Grine FE, Pearson OM (2007) Robusticity and sexual dimorphism in the postcranium of modern hunter-gatherers from Australia. Am J Phys Anthropol 134:9–23

Carlson KJ, Lublinsky S, Judex S (2008a) Do different locomotor modes during growth modulate trabecular architecture in the murine hind limb? Integr Comp Biol 48:385–393

Carlson KJ, Sumner DR, Morbeck ME, Nishida T, Yamanaka A, Boesch C (2008b) Role of non-behavioral factors in adjusting long bone diaphyseal structure in free-ranging *Pan troglodytes*. Int J Primatol 29:1401–1420

Daley MA, Biewener AA (2006) Running over rough terrain reveals limb control for intrinsic stability. Proc Natl Acad Sci U S A 103:15681–15686

Daley MA, Usherwood JR, Felix G, Biewener AA (2006) Running over rough terrain: guinea fowl maintain dynamic stability despite a large unexpected change in substrate height. J Exp Biol 209:171–187

Demes B, Stern JT, Hausman MR, Larson SG, McLeod KJ, Rubin CT (1998) Patterns of strain in the macaque ulna during functional activity. Am J Phys Anthropol 106:87–100

Demes B, Qin Y, Stern JT Jr, Larson SG, Rubin CT (2001) Patterns of strain in the macaque tibia during functional activity. Am J Phys Anthropol 116:257–265

Demes B, Carlson KJ, Franz TM (2006) Cutting corners: the dynamics of turning behaviors in two primate species. J Exp Biol 209:927–937

Flurkey K, Currer JM, Harrison DE (2007) Mouse models in aging research. In: Fox JG, Davisson MT, Quimby FW, Barthold SW, Newcomer CE, Smith AL (eds) The mouse in biomedical research. Academic, Amsterdam, pp 637–672

Green DJ, Hamrick MW, Richmond BG (2011) The effects of hypermuscularity on shoulder morphology in myostatin-deficient mice. J Anat 218:544–557

Higgins RW (2014) The effects of terrain on long bone robusticity and cross-sectional shape in the lower limb bones of bovids, Neanderthals, and Upper Paleolithic modern humans. In: Carlson KJ, Marchi D (eds) Reconstructing mobility: environmental, behavioral, and morphological determinants. Springer Press, New York

Hunt KD, Cant JGH, Gebo DL, Rose MD, Walker SE (1996) Standardized descriptions of primate locomotor and postural modes. Primates 37:363–387

Judex S, Carlson KJ (2009) Is bone's response to mechanical signals dominated by gravitational loading? Med Sci Sports Exerc 41:2037–2043

Judex S, Donahue LR, Rubin C (2002) Genetic predisposition to low bone mass is paralleled by an enhanced sensitivity to signals anabolic to the skeleton. FASEB J 16:1280–1282

Kelly RL (1995) The foraging spectrum—diversity in hunter-gatherer lifeways. Smithsonian Institution Press, Washington

Lieberman DE, Polk JD, Demes B (2004) Predicting long bone loading from cross-sectional geometry. Am J Phys Anthropol 123:156–171

Marchi D (2008) Relationships between lower limb cross-sectional geometry and mobility: the case of a Neolithic sample from Italy. Am J Phys Anthropol 137:188–200

Marchi D, Shaw CN (2011) Variation in fibular robusticity reflects variation in mobility patterns. J Hum Evol 61:609–616

Marchi D, Sparacello V, Shaw CN (2011) Mobility and lower limb robusticity of a pastoralist Neolithic population from north-western Italy. In: Pinhasi R, Stock JT (eds) Human bioarchaeology of the transition to agriculture. Wiley-Blackwell, Oxford, pp 317–346

Milgrom C, Finestone A, Levi Y, Simkin A, Ekenman I, Mendelson S, Millgram M, Nyska M, Benjuya N, Burr D (2000) Do high impact exercises produce higher tibial strains than running? Br J Sports Med 34:195–199

Moreno CA, Main RP, Biewener AA (2008) Variability in forelimb bone strains during non-steady locomotor activities in goats. J Exp Biol 211:1148–1162

Nagurka ML, Hayes WC (1980) An interactive graphics package for calculating cross-sectional properties of complex shapes. J Biomech 13:59–64

Parfitt AM, Drezner MK, Glorieux FH, Kanis JA, Malluche H, Meunier PJ, Ott SM, Recker RR (1987) Bone histomorphometry: standardization of nomenclature, symbols, and units. Report of the ASBMR Histomorphometry Nomenclature Committee. J Bone Miner Res 2:595–610

Pearson OM, Lieberman DE (2004) The aging of Wolff's "Law": ontogeny and responses to mechanical loading in cortical bone. Yrbk Phys Anthropol 47:63–99

Prost JH (1965) A definitional system for the classification of primate locomotion. Am Anthropol 67:1198–1214

Ruff CB (1999) Skeletal structure and behavioral patterns of prehistoric Great Basin populations. In: Hemphill BE, Larsen CS (eds) Prehistoric lifeways in the Great Basin wetlands: bioarchaeological reconstruction and interpretation. University of Utah Press, Salt Lake City, pp 290–320

Ruff CB (2009) Relative limb strength and locomotion in *Homo habilis*. Am J Phys Anthropol 138:90–100

Ruff CB, Larsen CS (2014) Long bone structural analyses and the reconstruction of past mobility: a historical review. In: Carlson KJ, Marchi D (eds) Reconstructing mobility: environmental, behavioral, and morphological determinants. Springer, New York

Ruff CB, Runestad JA (2002) Primate limb bone structural adaptations. Annu Rev Anthropol 21:407–433

Ruff CB, Larsen CS, Hayes WC (1984) Structural changes in the femur with the transition to agriculture on the Georgia Coast. Am J Phys Anthropol 64:125–136

Ruff CB, Trinkaus E, Walker A, Larsen CS (1993) Postcranial robusticity in *Homo*. I. Temporal trends and mechanical interpretation. Am J Phys Anthropol 91:21–53

Ruff CB, Holt B, Trinkaus E (2006) Who's afraid of the big bad Wolff?: "Wolff's Law" and bone functional adaptation. Am J Phys Anthropol 129:484–498

Schmitt D, Zumwalt AC, Hamrick MW (2010) The relationship between bone mechanical properties and ground reaction forces in normal and hypermuscular mice. J Exp Zool A Ecol Genet Physiol 313:339–351

Shaw CN, Stock JT (2009a) Intensity, repetitiveness, and directionality of habitual adolescent mobility patterns influence the tibial diaphysis morphology of athletes. Am J Phys Anthropol 140:149–159

Shaw CN, Stock JT (2009b) Habitual throwing and swimming correspond with upper limb diaphyseal strength and shape in modern human athletes. Am J Phys Anthropol 140:160–172

Sparacello V, Marchi D (2008) Mobility and subsistence economy: a diachronic comparison between two groups settled in the same geographical area (Liguria, Italy). Am J Phys Anthropol 136:485–495

Sparacello VS, Marchi D, Shaw CN (2014) The importance of considering fibular robusticity when inferring the mobility patterns of past populations. In: Carlson KJ, Marchi D (eds) Reconstructing mobility: environmental, behavioral, and morphological determinants

Stock JT, Pfeiffer SK (2001) Linking structural variability in long bone diaphyses to habitual behaviours: foragers from the Southern African Later Stone Age and the Andaman Islands. Am J Phys Anthropol 115:337–348

Stock JT, Pfeiffer SK (2004) Long bone robusticity and subsistence behaviour among Later Stone Age foragers of the forest and fynbos biomes of South Africa. J Archaeol Sci 31:999–1013

Sugiyama T, Price JS, Lanyon LS (2010) Functional adaptation to mechanical loading in both cortical and cancellous bone is controlled locally and is confined to the loaded bones. Bone 46:314–321

Trinkaus E, Stringer CB, Ruff CB, Hennessy RJ, Roberts MB, Parfitt SA (1999) Diaphyseal cross-sectional geometry of the Boxgrove I Middle Pleistocene human tibia. J Hum Evol 37:1–25

Voloshina AS, Kuo AD, Daley MA, Ferris DP (2013) Biomechanics and energetics of walking on uneven terrain. J Exp Biol 216:3963–3970

Wallace IJ, Tommasini SM, Judex S, Garland T Jr, Demes B (2012) Genetic variations and physical activity as determinants of limb bone morphology: an experimental approach using a mouse model. Am J Phys Anthropol 148:24–35

Wallace IJ, Kwaczala AT, Judex S, Demes B, Carlson KJ (2013) Physical activity engendering loads from diverse directions augments the growing skeleton. J Musculoskel Neuron Interact 13:245–250

Wergedal JE, Sheng MHC, Ackert-Bicknell CL, Beamer WG, Baylink DJ (2005) Genetic variation in femur extrinsic strength in 29 different inbred strains of mice is dependent on variations in femur cross-sectional geometry and bone density. Bone 36:111–122

Chapter 15
Femoral Mechanics, Mobility, and Finite Element Analysis

Kelli H. Tamvada

Abstract Traditional analyses of long bone morphology, e.g., applying beam theory to imaged cross sections of bone or investigating diaphyseal curvature, examine the effect of skeletal variables on structural integrity separately, an approach that does not incorporate information on the entire bone. Finite element analysis allows exploration of the structural integrity of complete bones under specific loading conditions, providing a more detailed picture of precisely how morphological differences affect a bone's strength and patterns of stress and strain. Finite element analysis also allows complex variables such as differences in joint configurations between species to be modeled. Finite element models further allow the examination of how bones behave during simulations of particular activities, at various magnitudes of loading, and at different angles of excursion. Here I provide an overview of finite element analysis and examine how it contributes to studies of mobility using a case study of a human femur.

Keywords FEA • Femur • Biomechanics • Stress • Strain

Bony responses to mechanical loading, particularly the rate and frequency of loading, are well-documented (Goodship et al. 1979; Hert et al. 1969, 1971, 1972; Jones et al. 1977; Krolner and Toft 1963; Lanyon 1987; Lanyon and Bourn 1979; Lanyon et al. 1979, 1982; Nordstrom et al. 1996; Paul 1971; Ruff 2005; Ruff et al. 2006; Skerry 2000; Taylor et al. 1996; Tilton et al. 1980; Woo 1981). For this reason, bones are thought to be useful sources for understanding activity in populations or organisms whose activity cannot be directly observed. One aspect of activity is mobility, defined here as linear movement across the landscape (Carlson et al. 2007;

K.H. Tamvada (✉)
Department of Anthropology, University at Albany, State University of New York, 1400 Washington Avenue, Albany, NY 12222, USA
e-mail: KTamvada@albany.edu

K.J. Carlson and D. Marchi (eds.), *Reconstructing Mobility: Environmental, Behavioral, and Morphological Determinants*, DOI 10.1007/978-1-4899-7460-0_15,

Kelly 1995), which is often quantified as the ecological variable day range. Day range is the distance an animal or focal group typically travels in the pursuit of resources in the course of one day. Mobility patterns elucidate interesting aspects of culture in prehistoric societies such as subsistence strategies, hunting techniques, seasonal activity levels, home range size, resource availability, and other behavioral variables (Larsen 1987). In order to infer clues about mobility from bones, we must first identify which aspects of bony morphology are important in reconstructing mobility and why.

There are several characteristics of the human femur that are likely to be related to activity levels. These include neck-shaft angle (Trinkaus 1993), diaphyseal cross-sectional morphology and robusticity (i.e. relative biomechanical strength) (Cowgill, 2014; Ruff et al. 1993; Stock and Shaw 2007; Lieberman et al. 2003; Trinkaus and Ruff 1999; Trinkaus et al. 1999), and diaphyseal curvature (Bertram and Biewener 1988; Ruff 1995; Shackelford and Trinkaus 2002). Human infants are born with high neck-shaft angles, but as load-bearing begins, this angle decreases. Given the plastic nature of these traits in sub-adults, they may be indicative of activity levels during development (Cowgill 2010, 2014; Cowgill et al. 2010; Trinkaus 1993). Clearly, diaphyseal robusticity is related to activity levels in that frequent loading induces bony changes meant to reinforce the strength of the bone, usually through bone deposition on the periosteal surface, e.g., (Goodship et al. 1979; Hert et al. 1971, 1972; Lanyon and Baggott 1976; Lanyon et al. 1979). When elevated activity levels are a consequence of locomotion, then those activity levels may be associated with increased mobility. Aspects of shape, such as cross-sectional geometry and longitudinal curvature, serve to elevate and influence the predictability of stress transmission through the shaft (Bertram and Biewener 1988). Predictability of stress transmission may be an important adaptation to resisting eccentrically-directed loads (Bertram and Biewener 1988; Biewener et al. 1983). Thus, there are two means by which a bone may reinforce itself: size and shape, both of which must be considered when studying mobility. Each of these morphological traits is worthy of investigation, but the femur, like any other bone, is an integrated structure (Bertram and Biewener 1988; Currey 2002) rather than discrete characteristics (e.g., diaphyseal curvature, cross-sectional geometry) independently grouped together. Finite element models (FEMs) provide an advantage relative to two-dimensional analyses [such as applying beam analysis to variously imaged cross sections of bone (Ruff 1989)] in that they can potentially provide a more complete understanding of bone behavior under various, specified loading environments. For this reason, finite element analysis (FEA) has a promising role to play in mobility studies aimed at deciphering the effect of long bone morphology on bone behavior.

This method is particularly powerful in that, once a model is created and validated, an array of modeling experiments can be performed to test various questions regarding the modeled structure. In principle, such experiments are limited only by the accuracy of input data, such as geometry, material properties, or muscle force magnitudes and directions. Note, however, that FEA does not provide direct information about mobility patterns or ranging behavior. Rather, FEA provides a means of testing how well structures "perform" mechanically under specific loading conditions that may simulate those experienced by an organism during particular behaviors. If these hypothesized conditions reflect those experimentally determined to be adaptive in organisms that, for example, range over long distances, then it is possible to test whether or not expressed morphologies of organisms confer a biomechanical advantage compared to alternative morphologies (e.g., structures of different shapes). For example, if it is hypothesized that a given bone is routinely loaded with high forces, or that it is loaded especially frequently, then one might hypothesize that the bone should exhibit a morphology that makes it structurally strong in the face of these loads. Alternatively, it might possess a shape that increases the predictability of its strain environment. In either case, these predictions are mechanical in nature, and importantly, they can be tested with FEA.

FEA is a remarkably powerful and flexible analytical tool. In principle, it should be possible to perform a series of experiments modeling the performance of a bony structure over the course of a given activity, e.g., a femur during running at different points of the gait cycle, or a humerus during the act of rowing a boat. Indeed, FEA can, in principle, be used to dynamically simulate complex behaviors. However, such applications would require detailed information about applied loads and kinematics that may not presently exist. It also would be interesting to examine the effect of bone remodeling, or changes in a structure's morphology, on its performance. In the case of remodeling, this could potentially be carried out by artificially altering the FEM so that periosteal deposition is simulated using Virtual Anthropology techniques (Weber and Bookstein 2011). Advances in the confluence of these two methods leave the field ripe for discovery (Weber et al. 2011).

The aim of this chapter is to give readers an overview of how FEA works, to illustrate potential research applications of FEMs in anthropology (focusing on mobility) with a human femur FEM test case, and to identify avenues of future research involving postcrania.

15.1 Finite Element Analysis: The Method

One purpose of FEA is to elucidate the manner or degree to which a structure responds to external loads. Key outputs of FEA include information about stress (force per unit area) and strain (change in length divided by original length) (Richmond et al. 2005) experienced by a loaded object. Two applications of FEA are of particular interest in this chapter. In the first application, biologically realistic models are created for use in experiments aimed at better understanding how the bone (or structure of interest) behaves under specific loading conditions. For this application, a well-validated model (discussed below) is of the utmost importance. The second application involves comparisons between similar structures, as may occur when comparing different fossil taxa. In other words, how do different skeletal designs compare mechanically to one another? For example, how do differences in femoral shape and size between the Neanderthal and modern human lower limb affect each system's ability to withstand loads associated with walking? Since muscle force data, and to some degree, body mass data, are unknowable for extinct taxa, this application is better employed when investigating relative abilities of structures to resist loads.

There are four main steps involved in FEA: model creation, model solving, and validation followed by interpretation. The first step, model creation, is often the most time-consuming process. During model creation, the investigator makes decisions regarding the geometric design of the structure of interest, boundary conditions, material properties, and the loads that will be applied to the model. Once a model is created, it is solved by computer hardware and software capable of performing a vast number of mathematical equations that result in stress, strain, and displacement calculations for the entire structure. Afterwards, the really interesting questions can be asked. For example, are the results realistic, and what do they mean in a biological context?

Finite element models of skeletal structures are typically created from serial computed tomography (CT) scans so that both external and internal geometry can be modeled. Tessellated surface models, which are composed of hundreds of thousands of geometrically simple surfaces (such as triangles) arranged in a mosaic pattern and enclosing volumes representing bone, are generated using medical imaging software in a multi-step process. These software programs typically require the use of a combination of manual and automatic thresholding techniques to separate trabecular bone from cortical bone, and bone from air. This procedure can be quite time consuming, but long bones, particularly the diaphysis, that are key components of mobility studies, have relatively simple geometries and thus are less difficult to model than skeletal structures like crania. Once separate volumes of bone are created, they are divided into a large, but finite number of elements of a simple shape, joined together at vertices called nodes. These simple shapes collectively create the mesh that comprises a model. Depending on the software being used, these shapes may be tetrahedra or "bricks" with a varying number of nodes and sides. As the number of nodes and/or elements increases, the accuracy of the model should increase, but a trade-off is incurred since more computational power is needed to solve the model (Richmond et al. 2005).

Following creation of a FEM, it must be assigned material properties. In the case of a femur, the relevant material is bone. Key properties include the elastic modulus and Poisson's ratio. The elastic modulus, E, describes how much strain a structure will experience in response to a given stress when the object is loaded axially. More specifically, it represents the slope of the linear (elastic) portion of the stress–strain curve for a given material. This describes the stiffness of the object during tension or compression. Poisson's ratio (ν=lateral strain/axial strain) describes how much the sides of an object will contract or expand laterally during tensile or compressive axial loads, respectively. If a material is isotropic, then its material properties are the same in all directions at any given point, and thus the elastic modulus and Poisson's ratio are the only two properties that need to be specified. Most FEA studies assume that cortical bone is isotropic, but this is typically not the case in life. Rather, bone tends to range between being roughly transversely isotropic (i.e., material properties in the axial direction of a long bone may differ from those in the cross section of the bone) to being orthotropic (material properties vary in each of three orthogonal directions). Moreover, most FEA studies assume that the material properties of cortical bone are spatially homogenous (i.e., they are the same in all regions of the bone), when in fact those properties may be heterogeneously distributed [i.e., they may vary from region to region (Bozanich et al. 2009; Wang et al. 2006)]. Finally, cortical and trabecular bone have different elastic moduli; cortical bone is much stiffer than trabecular bone (Currey 2002).

Constraints and applied forces are referred to as boundary conditions. It is necessary to constrain the model from moving in at least some fashion, although it is also important not to overly constrain it as that may result in unrealistic stresses and/or strains throughout the model (Richmond et al. 2005). The application of constraints ensures that models resist translational or rotational movement; it anchors them in three-dimensional space and ensures that the applied forces will cause deformations in the model. Constraints are typically chosen in locations imitating biological constraints, such as ligaments, or contact between bones. For instance, when modeling a femur, one might choose to apply constraints at the fovea capitis on the femoral head and on the distal-most surface of the epiphyses to simulate contact with the tibial plateau. Because the selected nodes are not allowed to move, strain will be concentrated at and around those locations, possibly producing unrealistic local strains. Therefore, if possible, it is best to analyze strain at locations away from the constraints so as not to bias the results of the experiment.

Muscle forces can be applied to the model as vectors running from the origin of a muscle towards its insertion. Muscles with multiple compartments that may not all act simultaneously are best modeled with multiple origins, or as separate muscles. Ideally, surface models of bones articulating with the bone of interest will be positioned such that they can serve as origin and insertion points for the muscles. For instance, surface models of the pelvis, tibia, and fibula may be necessary to apply muscle forces to a femur FEM during simulation of bipedal walking insofar as many muscles active during walking either arise from or insert on one of these surrounding bones.

Once a model has been created, volumes have been assigned material properties, and boundary conditions have been applied, it is possible to solve the model and interpret results. Computer software solves the model by calculating nodal displacements due to applied forces, and the stresses and strains corresponding to these nodal displacements (Zienkiewicz et al. 2005).

Once a model has been solved, there is not yet reason to be confident that the model accurately depicts what happens in a real biological system. In order to know this, the investigator must validate the model. Preferably, this would mean comparing strain data obtained from the FEM of a bone to strain data obtained from in vivo measurements by strain gages affixed to the same bone during the same loading scenario as was applied to the FE model of the bone. However, this is not always possible due to both practical and ethical reasons (in the case of humans and animals, respectively) as the procedure is highly surgically invasive, and is not even the norm, especially for experiments focusing on human subjects. In cases where in vivo strain gage measurements are impossible to obtain, in vitro cadaveric experiments are a reasonable alternative. However, in vivo and in vitro validation experiments measure different things. Generally speaking, in vitro bone strain experiments entail the application of forces that only coarsely approximate those used in actual behaviors. However, an advantage of such studies is that it is generally relatively straightforward to simulate those loads (as well as constraints) in FEA. Thus, in vitro validation is most useful in assessing the validity of the geometry and material properties of a bony structure. In contrast, in vivo validation experiments examine the degree to which all of the assumptions incorporated into the simulation of a behavior (e.g., geometry, loads, constraints, material properties) are collectively valid. In a perfect scenario, FEMs would be validated using both in vivo and in vitro data, although this is not typically done. Regardless, a well-validated model is essential if the purpose is to realistically depict the performance of a structure in a biological context. Once it is reasonably certain that the FEM behaves in a biologically realistic manner, loads or other input variables can be changed to reflect those obtained from in vivo experiments, and interpretation of the results may proceed with a level of confidence equal to the rigor of the validation test.

Examination of the patterns of stress or strain due to specific loads allows an investigator to identify weak points in the structure, the overall pattern of deformation, or how each set of loading conditions affects the behavior of the model. Applying the same loads to different models shows how size *and* shape differences in the structures affect each structure's ability to resist loads. However, when comparing bones of different morphology, an investigator may want to know what effect shape alone has on stress and strain. By scaling magnitudes of forces applied to a FEM by the volume of a model raised to the 2/3 power, one can remove size as a factor in FEA experiments and simply compare the effects of scale-free shape differences (Dumont et al., 2009).

15.2 Finite Element Analysis of a Human Femur

15.2.1 Model Creation

Serial computed tomography (CT) scans of a modern human femur were first imported as TIFF files and processed in the computer software program Mimics v13 (Materialise, Ann Arbor, MI, USA), in which surface meshes composed of triangles were created. An automatic thresholding algorithm was used to separate bone from empty space. Then, through manual slice-by-slice segmentation, three separate surfaces were generated representing the outer layer of cortical bone, and two volumes of trabecular bone, one each in the proximal and distal ends of the bone. The medullary cavity was modeled as an empty space (Fig. 15.1).

These surfaces were exported into the surface editing program Geomagic Studio v12 (Research Triangle Park, NC, USA) as binary STL files. In Geomagic Studio, surfaces were rid of imperfections such as holes, overlapping triangles, spikes, and other abnormalities or distortions created during the manual segmentation process.

Fig. 15.1 The human FEM displayed transparently in the surface editing program Geomagic Studio, with solid black lines marking divisions between proximal and distal volumes of trabecular bone and the central medullary cavity

Once the geometry was judged to be clean, surfaces were re-imported into Mimics where they were once again meshed to check for overlapping triangles. If no intersections were found, surfaces were volume meshed to create watertight solid volumes composed of thousands of tiny tetrahedral elements connected by nodes, rather than simple surfaces. The end result of this process was four mutually exclusive volumes: outer cortical bone, inner medullary cavity, proximal and distal trabecular bone.

Each volume was imported into the Strand7 Finite Element Analysis Software System (Strand7 Pty Ltd, Sydney, NSW) as a NASTRAN file. Strand7 allows the application of various material properties, constraints, and force loads. In this model, the medullary cavity volume was deleted, leaving an empty space inside the volume of cortical bone, which separated each trabecular bone volume (Fig. 15.1). Volumes were assigned isotropic material properties. Cortical bone was given an elastic modulus (E) of 20 gigapascals (GPa), and a Poisson's ratio (v) of 0.3 (Currey and Butler 1975). Trabecular bone was modeled as a solid rather than as individual trabeculae due to the prohibitively time-consuming nature of the task. Trabecular bone was assigned $E=749$ megapascals (MPa) and $\nu=0.3$ (Kaneko et al. 2004; Strait et al. 2005).

15.2.2 Constraints

Constraints were applied at seven locations on the femur (Fig. 15.2). One node within the fovea capitis was constrained from anteroposterior (AP) and mediolateral (ML) movement to simulate ligamentum teres. One node on the inferior most surface of each femoral condyle was constrained from moving in the vertical direction. This simulates contact between the femur and the tibia. One node on the inner surface of each condyle within the intercondylar groove was constrained from moving

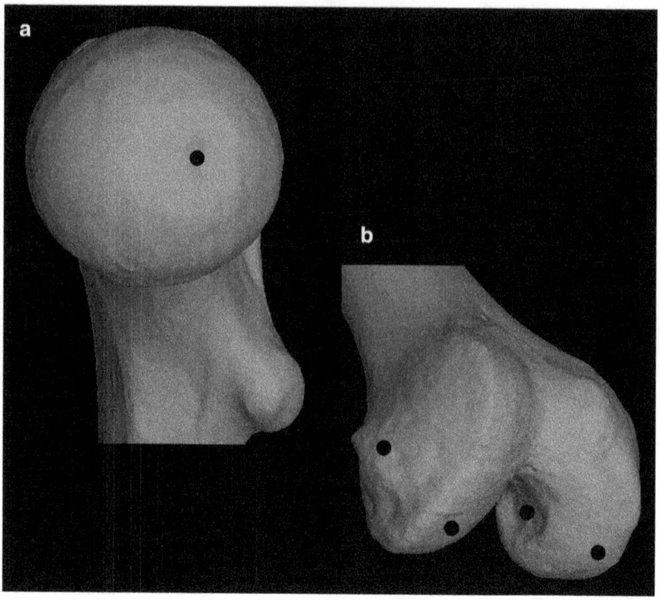

Fig. 15.2 Constraints, marked with black dots, were applied to one location on the femoral head (**a**) and six nodes on the distal end (**b**) of the FEM. Constraints represent the effect of contact with ligamentum teres, the tibial plateau, the cruciate ligaments, and collateral ligaments. Note that constraints representing the medial collateral ligament and the lateral attachment of the cruciate ligaments are not shown, but mirror their counterparts

in the AP direction, simulating the effect of the cruciate ligaments. Finally, one node was constrained from moving in the ML direction on the lateral surfaces of each epicondyle, corresponding to the collateral ligaments. It is important not to over-constrain the FEM, as would be the case if a region of nodes corresponding to the cross-sectional area of each ligament was constrained, since this can have an adverse effect on results (Haut Donahue et al. 2002).

15.2.3 Validation

This FEM was validated by replicating a cadaveric experiment conducted by Huiskes (1982). In that experiment, an embalmed human femur, dissected free of all soft tissue, was secured in a laboratory setting, and loaded with strain gages at seven horizontal levels along the diaphysis (Fig. 15.3). At each of these horizontal levels, seven strain gages were applied to the circumference of the diaphysis to measure maximum and minimum principal stress. Ten thousand Newton millimeters (Nmm) of torque were applied to the femoral head; the resulting principal stresses were measured at 49 locations on the diaphysis of the femur. In order to recreate the loading regime of the cadaveric femur used by Huiskes (1982) for the femur FEM, it

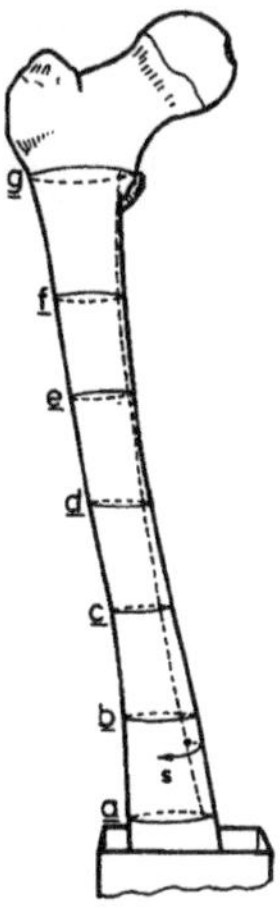

Fig. 15.3 Location of the seven horizontal levels around the femoral diaphysis indicating areas where strain gages were affixed and stress was measured [modified from Huiskes (1982)]. The first location at each level begins at the dotted vertical line along the posterior diaphysis and subsequent points of stress are measured at even intervals proceeding medially as indicated by the *arrow* and letter "s"

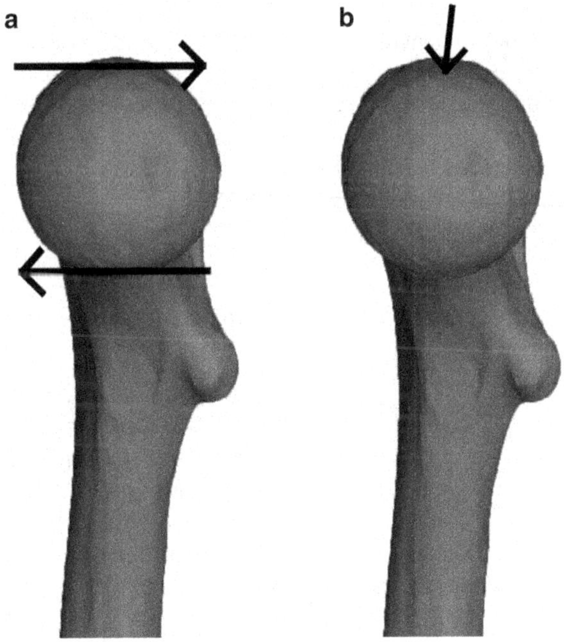

Fig. 15.4 Medial view of forces applied to the femoral head in the validation study (**a**) and simulation of loading during gait (**b**). In the validation study (**a**), 227.27 N were directed anteriorly and posteriorly as indicated by the solid *black arrows* inferiorly and superiorly, respectively. In the simulation of loading during gait, a resultant force of 1,951 N was applied to the superior portion of the femoral head (**b**). Solid *black arrow* indicates direction of applied force

was determined that the force couple producing 10,000 Nmm torque in the human cadaveric femur was 454.54 N for a femoral head with a radius of 22 millimeters (mm), as in the FEM. In order to create a torque, two forces (one each directed posteriorly and anteriorly) were applied to the femoral head. So (10,000 Nmm/(22 mm))/ 2 = 227.27 N. Therefore a 227.27 N force directed posteriorly was applied to the superior portion of the femoral head, and a 227.27 N force directed anteriorly was applied to the inferior surface of the femoral head (Fig. 15.4a).

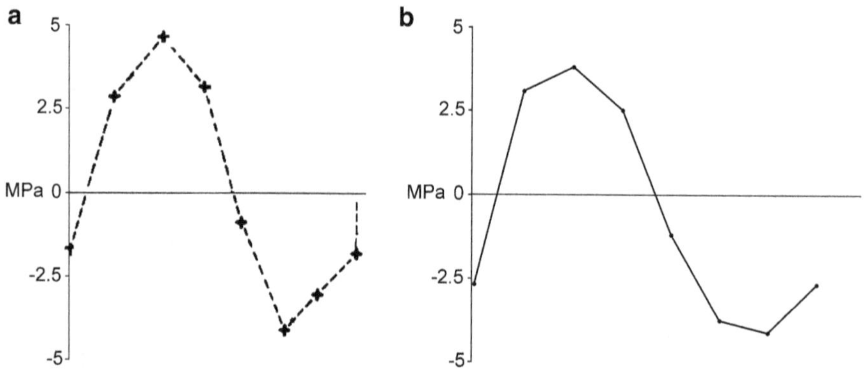

Fig. 15.5 Stress, recorded at seven locations along the midshaft circumference by strain gages affixed to a cadaveric human femur [modified from Huiskes (1982)] (**a**), is compared to stress measured along the midshaft diaphysis on the FEM (**b**). Location corresponds to level "d" in Fig. 15.3

Once the model was solved in Strand7, maximum and minimum principal stresses were measured at approximately the same locations as on the cadaveric femur (Fig. 15.3); the data show a close correspondence in both pattern and magnitude (Fig. 15.5). When subjected to a posterior bending moment, the femur experiences tension along the anterior portion of the diaphysis, and compression posteriorly, as does the femur FEM.

15.2.4 Simulation of Loading Conditions During Gait

Once the femur FEM is satisfactorily validated, it is used to simulate a more biologically interesting loading condition, namely the instant of peak acetabular force during push-off, directly prior to toe-off. In principle, one could model the femur dynamically as it is loaded throughout an entire gait cycle, but this introduces considerable complexity into the modeling procedure. As an alternative, the femur was modeled statically using the forces corresponding to the instant of peak acetabular force (although, there are many informative events within the gait cycle, such as at peak vertical substrate reaction force and peak acetabular force at heel strike). Pedersen et al. (1997) used a combination of kinematic and kinetic methods along with an optimization procedure to calculate acetabular force magnitude and direction, as well as the magnitudes of the forces of 22 hip and thigh muscles at 32 intervals during the gait cycle. At push-off, acetabular force was calculated to equal 314.8 % body weight, and was divided into three component directions (vertical, ML, and AP) with a resultant force of 1951 Newtons (N) (Table 15.1). These forces were applied to a rectangular selection of bricks on the femoral head of the model (Fig. 15.4b). Of the 22 muscles monitored by Pedersen et al. (1997), ten were active during the gait event (Table 15.1). Muscle forces were applied using the

Table 15.1 Forces applied to the human FEM

Axial force (N)	
Vertical	1,859
Mediolateral	555
Anteroposterior	214
Muscle forces (N)	
Gluteus maximus	420
Gluteus medius	286
Gluteus minimus	124
Biceps femoris	256
Psoas major	71
Iliacus	91
Gemellus superior	3
Gemellus inferior	7
Obturator internus	40
Piriformis	98

Source: Pedersen et al. (1997)

tangential-plus-normal loading procedure of Boneload (Grosse et al. 2007), a software package that interfaces with Strand7 allowing the modeling of complex muscle vectors that wrap around the surface of a model. Surface models of a pelvis and tibia belonging to the same individual as the femur from which the model was built were used to determine muscle attachment sites. Muscle forces were applied as the femur was in a slightly extended position relative to the pelvis, as would occur during push-off (Fig. 15.6). Some muscles, such as the adductors, with a linearly large attachment site were divided and measured in multiple components by Pedersen et al. (1997). We followed their procedure for dividing those muscles, and otherwise origin/attachment sites were directed from/to the center of the attachment site. Loads were applied to regions on the femur and were directed outward, either to the origin or insertion site, depending on the muscle. For example, gluteus medius originates from a large area on the ilium, and although it is a large muscle, it has a small insertion area on the greater trochanter of the femur. The entire insertion region received a load of 286 N (Table 15.1) divided evenly over its surface area and directed towards the center of origin on the ilium.

15.2.5 Results

Results show that a band of tension, as evidenced by maximum principal stress, begins on the lateral aspect of the greater trochanter and continues down and across the anterior diaphysis, ending on the anteromedial metaphysis (Fig. 15.7a). Similarly, but conversely, a band of compression, shown by minimum principal stress, originates on the posterior femoral head and continues down the femoral neck to the posterior diaphysis where it ends on the posterolateral metaphysis (Fig. 15.7b). Von Mises stress is most closely related to bone failure (Keyak and

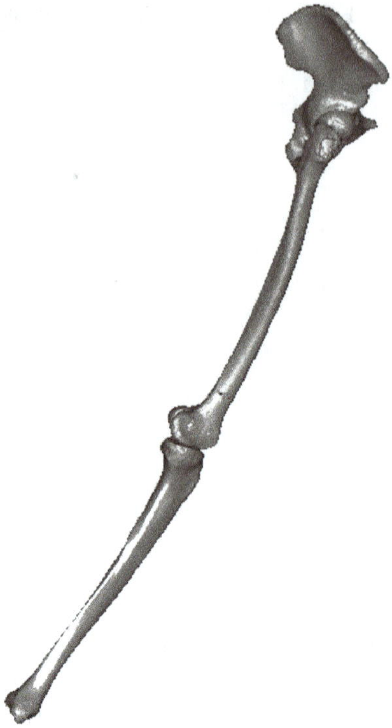

Fig. 15.6 The femur FEM is articulated with surface models of the pelvis and tibia in the approximate positioning of the lower limb during the instant of the gait cycle directly prior to toe-off. These positions were used to direct muscle vectors in the gait simulation experiment

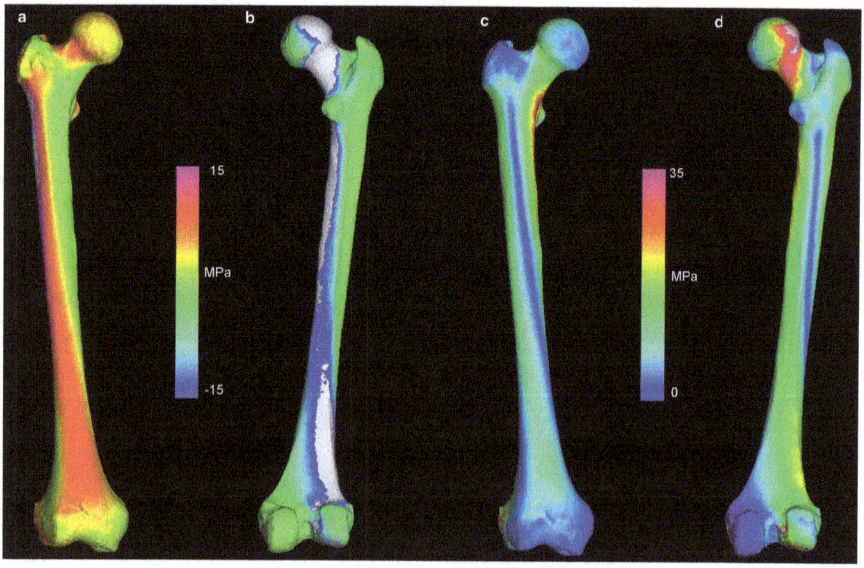

Fig. 15.7 Maximum principal stress (**a**) displayed on an anterior view of the FEM shows where the model experiences highest tensile stresses. Minimum principal stress (**b**) is shown on a posterior view of the human FEM. The posterior diaphysis and femoral neck experience compression in this loading environment. Von Mises stress patterns follow those of maximum and minimum principal stress, anteriorly (**c**) and posteriorly (**d**), respectively, but are highest on the posterior side where the model experiences compression. Units are in megapascals

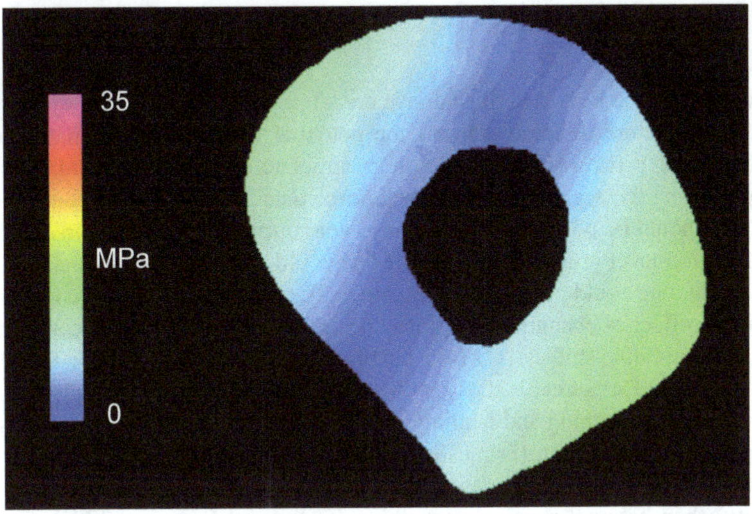

Fig. 15.8 Midshaft diaphyseal cross section showing von Mises stress. The area experiencing the least amount of stress is indicated in *blue*, showing the axis about which the bone bends

Rossi 2000); patterns of von Mises stress follow maximum and minimum principal stresses on the anterior (Fig. 15.7c) and posterior (Fig. 15.7d) diaphysis, respectively, but are higher posteriorly where the model experiences compression. Regions of highest von Mises stress produced by these loads are found on the femoral neck, posterolateral distal metaphysis, posterior diaphyseal midshaft, and lateral proximal diaphysis (in order of highest to lowest stress).

15.2.6 Implications

Two key biomechanical insights provided by this FEA are relevant to interpretations of mobility. First, as evidenced by analyses of diaphyseal cross sections, midshaft bending is primarily ML rather than AP in orientation (Fig. 15.8). If interpretations about mobility depend on interpretations of femoral strength, then it follows that the key measure of strength is ML bending strength. Although modern human femora are deeper anteroposteriorly than they are mediolaterally (Trinkaus and Ruff 1999), it nonetheless appears that ML strength is the key variable limiting bone failure, at least during normal walking on a level surface. Secondly, cross-sectional analyses of strength typically rely on assumptions about the location of the neutral axis and the bending direction of a bone; this study corroborates other work suggesting that such assumptions may not strictly apply (Lieberman et al. 2004).

15.3 Future Directions

Implications of the FEA described above are, at present, limited, but they point the way towards future research that has the potential to be more informative about mobility. First, it is possible in principle to simulate other loading conditions. For example, using the acetabular and muscle force data gathered from Pedersen et al. (1997), it should be possible to model any event during a gait cycle corresponding to walking on level ground. Moreover, if one could gather adequate kinematic and kinetic data, one could potentially simulate walking on sloped or uneven terrain as well as the effect of changing directions during locomotion, topics of considerable interest in mobility studies (Carlson, 2014; Carlson and Judex 2007; Daley and Biewener 2006; Demes et al. 2001, 2006). One could also model running on a variety of terrains, or leaping and landing in order to investigate the effect of substrate use on stress transmission (Demes et al. 1995). One could also model stumbling, which might represent a load case more likely to cause injury (and threaten bone integrity) than habitual running or walking (Daley and Biewener 2006). Specific hypotheses exist regarding trade-offs between high and low leg retraction velocities relating to injury and stumbling risk when running over uneven terrain (Daley and Usherwood 2010), hypotheses that might be testable using FEA. In addition, one could simulate traumatic blows. One might potentially find that femora that are strong under one loading regime might be weak under another, and this might lead researchers to a more nuanced understanding of the selection forces that may have influenced the evolution of femoral form. Furthermore, application of FEA is not limited to femora. Other postcranial elements that are the subject of studies of mobility include the fibula (Sparacello et al., 2014), humerus (Marchi et al. 2006), tibia (Demes et al. 2001), or multiple elements considered together (Sparacello and Marchi 2008; Stock 2006).

Major insights about mobility and femoral functional anatomy are likely to emerge through comparative biomechanics. For example, it has been hypothesized that Neanderthals ranged more widely than modern humans because metric analysis suggests that their femora were very strong. Comparisons of finite element models of modern human and Neanderthal femora could test this hypothesis with respect to specific loading conditions, and quantify the mechanical differences between them.

15.4 Summary

There are many hypotheses about activity (broadly) and mobility (in particular) that make predictions regarding long bone biomechanics. We want to understand the consequences of different activity levels and patterns of mobility on skeletal morphology, whether to reconstruct prehistoric lifeways, ranging behavior, or even to understand modern orthopedics. FEA is a versatile tool that allows the modeling of bone behavior under various loading environments. The results of FEA provide insights into how bones perform mechanically as whole structures. Although FEA is a time-intensive process, the rewards are potentially great as the versatility of FEMs allows many hypotheses to be tested in an integrative manner.

Acknowledgements This study was supported by NSF BCS 1060835 and NSF BCS 0725126. I am especially grateful to Dr. David Strait for his helpful reviews and comments.

References

Bertram JEA, Biewener AA (1988) Bone curvature: sacrificing strength for load predictability. J Theor Biol 131:75–92

Biewener AA, Thomason J, Goodship AE, Lanyon LE (1983) Bone stress in the horse forelimb during locomotion at different gaits: a comparison of two experimental methods. J Biomech 16:565–576

Bozanich J, Byron CD, Chalk J, Grosse IR, Lucas PW, Richmond BG, Ross CF, Slice DE, Smith AL, Spencer MA, et al. Density and material orientation in ape mandibular cortical bone; 2009 April 1–4, Miami.

Carlson KJ (2014) Linearity in the real world—an experimental assessment of non-linearity in terrestrial locomotion. In: Carlson KJ, Marchi D (eds) Reconstructing mobility: environmental, behavioral, and morphological determinants. Springer, New York

Carlson KJ, Grine FE, Pearson OM (2007) Robusticity and sexual dimorphism in the postcranium of modern hunter gatherers from Australia. Am J Phys Anthropol 134:9–23

Carlson KJ, Judex S (2007) Increased non-linear locomotion alters diaphyseal bone shape. J Exp Biol 210:3117–3125

Cowgill LW (2010) The ontogeny of Holocene and Late Pleistocene human postcranial strength. Am J Phys Anthropol 141:16–37

Cowgill LW (2014) Femoral diaphyseal shape and mobility: an ontogenetic perspective. In: Carlson KJ, Marchi D (eds) Reconstructing mobility: environmental, behavioral, and morphological determinants. Springer, New York

Cowgill LW, Warrener A, Pontzer H, Ocobock C (2010) Waddling and toddling: the biomechanical effects of an immature gait. Am J Phys Anthropol 143:52–61

Currey J (2002) Bones: structure and mechanics. Princeton University Press, Princeton

Currey JD, Butler G (1975) The mechanical properties of bone tissue in children. J Bone Joint Surg 57-A:810–814

Daley MA, Biewener AA (2006) Running over rough terrain reveals limb control for intrinsic stability. Proc Natl Acad Sci U S A 103(2):15681–15686

Daley MA, Usherwood JR (2010) Two explanations for the compliant running paradox: reduced work of bouncing viscera and increased stability in uneven terrain. Biol Lett 6:418–421

Demes B, Carlson KJ, Franz TM (2006) Cutting corners: the dynamics of turning behaviors in two primate species. J Exp Biol 209:927–937

Demes B, Jungers WL, Gross TS, Fleagle JG (1995) Kinetics of leaping primates: influence of substrate orientation and compliance. Am J Phys Anthropol 96(4):419–429

Demes B, Qin Y-X, Stern JT Jr, Larson SG, Rubin CT (2001) Patterns of strain in the macaque tibia during functional activity. Am J Phys Anthropol 116(4):257–265

Dumont E, Grosse I, Slater G (2009) Requirements for comparing the performance of finite element models of biological structures. J Theor Biol 256:96–103

Goodship AE, Lanyon LE, MacFie H (1979) Functional adaptation of bone to increased stress. J Bone Joint Surg 61-A:539–546

Grosse IR, Dumont ER, Coletta C, Tolleson A (2007) Techniques for modeling muscle-induced forces in finite element models of skeletal structures. Anat Rec 290(9):1069–1088
Haut Donahue TL, Hull ML, Rashid MM, Jacobs CR (2002) A finite element model of the human knee joint for the study of tibio-femoral contact. J Biomech Eng 124:273–280
Hert J, Liskova M, Landa J (1971) Reaction of bone to mechanical stimuli. Part I. Continuous and intermittent loading of tibia in rabbit. Folia Morphol 19:290–300
Hert J, Liskova M, Landgrot B (1969) Influence of the long-term continuous bending on the bone. An experimental study on the tibia of the rabbit. Folia Morphol 17:389–399
Hert J, Pribylova E, Liskova M (1972) Reaction of bone to mechanical stimuli. Part 3. Microstructure of compact bone of rabbit tibia after intermittent loading. Acta Anat 82:218–230
Huiskes R (1982) On the modelling of long bones in structural analyses. J Biomech 15(1):65–69
Jones HH, Priest JD, Hayes WC, Tichenor CC, Nagel DA (1977) Humeral hypertrophy in response to exercise. J Bone Joint Surg 59A:204–208
Kaneko TS, Bell JS, Pejcic MR, Tehranzadeh J, Keyak JH (2004) Mechanical properties, density, and quantitative CT scan data of trabecular bone with and without metastases. J Biomech 37: 523–530
Kelly RL (1995) The foraging spectrum-diversity in hunter-gatherer lifeways. Smithsonian Institution Press, Washington
Keyak JH, Rossi SA (2000) Prediction of femoral fracture load using finite element models: An examination of stress- and strain-based failure models. J Biomech 33:209–214
Krolner B, Toft B (1963) Vertebral bone loss, an unheeded side effect of therapeutic bed rest. Clin Sci 64:537–540
Lanyon LE (1987) Functional strain in bone tissue as an objective, and controlling stimulus for adaptive bone remodelling. J Biomech 20(11/12):1083–1093
Lanyon LE, Baggott DG (1976) Mechanical function as an influence on the structure and form of bone. J Bone Joint Surg 58-B(4):436–443
Lanyon LE, Bourn S (1979) The influence of mechanical function on the development and remodeling of the tibia: an experimental study in sheep. J Bone Joint Surg 61:263–273
Lanyon LE, Goodship AE, Pye CJ, MacFie H (1982) Mechanically adaptive bone remodeling. J Biomech 15:141–154
Lanyon LE, Magee PT, Baggot DG (1979) The relationship of functional stress and strain to the processes of bone remodeling. An experimental study on the sheep radius. J Biomech 12: 593–600
Larsen CS (1987) Bioarchaeological interpretations of subsistence economy and behavior from human skeletal remains. Adv Archaeol Method Theory 10:339–445
Lieberman DE, Pearson OM, Polk JD, Demes B, Crompton AW (2003) Optimization of bone growth and remodeling in response to loading in tapered mammalian limbs. J Exp Biol 206: 3125–3138
Lieberman DE, Polk JD, Demes B (2004) Predicting long bone loading from cross-sectional geometry. Am J Phys Anthropol 123:156–171
Marchi D, Sparacello VS, Holt BM, Formicola V (2006) Biomechanical approach to the reconstruction of activity patterns in Neolithic Western Liguria, Italy. Am J Phys Anthropol 131: 447–455
Nordstrom P, Thorsen K, Bergstrom E, Lorentzon R (1996) High bone mass and altered relationships between bone mass, muscle strength, and body constitution in adolescent boys on a high-level of physical-activity. Bone 19:189–195
Paul JP (1971) Load actions on the human femur in walking and some resultant stresses. Exp Mech 11:121–125
Pedersen DR, Brand RA, Davy DT (1997) Pelvic muscle and acetabular contact forces during gait. J Biomech 30(9):959–965
Richmond BG, Wright BW, Grosse I, Dechow PC, Ross CF, Spencer MA, Strait DS (2005) Finite element analysis in functional morphology. Anat Rec A Discov Mol Cell Evol Biol 283A: 259–274

Ruff C (1995) Biomechanics of the hip and birth in early *Homo*. Am J Phys Anthropol 98: 527–574

Ruff C (2005) Mechanical Determinants of Bone Form: Insights from Skeletal Remains. Muculoskelet Neutonal Interact 5(3):202–212

Ruff C, Holt B, Trinkaus E (2006) Who's afraid of the big bad Wolff?: "Wolff's Law" and bone functional adaptation. Am J Phys Anthropol 129:484–498

Ruff CB (1989) New approaches to structural evolution of limb bones in primates. Folia Primatol 53:142–159

Ruff CB, Trinkaus E, Walker A, Larsen CS (1993) Postcranial robusticity in *Homo*. I: Temporal trends and mechanical interpretation. Am J Phys Anthropol 91:21–53

Shackelford LL, Trinkaus E (2002) Late Pleistocene human femoral diaphyseal curvature. Am J Phys Anthropol 118:359–370

Skerry TM (2000) Biomechanical influences on skeletal growth and development. In: O'Higgins P, Cohn MJ (eds) Development growth and evolution: implications for the study of the hominid skeleton. Academic, San Diego, pp 29–39

Sparacello V, Marchi D (2008) Mobility and subsistence economy: a diachronic comparison between two groups settled in the same geographical area (Liguria, Italy). Am J Phys Anthropol 136:485–495

Sparacello V, Marchi D, Shaw C (2014) The importance of considering fibular robusticity when inferring the mobility patterns of past populations. In: Carlson KJ, Marchi D (eds) Reconstructing mobility: environmental, behavioral, and morphological determinants. Springer, New York

Stock J (2006) Hunter-gatherer postcranial robusticity relative to patterns of mobility, climatic adaptation, and selection for tissue economy. Am J Phys Anthropol 131(2):194–204

Stock J, Shaw C (2007) Which measures of diaphyseal robusticity are robust? A comparison of external methods of quantifying the strength of long bone diaphyses to cross-sectional geometric properties. Am J Phys Anthropol 134:412–423

Strait DS, Qang Q, Dechow PC, Ross CF, Richmond BG, Spencer MA, Patel BA (2005) Modeling elastic properties in finite-element analysis: How much precision is needed to produce an accurate model? Anat Rec A Discov Mol Cell Evol Biol 283A(2):275–287

Taylor ME, Tanner KE, Freeman MAR, Yettram AL (1996) Stress and strain distribution within the intact femur: compression or bending? Med Eng Phys 18:122–131

Tilton FE, Degioanni TTC, Schneider VS (1980) Long term follow up on Skylab bone demineralisation. Aviat Space Environ Med 51:209–213

Trinkaus E (1993) Femoral neck-shaft angles of the Qafzeh-Skhul early modern humans, and activity levels among immature Near Eastern Middle Paleolithic hominids. J Hum Evol 25: 393–416

Trinkaus E, Ruff CB (1999) Diaphyseal cross-sectional geometry of Near Eastern Middle Palaeolithic humans: the femur. J Archaeol Sci 26:409–424

Trinkaus E, Ruff CB, Conroy GC (1999) The anomalous archaic *Homo* femur from Berg Aukas, Namibia: A biomechanical assessment. Am J Phys Anthropol 110:379–391

Wang Q, Strait DS, Dechow PC (2006) A comparison of cortical elastic properties in the craniofacial skeletons of three primate species and its relevance to the study of human evolution. J Hum Evol 51(4):375–382

Weber GW, Bookstein FL (2011) Virtual anthropology: a guide to a new interdisciplinary field. Springer, New York

Weber GW, Bookstein FL, Strait DS (2011) Virtual anthropology meets biomechanics. J Biomech 44(8):1429–1432

Woo SL-Y (1981) The relationships of changes in stress levels on long bone remodeling, vol 45. American Society of Mechanical Engineers, Applied Mechanics Division, AMD, New York, pp 107–129

Zienkiewicz O, Taylor R, Zhu J (2005) The finite element method: its basis and fundaments. Elsevier, Oxford

Index

K.J. Carlson and D. Marchi (eds.), *Reconstructing Mobility: Environmental, Behavioral, and Morphological Determinants*, DOI 10.1007/978-1-4899-7460-0,

www.ingramcontent.com/pod-product-compliance
Lightning Source LLC
Chambersburg PA
CBHW060743150926
51501CB00003B/77
9781489974594